Lecture Notes in Information Systems and Organisation

Volume 43

Lecture Notes in Information Systems and Organization—LNISO—is a series of scientific books that explore the current scenario of information systems, in particular IS and organization. The focus on the relationship between IT, IS and organization is the common thread of this collection, which aspires to provide scholars across the world with a point of reference and comparison in the study and research of information systems and organization. LNISO is the publication forum for the community of scholars investigating behavioral and design aspects of IS and organization. The series offers an integrated publication platform for high-quality conferences, symposia and workshops in this field. Materials are published upon a strictly controlled double blind peer review evaluation made by selected reviewers.

LNISO is abstracted/indexed in Scopus

More information about this series at http://www.springer.com/series/11237

Fred D. Davis · René Riedl ·
Jan vom Brocke · Pierre-Majorique Léger ·
Adriane B. Randolph · Thomas Fischer
Editors

Information Systems and Neuroscience

NeuroIS Retreat 2020

Editors
Fred D. Davis
Information Systems and Quantitative Sciences (ISQS)
Texas Tech University
Lubbock, TX, USA

Jan vom Brocke
Department of Information Systems
University of Liechtenstein
Vaduz, Liechtenstein

Adriane B. Randolph
Department of Information Systems
Kennesaw State University
Kennesaw, GA, USA

René Riedl
University of Applied Sciences
Upper Austria
Steyr, Austria

Pierre-Majorique Léger
Department of Information Technology
HEC Montréal
Montreal, QC, Canada

Thomas Fischer
Department for Digital Business
University of Applied Sciences
Upper Austria
Steyr, Austria

ISSN 2195-4968 ISSN 2195-4976 (electronic)
Lecture Notes in Information Systems and Organisation
ISBN 978-3-030-60072-3 ISBN 978-3-030-60073-0 (eBook)
https://doi.org/10.1007/978-3-030-60073-0

This Springer imprint is published by the registered company Springer Nature Switzerland AG
The registered company address is: Gewerbestrasse 11, 6330 Cham, Switzerland

Preface

The proceedings contain papers presented at the 12th annual NeuroIS Retreat held on June 2–4, 2020. NeuroIS is a field in information systems (IS) that uses neuroscience and neurophysiological tools and knowledge to better understand the development, adoption, and impact of information and communication technologies (see http://www.neurois.org/).

The NeuroIS Retreat is a leading academic conference for presenting research and development projects at the nexus of IS and neurobiology. This annual conference promotes the development of the NeuroIS field with activities primarily delivered by and for academics, though works often have a professional orientation.

In 2009, the inaugural NeuroIS Retreat was held in Gmunden, Austria. Since then, the NeuroIS community has grown steadily, with subsequent annual Retreats in Gmunden from 2010–2017. Beginning in 2018, the conference is taking place in Vienna, Austria. Due to the Corona crisis, the organizers decided to host a virtual NeuroIS Retreat in 2020.

The NeuroIS Retreat provides a platform for scholars to discuss their studies and exchange ideas. A major goal is to provide feedback for scholars to advance their research papers toward high-quality journal publications. The organizing committee welcomes not only completed research, but also work in progress. The NeuroIS Retreat is known for its informal and constructive workshop atmosphere. Many NeuroIS presentations have evolved into publications in highly regarded academic journals.

This year is the sixth time that we publish the proceedings in the form of an edited volume. A total of 41 research papers are published in this volume, and we observe diversity in topics, theories, methods, and tools of the contributions in this book. The 2020 keynote presentation entitled "NeuroIS as Qualitative Research: Solving the Reverse Inference Problem" was given by Alan R. Dennis, current president of the Association for Information Systems (AIS) and Professor of Information Systems and John T. Chambers Chair of Internet Systems in the Kelley School of Business at Indiana University, USA. Moreover, Aaron Newman, Professor and Chair of the Department of Psychology and Neuroscience at

Dalhousie University, Canada, gave a hot topic talk entitled "A Critical View of Neuroimaging."

Altogether, we are happy to see the ongoing progress in the NeuroIS field. Also, we can report that the NeuroIS Society, established in 2018 as a non-profit organization, has been developing well. We foresee a prosperous development of NeuroIS.

June 2020

Fred D. Davis
René Riedl
Jan vom Brocke
Pierre-Majorique Léger
Adriane B. Randolph
Thomas Fischer

Organization

Conference Co-chairs

Fred D. Davis	Texas Tech University, Texas, USA
René Riedl	University of Applied Sciences Upper Austria, Steyr, Austria & Johannes Kepler University Linz, Linz, Austria

Programme Co-chairs

Jan vom Brocke	University of Liechtenstein, Vaduz, Liechtenstein
Pierre-Majorique Léger	HEC Montréal, Montreal, Canada
Adriane B. Randolph	Kennesaw State University, Kennesaw, USA

Programme Committee

Marc Adam	University of Newcastle, Callaghan, Australia
Shamel Addas	Queen's School of Business, Kingston, Canada
Bonnie Anderson	Brigham Young University, Utah, USA
Ricardo Buettner	Aalen University, Aalen, Germany
Colin Conrad	Dalhousie University, Halifax, Canada
Alan Dennis	Indiana University, Indiana, USA
Soussan Djamasbi	Worcester Polytechnic Institute, Massachusetts, USA
Rob Gleasure	Copenhagen Business School, Frederiksberg, Denmark
Jacek Gwizdka	University of Texas at Austin, Austin, Texas
Armin Heinzl	University of Mannheim, Mannheim, Germany
Alan Hevner	Muma College of Business, Florida, USA
Marco Hubert	University of Aarhus, Aarhus, Denmark

Peter Kenning	Heinrich-Heine-University Düsseldorf, Düsseldorf, Germany
Brock Kirwan	Brigham Young University, Utah, USA
Pierre-Majorique Léger	HEC Montréal, Montreal, Canada
Ting-Peng Liang	Sun Yat-sen University, Kaohsiung, Taiwan
Aleck Lin	National Dong Hwa University, Taiwan
Jan Mendling	Vienna University of Economics and Business, Vienna, Austria
Randall Minas	University of Hawai'i at Māona, Honolulu, Hawaii
Gernot Müller-Putz	Graz University of Technology, Graz, Austria
Fiona Nah	Missouri University of Science and Technology, Missouri, USA
Aaron Newman	Dalhousie University, Halifax, Canada
Isabella Seeber	University of Innsbruck, Innsbruck, Austria
Sylvan Sénécal	HEC Montréal, Montreal, Canada
Stefan Tams	HEC Montréal, Montreal, Canada
Lars Taxén	Linköping University, Linköping, Sweden
Ofir Turel	California State University, California, USA
Anthony Vance	Fox School of Business, Pennsylvania, USA
Eric Walden	Texas Tech University, Texas, USA
Barbara Weber	University of St. Gallen, St. Gallen, Switzerland
Robert West	DePauw University, Indiana, USA
Eoin Whelan	National University of Ireland Galway, Galway, Ireland
Selina Wriessnegger	Graz University of Technolog, Graz, Austria

Organization Support

Thomas Fischer	University of Applied Sciences Upper Austria, Steyr, Austria

Keynote Talks

NeuroIS as Qualitative Research: Solving the Reverse Inference Problem (Keynote)

Alan R. Dennis

NeuroIS research is commonly quantitative, driven by large amounts data and p-values. My own research using EEG, specifically event-related spectral perturbation (ERSP), is a good example. Yet, I have come to realize that research using ERSP (and other forms of NeuroIS) has much more in common with the style of qualitative research popularized by Allen Lee (former EIC of MIS Quarterly) than it does with quantitative research. A researcher may hypothesize that a treatment will lead to differences in activation in some brain regions but ERSP is a hypothesis-free analysis that examines activity across the entire surface area of the brain, so ERSP commonly finds activation in regions the researcher did not hypothesize about. Some neuroscientists have called this the "reverse inference problem" as it is impossible to use deductive reasoning to determine the cause because the regions are often associated with several functions. Neuroscientists who see this as a "problem" lack qualitative research training. For centuries, qualitative researchers have used abductive reasoning to draw conclusions in situations such as this, so for them, reverse inference is a valid, well-used, and well-understood research method, not "problem." The irony is that the same neuroscientists who label this as a problem use abductive reasoning when they build their hypotheses to test using other methods. So, it is not a question of whether or not to use abductive reasoning; it is a question of when—before or after building hypotheses. Like many issues in science, the answer is that both approaches are useful and valid.

A Critical View of Neuroimaging (Hot Topic Talk)

Aaron Newman

Neuroimaging has opened new doors for understanding the brain and, ultimately, using this knowledge to improve quality of life. The increasing availability of neuroimaging tools is accelerating this process, through an explosion of available data. This has also exposed the limitations of these relatively coarse physiological measurements—and the need for big data, data sharing, and meta-analytic techniques, as well as an emphasis on reproducibility. NeuroIS holds great potential as an applied discipline, but must always reflect critically on the limitations of the available techniques. I will outline challenges and opportunities for NeuroIS that can be learned from the evolution of cognitive neuroscience.

Contents

Why We Love Blue Hues on Websites: A fNIRS Investigation of Color and Its Impact on the Neural Processing of Ecommerce Websites

Anika Nissen(✉)

University of Duisburg-Essen, Duisburg, Germany
anika.nissen@icb.uni-due.de

Abstract. Blue of all colors seems to be generally preferred by humans and animals. Consequently, the use of this color in ecommerce context has several positive effects such as increased trustworthiness and aesthetic ratings. These effects are, in this study, hypothesized to be caused by specific neural processes in the prefrontal cortex of human decision makers. Consequently, this study tackles the research question whether there is a distinct neural activation pattern for blue websites that helps to explain why blue is often most favored. To investigate this, one website is designed and manipulated in color to which user reactions are measured by employing functional near-infrared spectroscopy (fNIRS). The results of this study show that bluc colored websites seem to require generally less processing power related to cognitive processing while revealing increases in brain structures related to processing pleasant and aesthetic stimuli.

Keywords: Color · Aesthetics · Websites · fNIRS · Decision making · Neural measurements

1 Introduction

Out of the blue, the true blue, being blue or feeling blue – blue has several associations and is consistently rated as the most famous color even across different cultures [1, 2]. Further, this trend does not only occur for humans, but seems to count for animals, too [3]. Consequently, it is not surprising that websites are often designed in decent blue hues (i.e. Facebook, Twitter, PayPal) [4]. One reason for blue's popularity might be its association (among others) with space, openness, and faithfulness [1] while there is no negative association with this color. Next to these associations, colors can also influence our emotions and therefore impact heavily the first impressions and evaluations of products and websites [5, 6]. It was found that the first impression which is influenced by the color scheme of websites dominates 67% of the purchase process [7]. Consequently, it is suggested, that blue colors raise the likelihood of purchase decisions on websites [8]. Further, for (physical) products 62–90% of the evaluations made about them stem from

F. D. Davis et al. (Eds.): NeuroIS 2020, LNISO 43, pp. 1–15, 2020.
https://doi.org/10.1007/978-3-030-60073-0_1

the color of their packaging [9], which makes the study of color inevitable for design and marketing. While blue seems to be the preferred color scheme for websites as it is consistently rated as more favorable and more trustworthy [10–13], we are wondering whether there is an impact of blue observable in the emotional and cognitive processing of ecommerce websites. So far, general neuroscientific studies that deal with color vision and perception primarily focus on what is processed in the visual cortex which is directly related to vision, and not what is processed in the prefrontal cortex (PFC) in which higher cognitive and emotional processing happens [14, 15]. However, as there have also been studies revealing the emotional and cognitive impact of color on humans, an investigation of the PFC in which such processes take place might be reasonable. Especially, when the context of the study is in online shopping and website perception. Consequently, in this work in progress, we are questioning if changes in the PFC are observable for blue designed ecommerce websites when they are compared to the same website, albeit in a different color scheme. Thus, we tackle the research question *how are blue colored ecommerce websites processed in the prefrontal cortex?* To give answers to this question, the remainder of the paper is structured as follows. First, related literature dealing with color theory and human color vision in general, as well as the effects of blue color in particular are reviewed. From the latter, two working hypotheses are derived which are further investigated in this study. After that, the method functional near-infrared spectroscopy (fNIRS) is described, as well as the study design and procedure. Finally, results are presented and discussed, and conclusions are drawn.

2 Related Literature

Color Theory and Human Color Vision

Colors have been in focus of several studies reaching from physics to artists to finally, computer engineers. Apparently, these investigations resulted in several different approaches of how to group and mix primary colors into further tones. Generally, it can be differentiated between additive and reductive color mixing. The former is used i.e. in the RGB color model (=red, green, blue). This model was derived from the ideas coming from the trichromatic theory of color vision [16] and the Grassman's Laws of mixing colored light [17]. This is also in accordance with how humans perceive colors with their eyes. That is, we physically perceive colors as light sent at different wavelengths that leads to a sensation in the eye and consequently, in the brain. The peaks of each of the three colors can be defined in nanometers (nm) – for instance, the blue color peaks 440 nm which is a short-wavelength, the color green peaks at 540 nm, which is a medium-wavelength, and red peaks at 580 nm, which is a long-wavelength in the color spectrum [18]. The different wavelengths of the light are perceived through cones in the human eye for which a different amount of cones is available for each of the three colors [19]. The number of cones determines the sensitivity we have for a specific color – as there are about 40-times more cones for red than for blue, the human eye is more sensitive to perceiving the color red than the color blue [18, 19].

Reductive color mixing resulted in what is now known as the CMY(K) color model (cyan, magenta, yellow, black value for printers). Subtracting primary colors was originally called RYB (red, yellow, blue) and was derived by taking the opposing and complementary hues from afterimages of the primary color. From this, several theories of colors were derived with one result of this being the color wheel originally invented by Isaac Newton [20]. Based on the color wheel, different harmonies according to the color's hue can be mixed. Early approaches to color harmonies and their psychological effects reach back to the 19th century [21, 22]. Another approach is to distinguish colors due to contrasts on three dimensions being 1) **value** (meaning light vs. Dark), 2) **chroma** (or saturation, intensity or purity), and 3) **hue** (that means color family, i.e. blue, green, red…) [10]. These dimensions have resulted in several color atlases that can also be understood by lay people such as the Munsell color system [23] which is also frequently applied in literature focusing on color [24–26].

In website design, however, the RGB model either in form of dedicated RGB values or via HEX codes is most frequently used. Studies that focus on color use on websites, have shown that the color scheme can influence emotions and the mood of customers which might further impact their decision making in ecommerce environments [27]. Additionally, color also impacts beauty evaluations of the website and thus, they are one major impact factor on the aesthetic attribution of websites [10, 28–30]. In some studies, it has been shown that for instance blue colors positively influence customers' mood while red colors negatively impact this [31]. Furthermore, another study revealed that blue also positively influences user performance while red negatively impacts it [32]. However, all these studies have primarily used behavioral measurements and did not employ psychophysiological measures. Although the perception and processing of color might also be culture-dependent, the favor of blue seems to be consistent even across different cultures [1, 2]. As both prior described cognitive and emotional responses are processed in the prefrontal cortex (PFC) of the human brain, we further regard it as reasonable to investigate this with a neuroimaging method.

The Perception and Effects of Blue Color

As stated, the human eye possesses cones which can be identified and related to different primary colors. When considering both cones and neural pathways through the visual cortex, it can be differentiated between a blue-yellow and red-green pathway – with the human eye being less sensible to blue, than to red and green [33, 34]. Consequently, blue colored stimuli might lead to specific, unique effects which also unconsciously influence users' perceptions of ecommerce websites. The main identified effects of blue color are that it can positively impact human health [35], which might be due to its relaxation effect [24]. Consequently, blue is suggested to increase performance in creative tasks and decrease it in tasks requiring high detail [36]. With special focus on color impact on attention, red seems to evoke a larger and earlier response in the brain in attention-related tasks [37], which further supports the relaxing effect of blue. In the context of ecommerce and websites, this effect can also be of advantage as studies have shown that blue positively signals approach behavior [36], and thus, it leads to higher pleasure and lower arousal ratings [38]. In general economic contexts, blue is suggested to signal brand competence when used in a company logo [39], and thus, it might lead to higher purchase rates [8]. However, it is also perceived as more trustworthy and more favorable when used on ecommerce websites [10–13]. Finally, and also of special interest for online environments, blue websites tend to be perceived to download faster than i.e. orange websites although their actual download times do not differ [24].

When summarizing these effects on a more abstract level, they might be differentiated in (a) *emotional processes* which include approach behavior, higher pleasure, less arousal, brand competence and trustworthiness, and in (b) *cognitive processes* such as relaxation, attention, user performance, and probably also trust. Cognitive and emotional processing in terms of the here named constructs is further observable in the prefrontal cortex of humans. Thus, blue colored websites might in fact be relatable to distinct cognitive and emotional processes reflected in the human brain. This leads us to the following two working hypotheses:

H1: Blue websites will reveal significant neural activity changes in PFC areas related to processing pleasant stimuli.

H2: Blue websites will reveal significant neural activity changes in PFC areas related to reduced cognitive processing.

Given that the PFC is primarily responsible for processing such higher, more complex (attribution to) stimuli, this study further uses one website which is manipulated in color and observes whether significant patterns can be observed in the PFC for the blue colored website. These patterns are further analyzed for blue in comparison to (1) no color at all, and blue in comparison to (2) green and orange, which both lie on the red-green pathway.

3 fNIRS Study for Neural Color Processing

Method

The visual appeal of the websites was measured using three scales from the Visual Aesthetics of Website Inventory (VisAWI) developed by Moshagen and Thielsch [40, 41]. In their questionnaire, they include the aspects of colorfulness, simplicity, and diversity which are each represented by 4–5 items, including reversed items. Simplicity (SIM) mainly comprises whether the content is easy to grasp and whether the overall layout appears harmonic. Diversity (DIV) describes the prior described visual complexity and variety which represent the originality of a website. Further, colorfulness (COL) considers whether the color scheme seems appealing and attractive or not. In addition to this, we wanted to assess whether the change in color has a significant impact on purchase intentions (PUI) and consequently, PUI was added as construct.

To complement the questionnaire data, we use fNIRS as second method to assess **neural activity** in the prefrontal cortex (PFC) of participants to receive deeper insights into cognitive and emotional appraisal processes for the blue colored website. The method fNIRS has been applied in the field of neuroscience, as well as cognate disciplines [42, 43]. Furthermore, in the context of human-computer interaction and interaction design, fNIRS has been shown to be a feasible method for measuring usability and user experiences of graphical user interfaces [44, 45]. fNIRS offers a lightweight and portable method which can be applied in real life contexts, while measuring the same biological processes as fMRI. Further, when compared to electroencephalography (EEG), fNIRS tends to be more user-friendly and most importantly, more robust to movement artefacts of the whole body and facial expressions when participants for instance talk during recordings [43, 45, 46]. In earlier studies, fNIRS data was found to highly correlate with the more mature fMRI signal [47, 48]. fNIRS operates with near-infrared light sent by sources that is absorbed, reflected or scattered by the human brain tissue, or more precisely by hemoglobin, and received by detectors. As a consequence, fNIRS takes advantage of this characteristic as the amount of oxygenated and deoxygenated hemoglobin (HbO and HbR, respectively) for a given brain region can be calculated with the received light by the detectors [49, 50]. This measurement therefore is an indirect parameter of neural cortical brain activity. Consequently, fNIRS yields potential for IS research, as it is a portable and lightweight technology that appears to be mostly robust to (movement) artefacts and can therefore be applied in practice-relevant scenarios [51, 52]. However, fNIRS has also some shortcomings that might be relevant for IS research. Most predominately, fNIRS has a limited spatial resolution, allowing to penetrate the human brain only on cortical structures [49, 53]. This, however, is sufficient to capture neural responses related to cognitive and emotional processing and decision making on blue ecommerce websites.

Sample and Study Design

Overall a **sample** size of $N = 24$ participants was recruited from the local university with 75% being male, 25% being female. From the sample, 87.5% were right-handed, 12.5% were left-handed. All participants had normal or corrected to normal sight, except one that had a red-green color blindness. Average age of the participants was $M = 26.33$ ($SD = 3.985$).

For the **study design**, four different color manipulated websites (blue, green, orange, black) are used as stimuli (Fig. 1) in connection with the short scales of the VisAWI [40, 41] and a purchase intention question. The questions and websites are then shown randomized in an experimental paradigm which showed the question first for 2s, then the website for 4s, then came the question again with a rating 5-point Likert scale and finally came a jitter with a fixation cross for the mean of 2s. Before the experiment started, participants were informed about their privacy and data protection rights, as well as the operating principle of fNIRS both in verbal and written form. After participants signed the informed consent, the mobile fNIRS headband was placed on the participants' head while taking the craniometric point of the nasion as reference to ensure comparability. To avoid data biases due to experiment equipment interference, several variables were controlled which can produce noise in the fNIRS signal [54–56]. The headband was calibrated for every individual participant through which data quality was assured. Having finished the calibration procedure, participants were instructed to start with the study. After participants finished the study, they were freed of the fNIRS headband and had to fill out a closing questionnaire including demographic questions.

Data Acquisition and Pre-processing

Physiological data was acquired using a continuous-wave NIRSport device developed by NIRx with a headband montage holding 8 sources, 7 long-distance detectors (average

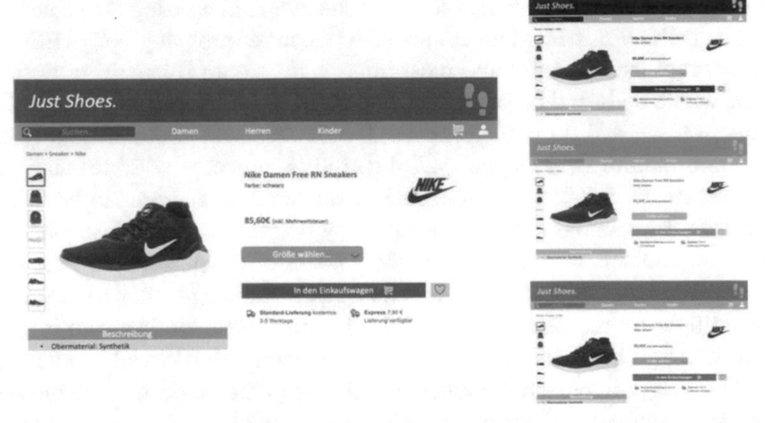

Fig. 1. Employed stimuli

distance 30 mm), and 8 short-distance detectors (average distance 8 mm). As commonly suggested by literature, short-distance measurements are crucial to filter out noise caused by extracerebral blood flow [57, 58] for which 8 mm provide an accurate distance [59, 60]. The wavelengths of the infrared light are 760 nm and 850 nm, the sampling frequency was 7.81 Hz which was resampled to 7 Hz in the pre-processing procedure. The raw fNIRS data was **processed** using the NIRS AnalyzIR toolbox [61]. At first, optical density was calculated, after which the data was bandpass filtered with a high cut-off frequency of 0.2 Hz and a low-cutoff frequency of 0.01 Hz which filters out artefacts due to heart rate and respiration [62, 63]. Further, short separation channel regression was applied as next pre-processing step by using the Linear Minimum Mean Square Estimations (LMMSE) [53, 64], which also heavily improves the hemodynamic response including when Mayer waves are present in the signal [65]. This is followed by calculating hemoglobin values by using the modified Beer-Lambert Law [66, 67]. Canonical hemodynamic response function (hrf) was used as baseline function for the general linear model (GLM). The questionnaire data was analyzed using an ANOVA with Tukey-HSD post-hoc tests. A threshold of $p < .05$ (Bonferroni corrected) was applied for the conducted group comparisons.

Results

The self-reported questionnaire results reflect what has been found in literature before. The ANOVA results were: $F_{Simplicity}(3, 188) = 3.045$ ($p < .03$), $F_{Diversity}(3, 188) = 3.916$ ($p < .01$), $F_{Colorfulness}(3, 188) = 3.730$ ($p < .012$), and $F_{Purchase}(3, 188) = 3.630$ ($p < .053$) which show statistical significance for all constructs except for purchase intention. For the other three constructs significant differences were found between the orange and the blue website with Bonferroni corrected p-values of $p = .018$ for simplicity, $p = .036$ for diversity, and most significantly $p = .006$ for colorfulness. As blue is in focus of this study, it provides the basis for the fNIRS group analysis in which it is compared to (1) an uncolored, black and white website, and (2) to both a green and an orange colored website. With this, the further focus on the data analysis and discussion are the prior named contrasts being (1) blue vs. non-colors and (2) blue versus other colors. For

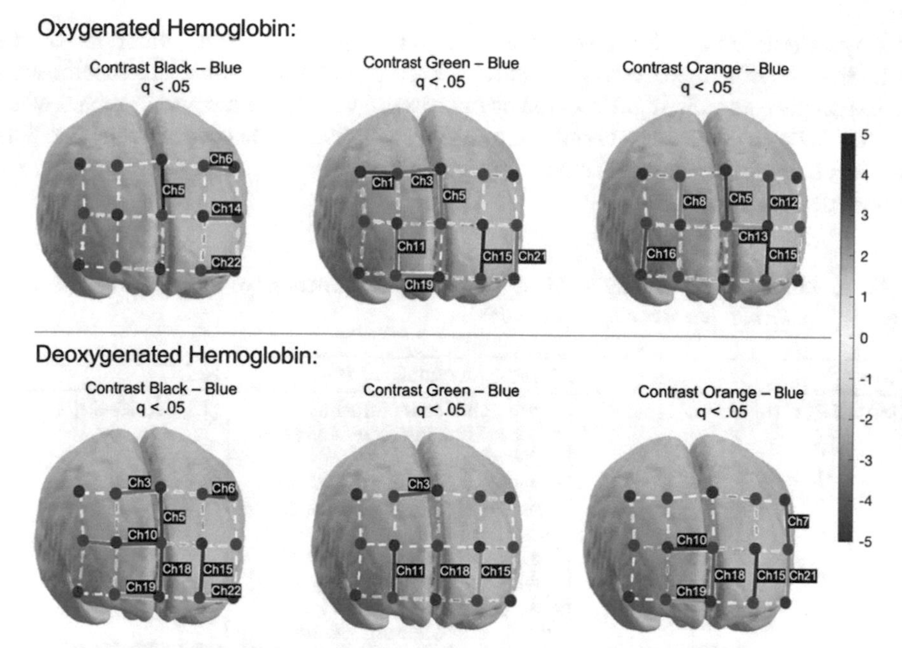

Fig. 2. **T-statistic contrasts between blue and other colored websites** (continuous and flagged lines represent significant channels).

each comparison, we applied paired t-tests with a false discovery rate (FDR) corrected threshold of $q < .05$ [68]. Figure 2 shows the t-statistic results of the contrasts between the blue and the other websites for HbO and HbR with the continuous lines representing significant channels (Ch) (with $q < .05$). When being compared to the uncolored website, decreased neural activity in the *left vlPFC* (Ch 22), and *right dmPFC* and *vmPFC* (Ch 3, 5, 10, 15, 18, 19) could be identified for the blue website. Further, this comparison also leads to significantly increased activity in the *left dlPFC* (Ch 6). When compared to both other colors (in this case orange and green), significant decreases on the blue website could be identified in the *dmPFC* (Ch 5) and the *left vmPFC* (Ch 15), as well as increases in the *right vmPFC* (Ch 11, 18, 19).

4 Discussion

In order to interpret our results and further analyze, whether the working hypotheses are supported by neural activity, we followingly present typical functions of the identified PFC regions which are related to decision making, together with the here found activity in these structures (Table 1). Firstly, and following our hypothesis H1, the left dlPFC seems to be related to processing pleasant stimuli which was identified increased for the blue website. Further, areas related to processing negative emotional stimuli such as the dmPFC and vlPFC were found to be decreased for the blue website which further supports our H1 which stated that blue websites will show significant differences in neural structures related to processing pleasant stimuli. Solely the vmPFC partly rejects this hypothesis, as it only showed increases on the right hemisphere, albeit not on the left. Given its role in processing pleasurable experiences, this hemisphere-specific activation seems contradicting. However, as increases in this region are also observed when confronted with the favorite brand of a product, activations in this region seem plausible for the blue website which was rated highest in the aesthetic scales. Consequently, H1 is regarded as supported.

Table 1. Identified activations in this study and typical functions of corresponding areas in the PFC (*italic points are used for discussion*)

Area	Typical functions	Ref.
(left) **dlPFC** (blue > non-color)	• performance monitoring • behavioral adjustment & error detection • reward evaluation & *linking sensory information to decision and action* • decreased when confronted with favorite brand • *left dlPFC increased for beautiful or pleasing stimuli*	[15, 42, 69–80]
dmPFC (blue < non-color & other-color)	• performance monitoring & adjusting behavior • forming and *processing first impressions* • *increased in attributing negative valence to stimuli* • emotion processing and *regulation* • activated in color discrimination and categorization	[69, 70, 73, 77, 81–85]
vmPFC (*left*: blue < non-color & other-color; *right*: blue > other color)	• increased for purchase intentions • *working memory* • necessary for rational decision making • *cognitive control of emotional processes* • *active when facing anticipation and uncertainty* • activated when facing one's first choice brand • increased for beautiful and pleasant stimuli	[15, 80, 86–94]
(left) **vlPFC** (blue < non-color)	• related to semantic processing and categorization processes • receives emotional and motivational information to process for decisions • controls hand and eye movements • *activated for facing negative, or unpleasant images*	[15, 83, 95–100]

Further, as already mentioned, the left vmPFC was decreased on the blue website compared to all three other websites. Therefore, in the given context this deactivation might reflect a lower impact of working memory on blue websites which would be in line with our second hypothesis H2. Cognitive deactivations were also observable in the dmPFC and vlPFC for the blue website, which both point to decreased effort to process the first impression, behavioral adjustments, and semantic processing – all of which can be regarded as more cognitive than emotional functions and thus, these findings support H2. However, the dlPFC which also incorporates cognitive processing through its role in performance monitoring, error detection, and behavior adjustment does not support this hypothesis. Given that this structure was only identified on the left hemisphere, and not bilaterally might support our assumption that this activation is more related to the emotional processing and not to the cognitive.

5 Conclusion

To conclude the findings of this work in progress, prior hypotheses which were derived from literature regarding the effects of the blue color on human perception and behavior were found to be supported in the neural activity in the prefrontal cortex. That is, blue websites seem to be characterized by a decrease in cognitive processing which becomes evident in areas of the vlPFC, vmPFC, and dmPFC. Further, blue websites also elicit activation patterns typically related to processing beautiful and thus, pleasant stimuli which further represents emotional processing and attribution of bluc websites when compared to other colors. These findings and conclusions further reveal the need to pursue **further research** in this direction in order to shed more light onto the role of the prefrontal cortex in website processing. Additionally, as mainly areas for decision making and processing emotional stimuli were activated, both processes seem to be tightly connected on ecommerce websites. Future research could therefore focus on how personality traits, gender-related color processing, or the impact of culture, as well as the use context, and website design influence these processes both on a behavioral and neural level and therefore, conduct between-group analyses. That is, the favorite color of a person might for instance influence their perception of the color use on a website, or specific meanings attached to color which are culture-dependent might impact its perception. Further, this study could also be reconducted with other methods such as eye tracking or EEG to better assess the visual pathways of users as well as get more timely neural information which may be crucial for decision making and color processing [101]. Although this paper provides a start into this research domain, it also comes with **limitations**. Among others, its major limitation is that participants only viewed screenshots of the colored websites and did not interact with them. The findings of this study thus are limited in their external validity and need to be further validated in real-life contexts where users actually interact with the websites – maybe even for longer time spans or repetitively. Further, the employed sample size included 75% males which might result in gender-related biased results, as men and women tend to processes information differently [102, 103]. Finally, the analysis and data interpretation solely focused on the blue website and did not focus on the other included colors and their unique neural patterns. Consequently, follow-up studies can further use these results and investigate different colors on websites and their impact on neural processing and consequently, on beauty perceptions and purchase intentions.

References

1. Heller, E.: Wie Farben auf Gefühl und Verstand wirken. Droemer, München (2000)
2. Granger, G.W.: Objectivity of colour preferences. Nature **170**, 778–780 (1952)
3. McManus, I.C., Jones, A.L., Cottrell, J.: The aesthetics of colour. Perception **10**, 651–666 (1981)
4. Herbert, P.: The Colors Used by the Ten Most Popular Sites. https://paulhebertdesigns.com/web_colors/
5. Bäumer, T., Leinberger, S., Beck, K., Kolb, F., Pfeifer, A.: Wahl ohne Qual – Wie Farben unsere Entscheidungen färben. Wirtschaftspsychologie **2019**, 108–118 (2019)
6. Bonnardel, N., Piolat, A., Le Bigot, L.: The impact of colour on Website appeal and users' cognitive processes. Displays **32**, 69–80 (2011)
7. Chang, W., Lin, H.: The impact of color traits on corporate branding. Afr. J. Bus. Manag. **4**, 3344–3355 (2010)
8. Becker, S.A.: An exploratory study on Web usability and the internationalization of US e-businesses. J. Electron. Commer. Res. **3**, 265–278 (2002)
9. Singh, S.: Impact of color on marketing. Manag. Decis. **44**, 783–789 (2006)
10. Cyr, D., Head, M., Larios, H.: Colour appeal in website design within and across cultures: a multi-method evaluation. Int. J. Hum. Comput. Stud. **68**, 1–21 (2010)
11. Seckler, M., Opwis, K., Tuch, A.N.: Linking objective design factors with subjective aesthetics: an experimental study on how structure and color of websites affect the facets of users' visual aesthetic perception. Comput. Human Behav. **49**, 375–389 (2015)
12. Fortmann-Roe, S.: Effects of hue, saturation, and brightness on color preference in social networks: gender-based color preference on the social networking site Twitter. Color Res. Appl. **38**, 196–202 (2013)
13. Palmer, S.E., Schloss, K.B.: An ecological valence theory of human color preference. Proc. Natl. Acad. Sci. U. S. A. **107**, 8877–8882 (2010)
14. Liu, X., Hong, K.S.: Detection of primary RGB colors projected on a screen using fNIRS. J. Innov. Opt. Health Sci. **10**, 1–11 (2017)
15. Zeki, S., Marini, L.: Three cortical stages of colour processing in the human brain. Brain **121**, 1669–1685 (1998)
16. Young, T.: The bakerian lecture: on the theory of light and colours. Philos. Trans. R. Soc. Lond. **92**, 12–48 (1802)
17. Grassmann, H.: Zur Theorie der Farbenmischung. Ann. der Phys. und Chemie. **165**, 69–84 (1853)
18. MacDonald, L.W.: Using color effectively in computer graphics. IEEE Comput. Graph. Appl. 20–35 (1999)
19. Hunt, R.W.G.: Measuring Colour. Fountain Press, UK (1998)
20. Newton, I.: Opticks (1704)
21. von Goethe, J.W.: Zur Farbenlehre. Tübingen (1810)
22. Chevreul, M.E., Martel, C.: The Principles of Harmony and Contrast of Colours, and Their Applications to the Arts. Longman, Brown, Green, and Longmans, Harlow (1855)
23. Munsell, A.H.: a pigment color system and notation. Am. J. Psychol. **23**, 236 (1912)
24. Gorn, G.J., Chattopadhyay, A., Sengupta, J., Tripathi, S.: Waiting for the web: how screen color affects time perception. J. Mark. Res. **41**, 215–225 (2004)

25. Elliot, A.J.: Color and psychological functioning: a review of theoretical and empirical work. Front. Psychol. **6**, 1–8 (2015)
26. Pridmore, R.W.: Chromatic induction: opponent color or complementary color process? Color Res. Appl. **33**, 77–81 (2008)
27. Patil, D.: Coloring consumer's psychology using different shades the role of perception of colors by consumers in consumer decision making process: a micro study of select departmental stores in Mumbai city, India. J. Bus. Retail Manag. Res. **7**, 60–74 (2012)
28. Pandir, M., Knight, J.: Homepage aesthetics: the search for preference factors and the challenges of subjectivity. Interact. Comput. **18**, 1351–1370 (2006)
29. Tuch, A.N., Bargas-Avila, J.A., Opwis, K., Wilhelm, F.H.: Visual complexity of websites: effects on users' experience, physiology, performance, and memory. Int. J. Hum. Comput. Stud. **67**, 703–715 (2009)
30. Zheng, X.S., Chakraborty, I., Lin, J.J.W., Rauschenberger, R.: Correlating low-level image statistics with users' rapid aesthetic and affective judgments of web pages. In: Conference on Human Factors in Computing Systems – Proceedings, pp. 1–10 (2009)
31. Abegaz, T., Dillon, E., Gilbert, J.E.: Exploring affective reaction during user interaction with colors and shapes. Procedia Manuf. **3**, 5253–5260 (2015)
32. Soldat, A.S., Sinclair, R.C., Mark, M.M.: Color as an environmental processing cue: external affective cues can directly affect processing strategy without affecting mood. Soc. Cogn. **15**, 55–71 (1997)
33. Engel, S., Zhang, X., Wandell, B.: Colour tuning in human visual cortex measured with functional magnetic resonance imaging. Nature **388**, 68–71 (1997)
34. Nathans, J., Thomas, D., Hogness, D.S.: Molecular genetics of human color vision: the genes encodiung blue, green, and red pigments. Science (80-.) **232**, 193–202 (1986)
35. Holzmann, D.C.: What's in a color? The unique human health effects of blue light. Environ. Health Perspect. **118**, A22–A27 (2010)
36. Mehta, R., Zhu, R.J.: Blue or red? Exploring the effect of color on cognitive task performances. Science (80-.) **323**, 1226–1229 (2009)
37. Anllo-Vento, L., Luck, S.J., Hillyard, S.A.: Spatio-temporal dynamics of attention to color: evidence from human electrophysiology. Hum. Brain Mapp. **6**, 216–238 (1998)
38. Valdez, P., Mehrabian, A.: Effects of color on emotions. J. Exp. Psychol. Gen. **123**, 394–409 (1994)
39. Labrecque, L.I., Milne, G.R.: Exciting red and competent blue: the importance of color in marketing. J. Acad. Mark. Sci. **40**, 711–727 (2012)
40. Moshagen, M., Thielsch, M.T.: Facets of visual aesthetics. Int. J. Hum. Comput. Stud. **68**, 689–709 (2010)
41. Moshagen, M., Thielsch, M.T.: VisAWI Manual (Visual Aesthetics of Websites Inventory) (2013)
42. Krampe, C., Gier, N., Kenning, P.: The application of mobile fNIRS in marketing research – detecting the 'first-choice-brand' effect. Front. Hum. Neurosci. **12**, 433 (2018)
43. Kim, H.Y., Seo, K., Jeon, H.J., Lee, U., Lee, H.: Application of functional near-infrared spectroscopy to the study of brain function in humans and animal models. Mol. Cells **40**, 523–532 (2017)
44. Pollmann, K., Vukelić, M., Birbaumer, N., Peissner, M., Bauer, W., Kim, S.: fNIRS as a method to capture the emotional user experience: a feasibility study. In: Kurosu, M. (ed.) HCI 2016, Part III. LCNS, vol. 9733, pp. 37–47. Springer, Cham (2016)
45. Hill, A.P., Bohil, C.J.: Applications of optical neuroimaging in usability research. Ergon. Des. **24**, 4–9 (2016)
46. Irani, F., Platek, S.M., Bunce, S., Ruocco, A.C., Chute, D.: Functional near infrared spectroscopy (fNIRS): an emerging neuroimaging technology with important applications for the study of brain disorders. Clin. Neuropsychol. **21**, 9–37 (2007)

47. Huppert, T.J., Hoge, R.D., Diamond, S.G., Franceschini, M.A., Boas, D.A.: A temporal comparison of BOLD, ASL, and NIRS hemodynamic responses to motor stimuli in adult humans. Neuroimage **29**, 368–382 (2006)
48. Strangman, G., Culver, J.P., Thompson, J.H., Boas, D.A.: A quantitative comparison of simultaneous BOLD fMRI and NIRS recordings during functional brain activation. Neuroimage **17**, 719–731 (2002)
49. Ferrari, M., Quaresima, V.: A brief review on the history of human functional near-infrared spectroscopy (fNIRS) development and fields of application. Neuroimage **63**, 921–935 (2012)
50. Funane, T., Atsumori, H., Katura, T., Obata, A.N., Sato, H., Tanikawa, Y., Okada, E., Kiguchi, M.: Quantitative evaluation of deep and shallow tissue layers' contribution to fNIRS signal using multi-distance optodes and independent component analysis. Neuroimage **85**, 150–165 (2014)
51. Brigadoi, S., Ceccherini, L., Cutini, S., Scarpa, F., Scatturin, P., Selb, J., Gagnon, L., Boas, D.A., Cooper, R.J.: Motion artifacts in functional near-infrared spectroscopy: a comparison of motion correction techniques applied to real cognitive data. Neuroimage **85**, 181–191 (2014)
52. Leff, D.R., Orihuela-Espina, F., Elwell, C.E., Athanasiou, T., Delpy, D.T., Darzi, A.W., Yang, G.Z.: Assessment of the cerebral cortex during motor task behaviours in adults: a systematic review of functional near infrared spectroscopy (fNIRS) studies. Neuroimage **54**, 2922–2936 (2011)
53. Scholkmann, F., Kleiser, S., Metz, A.J., Zimmermann, R., Mata Pavia, J., Wolf, U., Wolf, M.: A review on continuous wave functional near-infrared spectroscopy and imaging instrumentation and methodology. Neuroimage **85**, 6–27 (2014)
54. Gefen, D., Ayaz, H., Onaral, B.: Applying functional near infrared (fNIR) spectroscopy to enhance MIS research. AIS Trans. Hum. Comput. Interact. **6**, 55–73 (2014)
55. Cui, X., Baker, J.M., Liu, N., Reiss, A.L.: Sensitivity of fNIRS measurement to head motion: an applied use of smartphones in the lab. J. Neurosci. Methods **245**, 37–43 (2015)
56. Zhao, H., Cooper, R.J.: Review of recent progress toward a fiberless, whole-scalp diffuse optical tomography system. Neurophotonics **5**(1), 011012 (2017)
57. Tak, S., Ye, J.C.: Statistical analysis of fNIRS data: a comprehensive review. Neuroimage. **85**, 72–91 (2014)
58. Tachtsidis, I., Scholkmann, F.: False positives and false negatives in functional near-infrared spectroscopy: issues, challenges, and the way forward. Neurophotonics **3**, 039801 (2016)
59. Brigadoi, S., Cooper, R.J.: How short is short? Optimum source–detector distance for short-separation channels in functional near-infrared spectroscopy. Neurophotonics **2**, 1–9 (2015)
60. Goodwin, J.R., Gaudet, C.R., Berger, A.J.: Short-channel functional near-infrared spectroscopy regressions improve when source-detector separation is reduced. Neurophotonics **1**, 015002 (2014)
61. Santosa, H., Zhai, X., Fishburn, F., Huppert, T.: The NIRS Brain AnalyzIR toolbox. Algorithms **11**, 73 (2018)
62. Zhang, D., Zhou, Y., Hou, X., Cui, Y., Zhou, C.: Discrimination of emotional prosodies in human neonates: A pilot fNIRS study. Neurosci. Lett. **658**, 62–66 (2017)
63. Pinti, P., Scholkmann, F., Hamilton, A., Burgess, P., Tachtsidis, I.: Current status and issues regarding pre-processing of fNIRS neuroimaging data: an investigation of diverse signal filtering methods within a general linear model framework. Front. Hum. Neurosci. **12**, 1–21 (2019)
64. Saager, R.B., Berger, A.J.: Direct characterization and removal of interfering absorption trends in two-layer turbid media. J. Opt. Soc. Am. A. **22**, 1874 (2005)

65. Yücel, M.A., Selb, J., Aasted, C.M., Lin, P.Y., Borsook, D., Becerra, L., Boas, D.A.: Mayer waves reduce the accuracy of estimated hemodynamic response functions in functional near-infrared spectroscopy. Biomed. Opt. Express **7**, 3078 (2016)
66. Delpy, D.T., Cope, M., van der Zee, P., Arridge, S., Wray, S., Wyatt, J.: Estimation of optical pathlength through tissue from direct time of flight measurement. Phys. Med. Biol. **33**, 1433–1442 (1988)
67. Kocsis, L., Herman, P., Eke, A.: The modified Beer-Lambert law revisited. Phys. Med. Biol. **51**, N91–N98 (2006)
68. Benjamini, Y., Hochberg, Y.: Controlling the false discovery rate: a practical and powerful approach to multiple testing. J. R. Stat. Soc. Ser. B. **57**, 289–300 (1995)
69. Bird, C.M., Berens, S.C., Horner, A.J., Franklin, A.: Categorical encoding of color in the brain. Proc. Natl. Acad. Sci. U. S. A. **111**, 4590–4595 (2014)
70. Siok, W.T., Kay, P., Wang, W.S.Y., Chan, A.H.D., Chen, L., Luke, K.K., Tan, L.H.: Language regions of brain are operative in color perception. Proc. Natl. Acad. Sci. U. S. A. **106**, 8140–8145 (2009)
71. Cela-Conde, C.J., Marty, G., Maestú, F., Ortiz, T., Munar, E., Fernández, A., Roca, M., Rosselló, J., Quesney, F.: Activation of the prefrontal cortex in the human visual aesthetic perception. Proc. Natl. Acad. Sci. U. S. A. **101**, 6321–6325 (2004)
72. Wang, M.Y., Lu, F.M., Hu, Z., Zhang, J., Yuan, Z.: Optical mapping of prefrontal brain connectivity and activation during emotion anticipation. Behav. Brain Res. **350**, 122–128 (2018)
73. Taren, A.A., Venkatraman, V., Huettel, S.A.: A parallel functional topography between medial and lateral prefrontal cortex: evidence and implications for cognitive control. J. Neurosci. **31**, 5026–5031 (2011)
74. Hutcherson, C.A., Plassmann, H., Gross, J.J., Rangel, A.: Cognitive regulation during decision making shifts behavioral control between ventromedial and dorsolateral prefrontal value systems. J. Neurosci. **32**, 13543–13554 (2012)
75. Chen, M.Y., Jimura, K., White, C.N., Todd Maddox, W., Poldrack, R.A.: Multiple brain networks contribute to the acquisition of bias in perceptual decision-making. Front. Neurosci. **9**, 1–13 (2015)
76. Greening, S.G., Finger, E.C., Mitchell, D.G.V.: Parsing decision making processes in prefrontal cortex: response inhibition, overcoming learned avoidance, and reversal learning. Neuroimage **54**, 1432–1441 (2011)
77. Mitchell, D.G.V., Luo, Q., Avny, S.B., Kasprzycki, T., Gupta, K., Chen, G., Finger, E.C., Blair, R.J.R.: Adapting to dynamic stimulus-response values: differential contributions of inferior frontal, dorsomedial, and dorsolateral regions of prefrontal cortex to decision making. J. Neurosci. **29**, 10827–10834 (2009)
78. Heekeren, H.R., Marrett, S., Ruff, D.A., Bandettini, P.A., Ungerleider, L.G.: Involvement of human left dorsolateral prefrontal cortex in perceptual decision making is independent of response modality. Proc. Natl. Acad. Sci. U. S. A. **103**, 10023–10028 (2006)
79. Sanfey, A.G., Rilling, J.K., Aronson, J.A., Nystrom, L.E., Cohen, J.D.: The neural basis of economic decision-making in the Ultimatum Game. Science (80-.) **300**, 1755–1758 (2003)
80. Deppe, M., Schwindt, W., Kugel, H., Plaßmann, H., Kenning, P.: Nonlinear responses within the medial prefrontal cortex reveal when specific implicit information influences economic decision making. J. Neuroimaging **15**, 171–182 (2005)
81. Gilron, R., Gutchess, A.H.: Remembering first impressions: Effects of intentionality and diagnosticity on subsequent memory. Cogn. Affect. Behav. Neurosci. **12**, 85–98 (2012)
82. Dolcos, F., Iordan, A.D., Dolcos, S.: Neural correlates of emotion - cognition interactions: a review of evidence from brain imaging investigations. J. Cogn. Psychol. **23**, 669–694 (2011)

83. Ellard, K.K., Barlow, D.H., Whitfield-Gabrieli, S., Gabrieli, J.D.E., Deckersbach, T.: Neural correlates of emotion acceptance vs worry or suppression in generalized anxiety disorder. Soc. Cogn. Affect. Neurosci. **12**, 1009–1021 (2017)
84. Britton, J.C., Phan, K.L., Taylor, S.F., Welsh, R.C., Berridge, K.C., Liberzon, I.: Neural correlates of social and nonsocial emotions: An fMRI study. Neuroimage **31**, 397–409 (2006)
85. Etkin, A., Egner, T., Kalisch, R.: Emotional processing in anterior cingulate and medial prefrontal cortex. Trends Cogn. Sci. **15**, 85–93 (2011)
86. Cela-Conde, C.J., Garcia-Prieto, J., Ramasco, J.J., Mirasso, C.R., Bajo, R., Munar, E., Flexas, A., del-Pozo, F., Maestu, F.: Dynamics of brain networks in the aesthetic appreciation. Proc. Natl. Acad. Sci. **110**, 10454–10461 (2013)
87. Koenigs, M., Tranel, D.: Irrational economic decision-making after ventromedial prefrontal damage: evidence from the ultimatum game. J. Neurosci. **27**, 951–956 (2007)
88. Naqvi, N., Shiv, B., Bechara, A.: The role of emotion in decision making: a cognitive neuroscience perspective. Curr. Dir. Psychol. Sci. **15**, 260–264 (2006)
89. Delli Pizzi, S., Chiacchiaretta, P., Mantini, D., Bubbico, G., Ferretti, A., Edden, R.A., Di Giulio, C., Onofrj, M., Bonanni, L.: Functional and neurochemical interactions within the amygdala–medial prefrontal cortex circuit and their relevance to emotional processing. Brain Struct. Funct. **222**, 1267–1279 (2017)
90. Buhle, J.T., Silvers, J.A., Wage, T.D., Lopez, R., Onyemekwu, C., Kober, H., Webe, J., Ochsner, K.N.: Cognitive reappraisal of emotion: a meta-analysis of human neuroimaging studies. Cereb. Cortex. **24**, 2981–2990 (2014)
91. Motzkin, J.C., Philippi, C.L., Wolf, R.C., Baskaya, M.K., Koenigs, M.: Ventromedial prefrontal cortex lesions alter neural and physiological correlates of anticipation. J. Neurosci. **34**, 10430–10437 (2014)
92. Doi, H., Nishitani, S., Shinohara, K.: NIRS as a tool for assaying emotional function in the prefrontal cortex. Front. Hum. Neurosci. **7**, 1–6 (2013)
93. Plassmann, H., O'Doherty, J., Rangel, A.: Orbitofrontal cortex encodes willingness to pay in everyday economic transactions. J. Neurosci. **27**, 9984–9988 (2007)
94. Brown, S., Gao, X., Tisdelle, L., Eickhoff, S.B., Liotti, M.: Naturalizing aesthetics: brain areas for aesthetic appraisal across sensory modalities. Neuroimage **58**, 250–258 (2011)
95. Snyder, H.R., Banich, M.T., Munakata, Y.: Choosing our words: retrieval and selection processes recruit shared neural substrates in left ventrolateral prefrontal cortex. J. Cogn. Neurosci. **23**, 3470–3482 (2011)
96. Sakagami, M., Pan, X.: Functional role of the ventrolateral prefrontal cortex in decision making. Curr. Opin. Neurobiol. **17**, 228–233 (2007)
97. Wager, T.D., Davidson, M.L., Hughes, B.L., Lindquist, M.A., Ochsner, K.N.: Prefrontal-subcortical pathways mediating successful emotion regulation. Neuron **59**, 1037–1050 (2008)
98. Leung, H.C., Cai, W.: Common and differential ventrolateral prefrontal activity during inhibition of hand and eye movements. J. Neurosci. **27**, 9893–9900 (2007)
99. Heinen, S.J., Rowland, J., Lee, B.T., Wade, A.R.: An oculomotor decision process revealed by functional magnetic resonance imaging. J. Neurosci. **26**, 13515–13522 (2006)
100. Hoshi, Y., Huang, J., Kohri, S., Iguchi, Y., Naya, M., Okamoto, T., Ono, S.: Recognition of human emotions from cerebral blood flow changes in the frontal region: a study with event-related near-infrared spectroscopy. J. Neuroimaging **21**, 94–101 (2011)
101. Müller-Putz, G.R., Riedl, R., Wriessnegger, S.C.: Electroencephalography (EEG) as a research tool in the information systems discipline: foundations, measurement, and applications. Commun. Assoc. Inf. Syst. **37**, 911–948 (2015)

102. Darley, W.K., Smith, R.E.: Gender differences in information processing strategies: an empirical test of the selectivity model in advertising response. J. Advert. **24**, 41–56 (1995)
103. Putrevu, S.: Exploring the origins and information processing differences between men and women: implications for advertisers. Acad. Mark. Sci. Rev. **10**, 1–16 (2003)

Identifying Linguistic Cues of Fake News Associated with Cognitive and Affective Processing: Evidence from NeuroIS

Bernhard Lutz[1](✉), Marc T. P. Adam[2], Stefan Feuerriegel[3], Nicolas Pröllochs[4], and Dirk Neumann[1]

[1] University of Freiburg, Freiburg, Germany
{bernhard.lutz,dirk.neumann}@is.uni-freiburg.de
[2] University of Newcastle, Newcastle, Australia
marc.adam@newcastle.edu.au
[3] ETH Zurich, Zurich, Switzerland
sfeuerriegel@ethz.ch
[4] University of Giessen, Giessen, Germany
nicolas.proellochs@wi.jlug.de

Abstract. False information such as "fake news" is widely believed to influence the opinions of individuals. So far, information systems (IS) literature is lacking a theoretical understanding of how users react and respond to fake news. In this study, we analyze drivers of cognitive and affective processing in terms of linguistic cues. For this purpose, we performed a NeuroIS experiment that involved N = 42 subjects with both eye tracking and heart rate measurements. We find that users spend more cognitive effort (more eye fixations) in assessing the veracity of fake news when it is characterized by better readability and less affective words. In addition, we find that fake news is more likely to trigger affective responses (lower heart rate variability) when it is characterized by a higher degree of analytic writing. Our findings contribute to IS theory by disentangling linguistic cues that help to explain how fake news is processed. The insights can aid researchers and practitioners in designing IS to better counter fake news.

Keywords: Fake news · Linguistic cues · Information processing · Eye tracking · ECG

1 Introduction

The term "fake news" refers to fabricated news articles that were created with the intention of manipulating public opinion [1]. Although news has always been questioned in its veracity [2], the rise of social media allows any user to create and disseminate fake content with little to no barriers [3]. Fake news receives more attention than real news and reaches a wider range of users in very little time [4]. In fact, it was estimated that the average US citizen consumed and remembered between one and three fake news

F. D. Davis et al. (Eds.): NeuroIS 2020, LNISO 43, pp. 16–23, 2020.
https://doi.org/10.1007/978-3-030-60073-0_2

articles prior to the 2016 US presidential election, which may have had an influence on the result [1]. Understanding the phenomenon of fake news thus provides a considerable challenge for research and practice.

Earlier IS research studied the processing of fake news in regard to prior exposure [5], cognitive reasoning [6], and affective processing [7]. In addition, researchers evaluated different approaches to making users think more critically about the content. This includes warning messages [3, 8] and alternative presentation formats [9], as well as user and source ratings [10]. However, it remains unclear what makes users think and respond to fake news, specifically in regard to linguistic cues embedded within the news body.

IS literature has extensively studied linguistic cues in the context of deception detection (e.g., [11–13]). The rationale is that textual content lacks facial expressions or gestures which could be evaluated to determine if the author is lying. Deceivers are, for instance, more likely to write their messages with more positive and negative emotive words [12] and with lower readability scores (i.e., more readable) [11]. Yet, these studies solely focused on the perspective of the deceivers. That is, they specifically analyzed linguistic cues that are used by deceivers, but without considering how these cues are processed by the recipients of deceptive content such as fake news.

In this paper, we address the question of how linguistic cues are associated with cognitive and affective processing. For this purpose, we conducted a NeuroIS experiment [14, 15] involving 42 subjects that were presented 40 news articles of different veracity (real or fake). During the experiment, we measured subjects' eye fixations and heart rate variability (HRV) to provide insight into their cognitive and affective processing. For each article, we calculated the linguistic cues of (i) readability, the fractions of (ii) positive and (iii) negative emotive words, and (iv) analytic writing using LIWC 2015 [16]. These cues are subsequently used to explain users' cognitive (eye fixations) and affective processing (HRV).

Our findings reveal that a higher fraction of positive and negative words and higher readability scores are associated with more cognitive effort (i.e., more eye fixations) in the processing of real news and with less cognitive effort in the processing of fake news. In addition, we find that a higher degree of analytic writing in fake news is associated with affective responses in terms of lower HRV. In general, users spend more cognitive effort when processing real news as compared to fake news. Similarly, users spend less cognitive when they perceive an article as fake after processing the headline.

To the best of our knowledge, we present the first NeuroIS study that investigates how linguistic cues of online news articles are associated with users' cognitive and affective processing. Our study thereby extends existing literature on the processing of fake news [3, 5, 6, 8] by identifying linguistic cues that distinguish fake news in terms of how much cognitive effort users spend in their veracity assessment. Based on the identified cues, IS practitioners can, for instance, rank news articles according to whether users will spend cognitive effort or whether users are likely to experience affective responses. This might be beneficial in allocating existing fact-checking resources to articles for which users require more cognitive effort or even experience affective responses.

2 Theoretical Background

Information processing theory as a model for human thinking and learning builds on the assumption that the human brain processes any given information [17]. However, information processing theory also suggests that humans continuously categorize and filter the given information [18]. This intermediate step becomes particularly relevant when considering the fact that human information processing involves cognition and affect [19]. One reason for filtering information is cognitive dissonance [20], an affective state of psychological discomfort that arises when humans are provided with two pieces of information that cannot both be true at the same time. To avoid cognitive dissonance, humans seek for information that is consistent with their beliefs [21, 22]. In addition, humans have a tendency to ignore information that challenges existing beliefs, which is also referred to as confirmation bias [23].

Previous research [3, 5, 6, 8–10] on the human processing of fake news is often based on the assumption that deliberate cognitive reasoning leads to correct categorization of news articles. To stimulate cognitive reasoning, these studies evaluated different approaches, including warning messages [3, 8] or user ratings [9]. This is motivated by the "classical reasoning account" which suggests that cognitive reasoning leads to correct judgments [24]. However, humans have limited cognitive capabilities [25] and tend to avoid cognitive effort [26].

Fake news can also be considered through the lens of deception detection. IS literature contains many studies that specifically analyzed linguistic cues that characterize deceptive messages (e.g., [11–13]). For instance, deceivers were shown to use a higher fraction of positive and negative emotive words [12], and more readable messages [11]. Although these studies only considered the perspective of deceivers, they still motivate our selection of linguistic cues.

3 Hypotheses Development

Previous studies identified multiple linguistic indicators of deceptive content. This includes the complexity, i.e., readability of the text [27]. A common measure of readability is given by the Gunning-Fog readability index [28], which is defined as a linear combination of the number of words per sentence and the fraction of words with three or more syllables. Lower readability scores indicate better readability (i.e., the text is easier to read), whereas higher readability scores indicate lower readability (i.e., the text is harder to read). There appear two ways in which readability has an effect on users' cognitive processing. Prior research [11] has shown that deceivers tend to communicate their messages in a more readable way as they are lacking knowledge or memory about specific events. Accordingly, one could expect that fake news is easier to recognize when it is characterized by less complex writing. As an alternative, users might spend less effort when fake news is hard to read as they generally avoid cognitive effort [26]. In order to disentangle these differential effects, we propose the following alternative hypotheses:

Hypothesis 1a (H1a). *Users spend less cognitive effort on assessing the veracity of fake news when the news body is characterized by lower readability scores.*

Hypothesis 1b (H1b). *Users spend more cognitive effort on assessing the veracity of fake news when the news body is characterized by lower readability scores.*

Another important aspect of news articles is the degree to which the author justifies their claims with reasonable proofs [29, 30]. This also referred to as "analytic writing." For instance, prestigious newspapers like the New York Times exhibit a high degree of analytic writing, whereas blog entries are written with a lower degree of analytic writing [16]. Accordingly, news that is characterized by a higher degree of analytic writing should be perceived as more professional and also more reliable. As a consequence, it is more difficult for users to identify unjustified claims or statements that are characteristic of fake news. Users should therefore require more cognitive effort to assess the veracity of fake news when it is characterized by a higher degree of analytic writing. We therefore propose:

Hypothesis 2 (H2). *Users spend more cognitive effort on assessing the veracity of fake news when the news body is characterized by a higher degree of analytic writing.*

Users can experience cognitive dissonance when assessing the veracity of a news article. That is, the article provides credible information that contradicts the user's existing beliefs. This seems more likely when fake news appears more reliable due to a higher degree of analytic writing. Fake news that is characterized by a higher degree of analytic writing should therefore be more likely to trigger cognitive dissonance, and hence, also affective responses [31]. We thus propose:

Hypothesis 3 (H3). *Users experience affective responses in categorizing fake news when the news body is characterized by a higher degree of analytic writing.*

4 Method

We performed a within-subject experiment, where each subject was shown the same dataset (20 real news, 20 fake news) in random order. Following previous research [1, 4, 32], we used the fact-checking website "Politifact" to create a dataset of real and fake news. On Politifact, trained journalists investigate political statements and news articles to assign a veracity label that ranges from "true" for real news to "pants on fire" for fake news [33].

In our experiment, each news article is presented in two steps and in a generic format using the software "Brownie" [34]. In the first step, only the headline is shown and, based on this, subjects state their initial belief about the veracity of the article (from (0) strongly real to (6) strongly fake). In the second step, subjects read the news body and categorize the article as either real or fake. Our experiment is fully incentivized, i.e., subjects earn €0.50 (approx. USD 0.60) for each correct news categorization. During the whole experiment, we employed eye tracking and heart rate measurements.

Based on the raw data from our measurement devices, we calculate the dependent variables as follows. Our eye tracking device measures the relative coordinates (x, y) for both eyes on the screen between 0 and 1. We calculate the number of eye fixations with the velocity-based algorithm suggested by Engbert and Kliegl [35]. To calculate HRV, we first perform QRS detection to identify the heartbeats in the ECG signal [36]. We then use the tool "cmetx" [37] to calculate HRV as standard deviation of normal-to-normal inter-beat intervals.

We estimate mixed-effects regression models with subjects' eye fixations and HRV as dependent variables. We control for the treatment ($0 =$ real, $1 =$ fake), the sequence number (0–39), the headline categorization (from 0 strongly real to 6 strongly fake), the fractions of positive and negative emotive words, and a subject-specific random intercept. The key explanatory variables in our model are readability, given by the Gunning-Fog readability index [28], and the degree of analytic writing as calculated by "LIWC 2015" [16]. In addition, we include the interaction terms of all linguistic cues with the treatment variable to distinguish the processing of real and fake news.

5 Results

We first test our hypotheses H1 and H2 that relate readability and analytic writing to cognitive processing. The coefficient of *Readability* $\times$ *Treatment* is negative and statistically significant ($\beta = -13.139$, $p < 0.001$), while the coefficient of *Readability* alone is positive and significant ($\beta = 7.178$, $p < 0.01$). That is, an increase of one standard deviation in the readability of a fake news article makes users perform 5.96 fewer eye fixations. Hence, we find support for H1b and reject H1a. Subsequently, we test H2. The coefficient of *Analytic writing*$\times$ *Treatment* is not significant at common statistical levels. Accordingly, H2 is rejected.

Our hypothesis H3 relates analytic writing to affective responses. The coefficient of *Analytic writing* $\times$ *Treatment* is negative and statistically significant ($\beta = -2.658$, $p < 0.01$). This means that in the processing of fake news, an increase of one standard deviation in the degree of analytic writing lowers users' HRV by 2.66 ms. Hence, H3 is supported.

Concerning our control variables, we find that the headline categorization is negatively associated with the subsequent cognitive processing of the news body. That is, if the headline categorization is one standard deviation more towards fake, users perform 4.40 fewer eye fixations. Similarly, we find that, when processing fake news, users perform 12.80 fewer eye fixations as compared to processing real news. In addition, we found evidence for habituation, such that users exhibit fewer eye fixations and higher HRV with higher sequence numbers.

6 Discussion

Our results show that the processing of fake news depends on the linguistic cues that characterize the textual content. We find that readability is negatively associated with cognitive processing, such that users spend more cognitive effort when the news body is characterized by a lower readability score. This indicates that complex writing makes it harder for users to discriminate between real and fake news. In addition, we find that a higher degree of analytic writing is associated with affective responses. This suggests that users are more likely to experience cognitive dissonance when the content of fake news is better justified.

Our study is limited by the selection of subjects as college students are not representative for the whole of society [38]. However, the task provided considerable challenges for our subjects in determining the veracity of the presented news articles. The overall accuracy in correctly categorizing the veracity amounts to a mere 73.46%. This reveals the difficulty – even for college students – in separating real from fake news and thus confirms the relevance of our research.

References

1. Allcott, H., Gentzkow, M.: Social media and fake news in the 2016 election. J. Econ. Perspect. **31**(2), 211–236 (2017)
2. Gaziano, C., McGrath, K.: Measuring the concept of credibility. Journal. Q. **63**(3), 451–462 (1986)
3. Moravec, P., Kim, A., Dennis, A., Minas, R.: Fake news on social media: People believe what they want to believe when it makes no sense at all. MIS Q. **43**(4), 1343–1360 (2019)
4. Vosoughi, S., Roy, D., Aral, S.: The spread of true and false news online. Science **359**(6380), 1146–1151 (2018)
5. Pennycook, G., Cannon, T.D., Rand, D.G.: Prior exposure increases perceived accuracy of fake news. J. Exp. Psychol. Gen. **147**(12), 1865–1880 (2018)
6. Pennycook, G., Rand, D.G.: Lazy, not biased: susceptibility to partisan fake news is better explained by lack of reasoning than by motivated reasoning. Cognition **188**, 39–50 (2018)
7. Lutz, B., Adam, M.T.P., Feuerriegel, S., Pröllochs, N., Neumann, D.: Affective information processing of fake news: Evidence from NeuroIS. In: Davis, F.D., Riedl, R., vom Brocke, J., (eds.) NeuroIS Retreat 2019, pp. 121–128. Springer, Heidelberg (2020)
8. Pennycook, G., Bear, A., Collins, E., Rand, D.G.: The implied truth effect: attaching warnings to a subset of fake news stories increases perceived accuracy of stories without warnings. Manag. Sci. (Forthcoming) (2020)
9. Kim, A., Moravec, P.L., Dennis, A.R.: Combating fake news on social media with source ratings: The effects of user and expert reputation ratings. J. Manag. Inf. Syst. **36**(3), 931–968 (2019)
10. Kim, A., Dennis, A.: Says who? The effects of presentation format and source rating on fake news in social media. MIS Q. **43**(3), 1025–1039 (2019)
11. Zhou, L., Burgoon, J.K., Nunamaker, J.F., Twitchell, D.: Automating linguistics-based cues for detecting deception in text-based asynchronous computer-mediated communications. Group Decis. Negot. **13**(1), 81–106 (2004)
12. Ho, S.M., Hancock, J.T., Booth, C., Liu, X.: Computer-mediated deception: Strategies revealed by language-action cues in spontaneous communication. J. Manag. Inf. Syst. **33**(2), 393–420 (2016)
13. Siering, M., Koch, J.A., Deokar, A.V.: Detecting fraudulent behavior on crowdfunding platforms: The role of linguistic and content-based cues in static and dynamic contexts. J. Manag. Inf. Syst. **33**(2), 421–455 (2016)
14. Dimoka, A., Davis, F.D., Gupta, A., Pavlou, P.A., Banker, R.D., Dennis, A.R., Ischebeck, A., Müller-Putz, G., Benbasat, I., Gefen, D., et al.: On the use of neurophysiological tools in IS research: Developing a research agenda for NeuroIS. MIS Q. **36**(3), 679–702 (2012)
15. vom Brocke, J., Hevner, A., Léger, P.M., Walla, P., Riedl, R.: Advancing a NeuroIS research agenda with four areas of societal contributions. Eur. J. Inf. Syst. (Forthcoming) (2020)

16. Pennebaker, J.W., Boyd, R.L., Jordan, K., Blackburn, K.: The Development and Psychometric Properties of LIWC2015. LIWC.net, Austin (2015)
17. Atkinson, R.C., Shiffrin, R.M.: Human memory: A proposed system and its control processes. Psychol. Learn. Motiv. **2**, 89–195 (1968)
18. Shiffrin, R.M., Schneider, W.: Controlled and automatic human information processing: Perceptual learning, automatic attending and a general theory. Psychol. Rev. **84**(2), 127–190 (1977)
19. Browne, G.J., Parsons, J.: More enduring questions in cognitive IS research. J. Assoc. Inf. Syst. **13**(12), 1000–1011 (2012)
20. Festinger, L.: A Theory of Cognitive Dissonance. Stanford University Press, Stanford (1957)
21. Cooper, J., Worchel, S.: Role of undesired consequences in arousing cognitive dissonance. J. Pers. Soc. Psychol. **16**(2), 199–206 (1970)
22. Jonas, E., Schulz-Hardt, S., Frey, D., Thelen, N.: Confirmation bias in sequential information search after preliminary decisions: An expansion of dissonance theoretical research on selective exposure to information. J. Pers. Soc. Psychol. **80**(4), 557–571 (2001)
23. Nickerson, R.S.: Confirmation bias: A ubiquitous phenomenon in many guises. Rev. Gen. Psychol. **2**(2), 175–220 (1998)
24. Kohlberg, L.: Stage and Sequence: The Cognitive-Developmental Approach to Socialization. In: Goslin, D.A. (ed.) Handbook of Socialization Theory and Research, pp. 347–480. Rand McNally, Chicago (1969)
25. Miller, G.A.: The magical number seven, plus or minus two: Some limits on our capacity for processing information. Psychol. Rev. **63**(2), 81–97 (1956)
26. Simon, H.A.: Motivational and emotional controls of cognition. Psychol. Rev. **74**(1), 29–39 (1967)
27. Newman, M.L., Pennebaker, J.W., Berry, D.S., Richards, J.M.: Lying words: Predicting deception from linguistic styles. Pers. Soc. Psychol. Bull. **29**(5), 665–675 (2003)
28. Gunning, R.: The Technique of Clear Writing. McGraw-Hill, McGraw-Hill (1968)
29. Correnti, R., Matsumura, L.C., Hamilton, L.S., Wang, E.: Combining multiple measures of students' opportunities to develop analytic. Text Based Writ. Skills. Educ. Assess. **17**(2–3), 132–161 (2012)
30. Pennebaker, J.W., Chung, C.K., Frazee, J., Lavergne, G.M., Beaver, D.I.: When small words foretell academic success: The case of college admissions essays. PLoS ONE **9**(12), e115844 (2014)
31. Harmon-Jones, E.: Cognitive dissonance and experienced negative affect: Evidence that dissonance increases experienced negative affect even in the absence of aversive consequences. Pers. Soc. Psychol. Bull. **26**(12), 1490–1501 (2000)
32. Lazer, D.M.J., Baum, M.A., Benkler, Y., Berinsky, A.J., Greenhill, K.M., Menczer, F., Metzger, M.J., Nyhan, B., Pennycook, G., Rothschild, D., et al.: The science of fake news. Science **359**(6380), 1094–1096 (2018)
33. Graves, L.: Boundaries not drawn: mapping the institutional roots of the global fact-checking movement. Journal. Stud. **19**(5), 613–631 (2016)
34. Hariharan, A., Adam, M.T.P., Lux, E., Pfeiffer, J., Dorner, V., Müller, M.B., Weinhardt, C.: Brownie: a platform for conducting NeuroIS experiments. J. Assoc. Inf. Syst. **18**(4), 264–296 (2017)
35. Engbert, R., Kliegl, R.: Microsaccades uncover the orientation of covert attention. Vis. Res. **43**(9), 1035–1045 (2003)
36. Astor, P.J., Adam, M.T.P., Jerčić, P., Schaaff, K., Weinhardt, C.: Integrating biosignals into information systems: A NeuroIS tool for improving emotion regulation. J. Manag. Inf. Syst. **30**(3), 247–278 (2013)

37. Allen, J.J.B., Chambers, A.S., Towers, D.N.: The many metrics of cardiac chronotropy: A pragmatic primer and a brief comparison of metrics. Biol. Psychol. **74**(2), 243–262 (2007)
38. Koriat, A., Lichtenstein, S., Fischhoff, B.: Reasons for confidence. J. Exp. Psychol. Hum. Learn. Mem. **6**(2), 107–118 (1980)

Think Outside the Box: Small, Enclosed Spaces Alter Brain Activity as Measured with Electroencephalography (EEG)

Alessandra Natascha Flöck[1] and Peter Walla[1,2(✉)]

[1] CanBeLab, Psychology Department, Webster Vienna Private University, Praterstrasse 23, 1020 Vienna, Austria
neuroconsult.pw@gmail.com
[2] School of Psychology, Newcastle University, Callaghan, Newcastle, NSW, Australia

Abstract. This within-subjects study investigated the effect of small enclosed spaces on human brain activation during a simple word encoding task. A small and movable wooden box was designed, inside of which participants were exposed to visually presented words while asked to decide whether or not the first and last letters of each word were in alphabetical order. Simultaneously, brain activity was recorded via EEG. Respective encoding-related brain potentials were contrasted to an open space condition with the same task instruction. Data processing revealed that brain potentials were significantly more negative going at left lateral-frontal electrode locations when participants were inside the box compared to outside.

First, we interpret this finding to show an increase in frontal brain activity reflecting higher amygdala activation while inside the box, i.e. a bottom-up process. Given the enclosed nature of this condition, one may assume fear-related brain responses to occur, reflecting projections from the amygdaloid complex to the frontal cortex. A second interpretation is that the increased lateral-frontal activity while inside the box stems from frontal regulation of negative affective responses, i.e. a top-down process, associated with the enclosed space. A third alternative interpretation is that attentional processes mediated by the anterior cingulate cortex (ACC) are active on higher levels while inside the box versus outside.

Under the assumption that the event-related potential (ERP) differences are indeed fear-related, we want to suggest a new psychological construct that is subsequently introduced as implicit claustrophobia, i.e. a non-conscious fear of enclosed spaces.

Keywords: EEG · Small enclosed spaces · Word encoding · NeuroIS · Implicit claustrophobia · Non-conscious processing · ERP · Neuroimaging

1 Introduction

The human nature has evolved from enjoyment of open-spaced areas to the imposed necessity of spending the majority of one's time in rather enclosed spaces, such as

F. D. Davis et al. (Eds.): NeuroIS 2020, LNISO 43, pp. 24–30, 2020.
https://doi.org/10.1007/978-3-030-60073-0_3

office cubicles, car cockpits, etc. According to statistics, 2.2% of the population experience symptoms of severe claustrophobia, resulting in an approximated number of 169.4 million people worldwide suffering from discomfort and uneasiness when situated in places others may perceive as "normal" [1]. These indicators, however, are solely cases identified by the concerned individual itself, i.e. self-reported states of feeling given the presence of certain conditions, or differently termed, subjective experience of a conscious nature.

The advent of neuroimaging techniques has made possible the profound exploration of the human mind, without being delimited by the realm of consciousness, the measures obtained allowing for objective assessments of conditions. As demonstrated by Rugg et al. [2] such measurements potentially demonstrate differentiated results when comparing objectively gathered data to data gathered through self-report [3–7]. Subsequently, referring back to the statistics concerning claustrophobia, one may assume a huge dark number of cases to exist, due to the possibility of non-conscious effects, i.e. not subjected to conscious access, and therefore not reportable by the individual. The effects of this non-conscious form of claustrophobia, even if not causing any obvious distress or discomfort, may after all affect overall functioning of the human mind. If the subsequently proposed construct does present itself to exist, implications would hold within themselves a wide variety of influences, such as negatively affected performance in the workplace, decreased productivity, overall negatively affected well-being, etc.

The present study was designed to investigate if a small enclosed space alters brain activity during a simple word encoding task in comparison to an open-spaced condition. The enclosed space condition was created by implementing a wooden box, measuring 130 × 110 × 150 cm, respectively for length, width and height, while the open space condition was constituted by being situated in an approximately 30 m^2 room. The aforementioned encoding task was chosen for its standardized simplicity, consequently minimizing cognitive effort, allowing for the imposed space-related condition differences to be more dominantly processed. If a small enclosed space indeed alters brain activity, as according to our hypothesis, then we propose the introduction of a new psychological construct titled "*implicit claustrophobia*". Given the lack of research done on this topic, the novel construct could surly be deemed as rather speculative. Through implementing the methodology of Electroencephalography (EEG), falling into the category formerly deemed as "objective measures", it is assumed that if implicit claustrophobia exists, then corresponding changes should be detectable when comparing enclosed-space ERPs to open-space ERPs.

2 Materials and Methods

2.1 Participants

In sum, 19 subjects partook in the present study (mean age = 20.4 (SD = 1.5); 6 males). All of the participants were right handed and reported normal or corrected-to-normal vision capacities. All of the participants were Viennese residents and either enrolled in secondary or tertiary education programs. None of them reported a history of psychopathology and none were prescribed any psychotropic or otherwise potentially interfering medication. The participants reported sufficient knowledge of the English language to partake in the present study, which was approved by the ethics committee of Webster University in Saint Louis, Missouri, USA.

2.2 Stimuli

The stimuli consisted of five to seven letter, meaningful words that were presented for 1000 ms in white font against a black background on a Dell E2214hb 21.5" widescreen LED LCD monitor. The experimental paradigm was created with the E-Prime 2.0® software package. The experiment consisted of two conditions (open space/enclosed space) with respectively two phases (encoding/recognition). During the encoding phase 50 words were presented in a random sequence, which then reappeared in the immediately following recognition phase, in addition to 50 novel word stimuli. Further, participants were instructed to indicate the order of the initiating and ending letter, i.e. if the letters appeared in alphabetical order the left button was to be pressed, if they did not, the right button was to be pressed. The alphabetical encoding was implemented so as to maximize the effects of the enclosed space itself (a wooden box measuring 130 × 110 × 150 cm, respectively for length, width and height), rather than distracting the subject from this unusual target condition. For each trial in both conditions (target and control) the presentation sequence remained unaltered with the exception of the words shown. One trial consisted of a blank screen (1 s) followed by a white fixation cross (1 s), another blank screen (1 s), the stimulus (1 s) and a third blank screen with indefinite duration time, solely dependent on the response time of the partaker.

2.3 Data Collection

Partakers' task responses were collected using the PST Serial Response Box™ and recorded via the E-Prime 2.0® software. Electrical brain activation across all participants was acquired using the Geodesic EEG™ System 400 with a HydroGel Geodesic Sensor Net of 64 electrodes while potential changes were perpetually sampled at a rate of 1000 Hz (EGI; Electrical Geodesics, Inc.) The continuous EEG data was recorded by EGI Net Station 5.4 software.

2.4 Procedure

Participants were invited to come to the CanBe Lab (Cognitive & Affective Neuroscience and Behavior Lab) at the campus of Webster Vienna Private University. After having arrived, they were provided with information concerning the methodology of EEG, the study itself and behavioral adequacy to minimize movement artefacts. They were asked about their vision and psychopathological history as well as their proficiency regarding the English language. Subsequently, the participants were seated while the EEG sensor net was applied before initiating the experiment either in the open or enclosed space condition – depending on the counterbalanced sequence selected. The EEG net was applied over the whole scalp, the electrodes connected to the ground and referenced to the Cz point with impedance kept below 50 kΩ.

2.5 Data Analysis

EEG signal processing and extraction were carried out with the EEG DISPLAY 6.4.9 software (created by Ross Fulham). An offline bandpass filter from 0.1 to 30 Hz was applied before defining epochs from 100 ms before stimulus onset (used as baseline) to 1000 ms after stimulus onset. Respective epochs were inspected for obvious visual artefacts and if occurring selected and discarded. As a final step, the created ERP averages of each dataset were rereferenced to the common average across all inspected electrode sites [8]. Mean amplitude values of 60 ms ERP time windows were then taken to calculate potential differences between word processing ERPs inside versus outside.

3 Results

3.1 Electroencephalography (EEG)

Visual inspection of grand-averaged ERPs revealed a clear focus of differences at left lateral-frontal electrode sites, where brain potentials were found to be more negative going when participants were inside the box versus outside. Table 1 shows significant ERP differences between the two conditions across consecutive time windows for seven selected electrode locations over the left lateral-frontal brain area. As a result, electrode 7 shows most pronounced differences as can also be seen in Fig. 1 displaying actual ERPs including topographical maps. Visual inspection revealed no such differences over corresponding right lateral-frontal areas and indeed statistical analysis resulted in a lack of significant p-values for those sites.

Table 1. P-values as a result of t-tests comparing mean ERP amplitudes of 60 ms time windows inside versus outside for consecutive times after stimulus onset (top line shows time intervals in milliseconds). The left column shows for which electrodes respective results are listed. Cells highlighted in blue color show significant p-values. Note that no significant effects were found for right lateral-frontal electrode locations that indeed do not show any ERP differences between inside and outside the wooden box.

Ele/ms	48	108	168	228	288	348	408	468	528	588	648	708	768	828	888	948
4	.409	.312	.159	.623	.382	**.009**	.202	.343	.117	.241	**.001**	.378	.318	.103	.281	**.001**
6	.461	.502	.133	.473	.222	**.020**	.101	.247	.058	.086	.137	.173	.398	**.027**	**.041**	.084
7	.558	**.031**	**.006**	.208	.074	**.001**	**.007**	**.031**	**.015**	**.007**	**.042**	.131	.057	**.026**	.099	**.045**
8	.575	.866	.189	.305	.323	.147	.333	.602	.185	.415	.761	.637	.987	.251	.355	.468
15	.271	**.035**	.062	.283	.450	**.006**	.105	.203	.078	.058	.178	.198	.167	**.041**	.111	.072
53	.41	.98	.985	.086	.765	.831	.594	.976	.783	.924	.643	.445	.268	.824	.932	.861
54	.221	.671	.672	.072	.843	.406	.533	.888	.805	.988	.764	.649	.231	.939	.993	.991

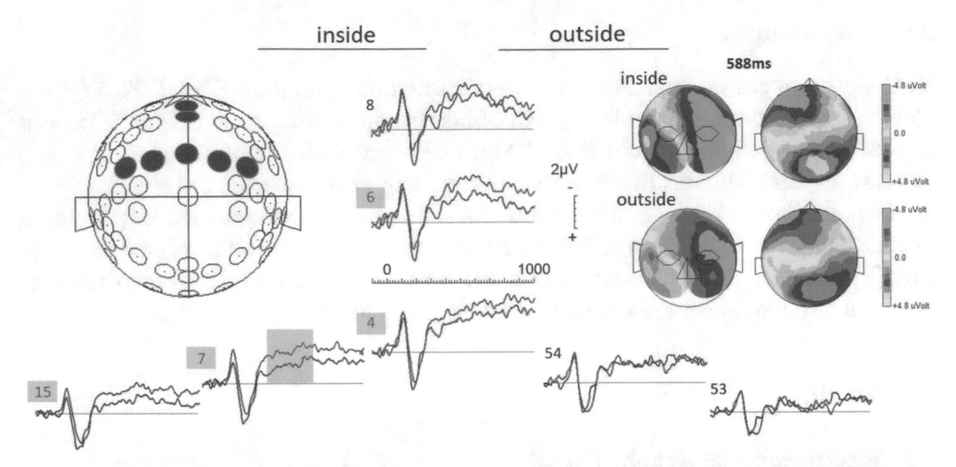

Fig. 1. Event-related potentials (ERPs) at the anterior frontal midline, left lateral-frontal and right lateral-frontal cortical areas. It is clearly visible that the midline and left lateral-frontal electrodes show enhanced negativity for the inside condition (red curve) compared to the outside condition. Blue boxes around electrode numbers mark locations with statistically significant effects. The highest significances were found at electrode number 7 (see time window marked in blue color; overlaid to ERPs). On the right top corner topogrphical maps are shown. The top two maps also demonstrate higher negativity for inside verus outside.

4 Discussion

The role of the frontal cortex in human behavior has been largely associated with monitoring, planning, organizing and other functions deemed to be unique human properties [9]. Damage to the aforementioned area has been proven to result in difficulties and general impairment of such activities [10]. The frontal cortex is assumed to receive input from structures embedded in the limbic system, which includes - among others - the amygdaloid complex.

The amygdaloid complex consists of several nuclei said to mediate responses to fear and threat-related stimuli, while both receiving and projecting information-laden signals from and to surrounding brain structures [11]. Damage to the amygdala has been shown to result in a far-reaching deactivation to fear eliciting stimulation, i.e. a decreased reactivity when confronted with dangerous situations. Therefore, the amygdala may well be considered an important constituent of the human's inherent vigilence system [12]. As an evolutionary evolved structure, the complex has been assumed to be especially responsive to stimuli of a highly ambiguous nature, which are still subjectively assumed to contain biological relevance for the organism. This assumption finds its roots in the nature of ambiguity, namely, a sense of uncertainty regarding both the origin and subsequent impact of the danger at hand.

Following the line of thought, the present study proposes a construct bearing a certain resemblance. It is assumed, that the enclosed space at hand – the wooden box – elicits similar activation. Therefore, the first conclusion drawn from the presented data is that the increased frontal activity, which was measured during the enclosed space condition, is a cortical representation of subcortical activation, stemming from the amygdala, hinting at the presence of a novel condition deemed as "implicit claustrophobia". This construct is proposed under the assumption that the human is an inherently open-spaced creature, the feeling of tightly enclosed spaces subsequently resulting in a nonconscious discomfort. However, as the methodology of EEG is widely known to have good temporal resolution, while lacking the same excellency when considering spatial properties, and is mainly sensitive to cortical neural activities (i.e. postsynaptic potentials of pyramidal cells in the cortex), processes evolving in subcortical structures can only be hypothesized rather than tested. However, due to strong connections between the amygdala and especially frontal cortical regions the proposed conclusion has been deemed as adequate.

As an alternative conclusion, we suggest one of the main properties of the frontal cortex, namely the process of monitoring to be the source of the observed effect. In the given context, this implicates frontal inhibition of amygdaloid complex activation. Ironside et al. [13] propose a connection between the prefrontal cortex and the amygdala in terms of a study reflecting that direct current stimulation of the former reduces threat-related activation of the latter, therefore illustrating a close link between the two structures. Within the realm of the proposed study, the enclosed space would substitute the original method of stimulation.

In summary, the two assumptions pose themselves as a dichotomy of thought, one reflecting a profoundly rooted bottom-up process, whereas the other makes obvious top-down influences. Firstly, the assumption is named that the measured frontal cortex negativity originates from the amygdaloid complex itself – the measurement at the site solely being a projection from the same. Taking into consideration the evolutionary imposed functions of the amygdala, one could consequently assume the enclosed space to elicit anxiety or discomfort. The second proposition, on the other hand, hints at a monitoring aspect stemming from the frontal cortex, controlling the drives arising from subcortical structures and subsequently reducing fear-related hyperactivity, resulting in ordinary cognitive functioning.

Proposed as an alternative hypothesis to the involvement of the amygdaloid complex, necessitated due to the spatial inaccuracy of EEG and the subsequent inability to locate the true origin of signals, is attention-related activation of the ACC. ACC neurons have been hypothesized to show increased activation when confronted with performance related tasks requiring attention [14]. Further, the ACC has been shown to direct attention to situationally relevant stimuli so as to limit the effects of distractors on cognitive processes [15]. This finding may be of high importance when reflecting on the results of the current study, consequently, the alterations in brain activity in the enclosed space condition may be the result of increased attention on the encoding paradigm the participants were presented with – the "relevant" task - while minimizing attention to the unusual setting of the wooden box – the "distraction".

References

1. Wardenaar, K.J., Lim, C., Al-Hamzawi, A.O., Alonso, J., Andrade, L.H., Benjet, C., Bunting, B., de Girolamo, G., Demyttenaere, K., Florescu, S.E., Gureje, O., Hisateru, T., Hu, C., Huang, Y., Karam, E., Kiejna, A., Lepine, J.P., Navarro-Mateu, F., Browne, M.O., Piazza, M., Posada-Villa, J., ten Have, M.L., Torres, Y., Xavier, M., Zarkov, Z., Kessler, R.C., Scott, K.M., de Jonge, P.: The cross-national epidemiology of specific phobia in the World Mental

Health Surveys. Psychol. Mèd. **47**(10), 1744–1760 (2017). https://doi.org/10.1017/S0033291717000174

2. Rugg, M.D., Mark, R.E., Walla, P., Schloerscheidt, A.M., Birch, C.S., Allan, K.: Dissociation of the neural correlates of implicit and explicit memory. Nature **392**(6676), 595–598 (1998). https://doi.org/10.1038/33396
3. Kunaharan, S., Halpin, S., Sitharthan, T., Bosshard, S., Walla, P.: Conscious and non-conscious measures of emotion: do they vary with frequency of pornography use? Appl. Sci. **7**, 493 (2017). 17p.
4. Walla, P., Koller, M., Brenner, G., Bosshard, S.: Evaluative conditioning of established brands: implicit measures reveal other effects than explicit measures. J. Neurosci. Psychol. Econ. **10**(1), 24–41 (2017)
5. Walla, P., Endl, W., Lindinger, G., Deecke, L., Lang, W.: Implicit memory within a word recognition task: an event-related potential study in human subjects. Neurosci. Lett. **269**(3), 129–132 (1999). https://doi.org/10.1016/S0304-3940(99)00430-9
6. Bosshard, S.S., Bourke, J.D., Kunaharan, S., Koller, M., Walla, P.: Established liked versus disliked brands: Brain activity, implicit associations and explicit responses. Cogent Psychol. **3**(1) (2016). https://doi.org/10.1080/23311908.2016.1176691
7. Geiser, M., Walla, P.: Objective measures of emotion during virtual walks through urban neighbour-hoods. Appl. Sci. **1**(1), 1–11 (2011). https://doi.org/10.3390/app1010001
8. Chang, M., Pavlevchev, S., Flöck, A.N., Walla, P.: The effect of body positions on word-recognition: a multi-methods NeuroIS study. In: Davis, F., Riedl, R., vom Brocke, J., Léger, P.M., Randolph, A., Fischer, T. (eds.) Information Systems and Neuroscience. Lecture Notes in Information Systems and Organisation, vol 32. Springer, Cham (2020). https://doi.org/10.1007/978-3-030-28144-1_36
9. Baumeister, R.F., Schmeichel, B.J., Vohs, K.D.: Self-regulation and the executive function: the self as controlling agent. In: Kruglanski, A., Higgins, E.T. (eds.) Social Psychology: Handbook of Basic Principles, 2nd edn., pp. 516–539. Guilford Press, New York (2007)
10. McCloskey, G., Perkins, L.: Essentials of Executive Functions Assessment. Wiley, Hoboken (2013)
11. Deussing, J.M., Wurst, W.: Amygdala and neocortex: common origins and shared mechanisms. Nat. Neurosci. **10**(9), 1081–1082 (2007). https://doi-org.library3.webster.edu/10.1038/nn0907-1081
12. Suslow, T., Ohrmann, P., Bauer, J., Rauch, A.V., Schwindt, W., Arolt, V., Heindel, W., Kugel, H.: Amygdala activation during masked presentation of emotional faces predicts conscious detection of threat-related faces. Brain Cogn. **61**, 243–248 (2006)
13. Ironside, M., Browning, M., Ansari, T.L., Harvey, C.J., Sekyi-Djan, M.N., Bishop, S.J., Harmer, C.J., O'Shea, J.: Effect of prefrontal cortex stimulation on regulation of amygdala response to threat in individuals with trait anxiety: a randomized clinical trial. JAMA Psychiatry **76**(1), 71–78 (2019). https://doi.org/10.1001/jamapsychiatry.2018.2172
14. Davis, K.D., Hutchinson, W.D., Lozano, A.M., Tasker, R.R., Dostrovsky, J.O.: Human anterior cingulate cortex neurons modulated by attention-demanding tasks. J. Neurophysiol. **83**(6), 3575–3577 (2000). https://doi.org/10.1152/jn.2000.83.6.3575
15. Weissman, D.H., Gopalakrishnan, A., Hazlett, C.J., Woldorff, M.G.: Dorsal anterior cingulate cortex resolves conflict from distracting stimuli by boosting attention toward relevant events. Cereb. Cortex **15**(2), 229–237 (2005). https://doi.org/10.1093/cercor/bhh125

Operationalization of Information Acquisition Switching Behavior in the Context of Idea Selection

Arnold Wibmer[1(✉)], Frederik Wiedmann[1], Isabella Seeber[1], and Ronald Maier[1,2]

[1] Department of Information Systems, Production and Logistics Management, University of Innsbruck, Innsbruck, Austria
{arnold.wibmer,frederik.wiedmann,isabella.seeber}@uibk.ac.at
[2] Vice Rectorate for Digitalisation and Knowledge Transfer, University of Vienna, Vienna, Austria
ronald.maier@univie.ac.at

Abstract. Decision makers ought to adapt their information acquisition (IA) contingent to the task, which has not yet been investigated in the context of idea selection. Therefore, this paper suggests an operationalization of IA switching behavior using eye-tracking data. A first data analysis indicates that raters switch between modes of high and low IA in an idea selection task. These modes of IA could be associated with compensatory and non-compensatory information integration. The extent of switches between IA modes seems to stay stable between the first and the second half of the task with a slight decreasing trend towards the end. Future research will add cognitive load to explain occurring switches between different IA modes and may allow to deduce recommendations for more efficient IT designs, preserving rater's cognitive resources.

Keywords: Clustering · Eye-tracking · Idea selection · Information acquisition

1 Introduction

Firms' open innovation initiatives have the potential to generate huge numbers of ideas, but can make it difficult for human raters to identify and select the most promising ideas [1]. Thus, raters require support when making decisions on the merit of ideas. IT tools can facilitate idea selection tasks [1], but their design requires a better understanding of the decision processes that raters' apply during idea selection [2]. Drawing on decision making research [3], raters go through information acquisition and evaluation processes before making a choice (action). Information acquisition (IA), defined as the process of information search and storage [3], determines the information that can be included for the ensuing evaluation of an idea. During evaluation, research in the context of idea selection has explored the role of formal criteria (e.g. explicit reasoning) and subjective criteria (e.g. gut feeling) [4], but has ignored how raters combine different pieces of information about ideas (idea attributes) in the IA phase. Idea attributes, such as the

F. D. Davis et al. (Eds.): NeuroIS 2020, LNISO 43, pp. 31–41, 2020.
https://doi.org/10.1007/978-3-030-60073-0_4

textual idea description [5] and machine- or community-generated feedback [6] have shown to be potential quality criteria. The integration of these idea attributes could either follow a compensatory [7, 8] or non-compensatory strategy [9, 10]. By applying a compensatory strategy, decision makers consider all relevant attributes of an idea (i.e. all available pieces of information about an idea) and allow trade-offs between values, which is a more effortful way of integrating information. In contrast, non-compensatory strategies allow ignoring idea attributes and excluding options from consideration, which is less effortful [11, 12]. Although the information integration strategy is difficult to observe directly [13], tracing the extent of IA allows us to make inferences on the applied information integration strategy since both are interdependent [3, 14]. Hence, measuring the IA of raters provides useful insights into how raters integrate idea attributes into their decisions at a later stage.

In addition, adaptive decision making theory [15] states that decision makers can switch strategies to adapt to the task environment and reduce effort, for example, if compensatory strategies would deplete their cognitive resources [16, 17]. When developing IT tools, we can make use of our understanding of this adaptive behavior [18] and design choice environments which influence raters' IA to support them in idea selection tasks. Moreover, eye-tracking methods enable to determine the IA during IT facilitated idea selection and to evaluate the design of the IT tool [19]. However, even though many studies explored IA based on single eye-tracking metrics for its direction, amount or depth [13, 20–22], we still lack a comprehensive measure for IA that considers different pieces of information in idea selection tasks. Consequently, such a measure needs to account for information like textual attributes (i.e. idea description), which require multiple fixations for full IA while reading, as well as short numeric attributes that could easily be acquired at a glance.

In the present paper, we suggest such an operationalization of IA in IT facilitated idea selection tasks. The proposed measure clusters constellations of IA as IA modes and shows the extent to which raters switched between IA modes. In addition, we applied our measure in a preliminary analysis to identify whether raters use different modes for IA and, if so, whether and how they switch between modes in the progress of the task.

RQ 1: "What modes of information acquisition exist in the context of idea selection?"
RQ 2: "To what extent and how do raters switch between modes of information acquisition during idea selection?"

2 Information Acquisition and Adaptive Decision Making

In the past decades, many studies have relied on eye-tracking to trace IA by observing the information search behavior of decision makers [23]. Insights about the IA of decision makers could then be used to infer the applied information integration strategy [24]. One way of examining IA is by its *direction.* Eye-tracking studies revealed that decision makers' scan paths during the search for information can follow either option-wise or attribute-wise patterns [12, 25]. By applying option-wise search, decision makers acquire information in a sequence, in which their fixations shift from one attribute of an option to another attribute of the same option. In contrast, decision makers follow an attribute-wise search if their fixations shift between the same attributes of different options. Option-wise search is typically linked to compensatory integration of the acquired information, while attribute-wise search is associated with a non-compensatory integration of attributes [11, 12]. However, knowing the direction of IA does not allow unambiguous interpretations of the applied information integration strategies because there is also evidence of a strategy that is characterized as attribute-wise and compensatory [13, 24]. Given that decision makers not always engage in an exhaustive search of all attributes [26], a more comprehensive view of IA needs to reflect the *amount of IA*. Constantly searching a high amount of information indicates compensatory information integration, while selective search of smaller amount is a sign of non-compensatory information integration [17, 27]. In the last decades, research has also looked at the *depth of IA* [28, 29]. Short single fixation durations (<150 ms) have been related to automatic and screening-oriented IA (non-compensatory), while long single fixation durations (>500 ms) seem to indicate a deliberate IA (compensatory) [21, 30]. Consequently, a comprehensive view of IA, which is missing so far, requires to include the three dimensions direction of IA, amount of IA and depth of IA. A measure of these dimensions could then allow inferences whether the IA reflects a compensatory or non-compensatory information integration strategy.

According to the theory of adaptive decision making, decision makers construct their strategies on the fly and adapt them contingent upon the task environment [15]. Hence, as the applied strategy changes, distinct modes of IA might be observed over the task progress. As an example, if task complexity is high due to more presented options and attributes, decision makers were found to engage in less IA. Conversely, in choice sets with low task complexity, decision makers acquire more information (i.e. higher amount of searched information) [13, 20, 31]. Indications for an adaption to the number of simultaneously presented options were already observed in the context of idea selection tasks [32]. A higher task complexity was also found to induce switching from option-wise to attribute-wise patterns [20, 31]. Eye-tracking also showed that these adaptations to incremental changes of the task complexity are quick [13]. Finally, an interesting finding regarding IA switching behavior comes from Shi et al. [33]. Although the task complexity was kept constant within the treatments, the study identified switching between option-wise and attribute-wise patterns. These switches might reflect changes between goals when making decisions. A cluster analysis based on a comprehensive measure of IA might provide more insights to the extent to which decision makers are variable in their IA even under constant task complexity.

3 Method

3.1 Experimental Procedure and Sample

We analyzed eye-tracking data from a prior laboratory experiment [32], which followed a two-factorial design by manipulating presentation mode and regulatory focus. In our analysis we are interested in information acquisition behavior independently form those manipulations. We included both treatments as control variables in our data analysis.

In the experiment, participants had to select the most promising ideas out of a pool of 32 ideas on gratitude at the workplace. With respect to presentation mode, we presented one group with 2 ideas on 16 screens, whereas the other group saw 4 ideas on 8 screens in a computer-aided randomized order. The regulatory focus priming was adapted from Chernev [34] and is intended to motivate participants to find the best ideas in the set (promotion focus) or to prevent bad ideas from being declared good (prevention focus). For more details about the priming procedure see [32].

Data was recorded using eye-tracking (Tobii Pro X3-120 eye-tracker with a sample rate of 120 Hz) and surveys. Prior to the task, an automatic 5-point calibration procedure was performed on each participant. The equipment ensured an error margin of 0.5 degree and was mounted on a 24-inch screen with a resolution of 1920 × 1080, presenting the stimuli. We defined non-overlapping Areas of Interest (AOIs), taking into account the error margin in the dimensioning, for the idea attribute "idea description" and each sub-attribute of "idea feedback", which are historical idea score (past success of the contributor), number of likes, creativity score (text mining score for creativity [35]), and tags. Participants were seated in a distance of 60–65 cm away from the screen and entered their responses using keyboard and mouse. In total, 63 graduate students took part in the experiment from May to July 2018 (31 participants) and from March to April 2019 (32 participants) at a European university.

Four cases had to be excluded from analysis due to measurement errors or inattentiveness of the participants to the task. Three of the four cases showed a shift of all fixations on every screen, caused by an inaccurate calibration. For the fourth, the insufficient processing time of the priming (below 1 min and therefore substantially less than the proposed 5 min) as well as the duration of the selection task (11.1 s per idea vs. a mean of 34.5 s per idea for the rest of the participants) were the reasons for exclusion. Finally, we ended up with in total 59 participants for analysis.

3.2 Eye-Tracking Measures

We operationalized IA behavior with its dimensions amount, depth and direction of search and built each of our measures on the well-established eye-tracking metrics fixation count, fixation duration [21] and transitions between AOIs [36]. We applied a minimum fixation duration of 60 ms to define fixations, meaning that we discarded all fixations with a fixation duration below this threshold from the analysis.

Amount of information acquisition refers to the quantity of task-relevant information searched. We measured the idea description separately from the idea feedback. The reason is that lengthy texts (idea descriptions were up to 130 words) require more fixations compared to idea feedback attributes that included icons (likes, creativity score) and tags (ranging from 5 to 9 words). Hence, we operationalized the quantity of IA on the idea description as *a) fixations description* measured as the mean fixation count on the AOI defined on the attribute idea description over all screens presented to each rater. The quantity of IA on idea feedback was operationalized as *b) feedback searched* measured as the number of idea feedback attributes fixated at least once on a screen divided by the total number of available idea feedback attributes per screen over all screens presented to each rater. For example, a value of 0.55 means that 55% of the feedback was visually searched and vice versa, 45% of the feedback was ignored on that screen.

Depth of information acquisition indicates the level of processing during IA and was measured using the average fixation duration of the single fixations. To avoid creating a biased measure by mixing text, which was found to have an average fixation duration of 225 to 250 ms [37], and non-text attributes, which were found to be linked with shorter (<150 ms) fixations [21], we distinguished between both. Consequently, we measured the a) *fixation duration description* by the average fixation duration in seconds on the AOI description and the b) *fixation duration feedback* by the average fixation duration in seconds on the AOIs of the idea feedback attributes. Each of these measures where then aggregated to the screens level for each rater.

Direction of information acquisition measures whether the rater followed an option-wise or attribute-wise search on the corresponding screen and is determined by the strategy index [27]. To calculate the strategy index for each screen, the number of attribute-wise transitions is subtracted from the number of option-wise transitions and is then divided by the total number of option-wise and attribute-wise transitions. A positive value indicates more option-wise search, whereas a negative value corresponds to more attribute-wise search [30].

3.3 Operationalization of Information Acquisition Switching Behavior

To determine the IA switching behavior, we propose following three consecutive steps as visualized in Fig. 1. The inputs for the operationalization are the measures fixations description, feedback searched, fixation duration description, fixation duration feedback, and strategy index.

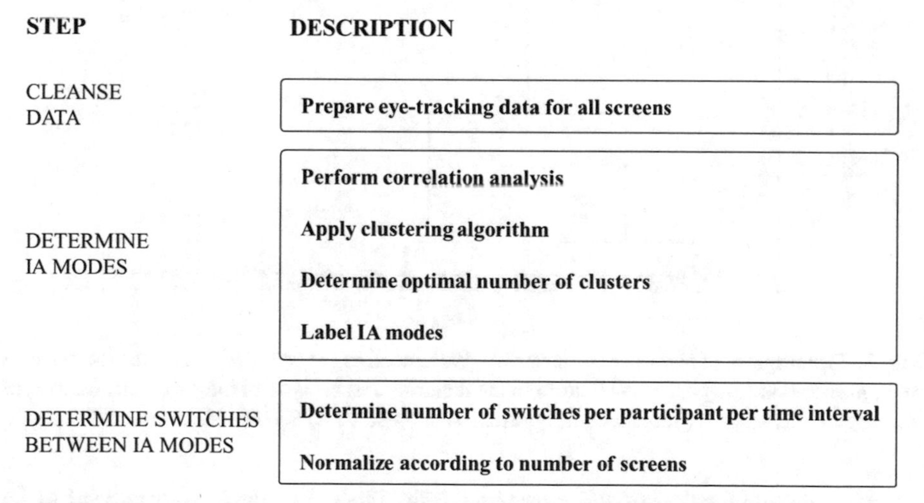

Fig. 1. Procedure for operationalization of information acquisition switching behavior (Source: own representation)

The purpose of the first step *cleanse data* is to prepare the data for clustering by dealing with missing as well as absent values. In our experiment, 2 out of 59 participants skipped one screen, so we excluded both screens from the set. In addition, absent values can occur for the mean fixation duration if participants do not fixate on any idea feedback on a screen. In our case, four participants had no fixations on idea feedback on 11 screens.

We decided to insert a 0 for the mean fixation duration for these absent values. Finally, we ended up with 702 screens as unit of analysis for in total 59 participants. Out of these 59 participants, 30 saw 8 screens (in total 238 screens) and 29 saw 16 screens (in total 464 screens).

In the second step, a cluster analysis is performed to *determine IA modes* on the measures of amount, depth, and direction of IA. Prior to the clustering analysis, it is necessary to examine the correlations between clusters to not overrepresent the same concepts and exclude variables with very high correlations above r = 0.8 from the clustering [38]. Thus, we tested correlations between the five variables of the measure of IA. Our cluster variables correlated between r = 0.114 and r = 0.558. Consequently, we included all variables in the clustering analysis. After the correlation analysis, we run the cluster algorithm. We applied a hierarchical clustering as the technique assists in the determination of how many clusters should be retained [39]. Therefore, the performed hierarchical clustering using Ward's criterion [40] for agglomerative cluster linkage is utilized to find clusters of IA modes in an exploratory way in our data [39]. The dendrogram in Fig. 2. visualizes the identified clusters at different levels of agglomeration.

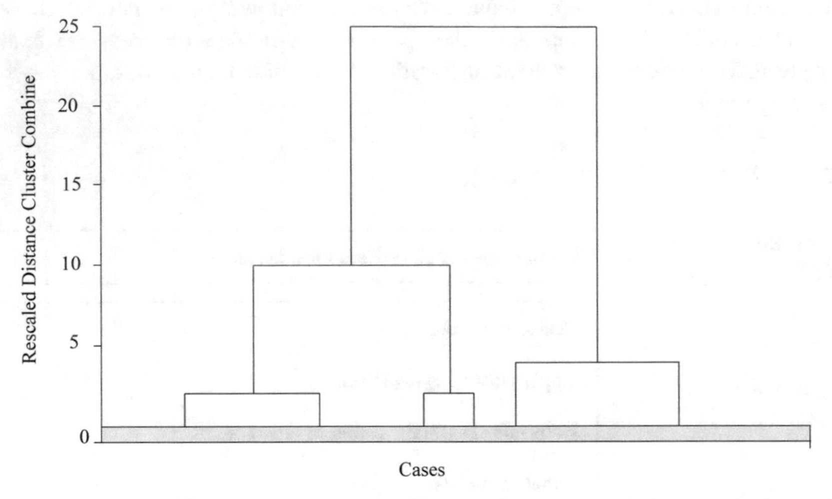

Fig. 2. Dendrogram of hierarchical clustering (We needed to grey out the cases at the lowest level of agglomeration (grey bar at the bottom in the dendrogram) because of the enormous number of branches at this stage.). (Source: own representation based on SPSS output)

After the application of the cluster analysis, these identified clusters need to be compared to derive the optimal number that best represents the structure of the data set. We determined the optimal number of clusters according to two criteria: the Calinski-Harabasz index [41] and the comparative length of the branches representing changes in the homogeneity measure between agglomeration steps [39] in the dendrogram. We compared the Calinski-Harabasz index between the longest (2 cluster solution) and the shorter branches (3 to 6 cluster solutions) of agglomeration in the dendrogram, where a higher index indicates a higher quality of the obtained clusters. However, as shown in Table 1, we could not derive the optimal number of clusters from the index as the criteria led to a monotonically increasing trend for a cluster solution from 2 to 7, converging to a solution where each cluster is one case. The index converges because no value exists for a solution with 702 clusters (it would mean that we divide by 0).

Table 1. Cluster quality determined with Calinski-Harabasz index.

Number of clusterS	Calinski-Harabasz index
2	957.95
3	1308.61
4	1421.85
5	1545.76
6	1920.597
7	2084.235
...	...
701	26779634

In contrast, a visual inspection of the dendrogram revealed two separate clusters as optimal number, since the longest branch between agglomeration levels is between four and two clusters. Due to the differentiation in the agglomeration of the branches in the dendrogram, we decided to continue the analysis with two clusters. Finally, the derived clusters are labeled as IA modes.

The purpose of the third step *determine switches between IA modes* is to study the IA behavior during the progress of the task based on participants' IA modes per screen. At this point, we changed the unit of analysis from screens to participants. We determined the number of switches for the first and second half of the task for each participant by counting how often a participant changed from one mode on a screen to a different mode on the next screen. This number needed to be normalized (divide number of switches by the maximum number of possible switches) as participants saw different numbers of screens (8 versus 16 screens in the two treatments on idea presentation mode). For example, the percentage switching value for a participant with 8 screens and 4 ideas per screen is calculated as follows for one half of the task: A maximum of 3 switches are possible on these 4 screens. Assuming the rater applied mode 1 on the first three screens and mode 2 on the last (fourth) screen in the first half, the resulting percentage switching value is 0.33 (1 actual switch/3 potential switches).

4 Preliminary Results

The preliminary application of our measure resulted in two distinct IA modes as presented in Table 2. Before interpreting the IA modes, we tested whether they were significantly different from each other on the variables subjected to clustering. As none of the variables followed a normal distribution, we employed a non-parametric test. A Mann-Whitney-U-test confirmed significant differences between the IA modes for all variables, that is for strategy index ($U = 48134.00$, $p < 0.05$), fixations description ($U = 0.000$, $p < 0.05$), feedback searched ($U = 42852.50$, $p < 0.05$), fixation duration description (U 43639.50, $p < 0.05$) as well as fixation duration feedback ($U = 51716.00$, $p < 0.05$).

Table 2. Descriptive statistics of information acquisition modes.

Dimension	Cluster variables	MODE 1: Less effortful IA (N = 318) Mean (SD)	MODE 2: More effortful IA (N = 384) Mean (SD)
Direction	Strategy index	0.2911 *(0.5015)*	0.4851 *(0.3459)*
Amount	Fixations description	81.13 *(17.60)*	137.36 *(28.15)*
	Feedback searched	0.5684 *(0.2889)*	0.7166 *(0.2386)*
Depth	Fixation duration description	0.2170 *(0.0320)*	0.2329 *(0.0286)*
	Fixation duration feedback	0.1616 *(0.0515)*	0.1740 *(0.0471)*

Mode 1 is characterized by less option-wise search (a lower strategy index), a smaller amount of acquired information (fewer fixations on the description and fewer searched feedback) and a shallower acquisition of information (shorter fixation durations on description and feedback). Therefore, we interpret this cluster as a rather non-compensatory information integration strategy, which is less effortful. Conversely, Mode 2 shows more option-wise search, a higher amount of acquired information and a deeper acquisition of idea attributes. We associate this IA mode to a more compensatory information integration strategy that is more effortful.

To understand whether and how the IA behavior changes over time, we tested the difference for percentage of switches in IA modes between the halves of the task progress in a repeated measures ANCOVA controlling for the presentation mode and regulatory focus treatments. In the results, the first half (mean = 0.2643; SD = 0.2657) and the second half (0.2373; SD = 0.2791) were not statistically different from each other ($F(1, 56) = 0.058$; $p = .811$; $\eta^2 = 0.001$). Neither the percent switches in the first half nor the percent switches in the second half followed a normal distribution, so we ran a Wilcoxon-test as robustness-check. Again, there were no significant differences between halves (mean rank first half = 21.33; mean rank second half = 17.47; $z = -.814$; $p = .415$). A manual inspection of the IA switching behavior provided additional insights. Forty-five out of 59 participants performed at least one switch between the identified IA modes for the whole task duration. In contrast, 14 participants showed no switches at all and thus remained stable in their IA. Another interesting fact is that one participant switched the IA mode in the first half of the task from screen to screen. This participant may have experienced difficulties in finding the appropriate IA mode for that kind of task.

5 Conclusion and Future Work

This paper contributes a measure to derive IA modes in the context of idea selection by clustering eye-tracking metrics of direction, amount and depth of IA. By calculating the number of switches for a certain task progress, the measure allows to examine the extent to which raters switch between different IA modes during the task. In an application of our method, we identified two distinct IA modes based on the eye-tracking data of a prior laboratory experiment. While the IA in the first IA mode is less option-wise and includes less information in less depth, the second identified IA mode is more option-wise and includes more information in more depth. Thus, the first IA mode indicates a more non-compensatory and the second IA mode a more compensatory information integration strategy.

Our preliminary findings also reveal the presence of switches between these IA modes since most raters changed their IA mode at least once during the task. Although our measure was built in the context of idea selection tasks, we assume that it might be useful for other decision tasks. In cases where it is not possible to derive an exact decision strategy [24] e.g. due to the presence of extensive textual attributes that require multiple fixations to acquire the full amount of information, our measure could be seen as a remedy to explore IA and make inferences on the information integration strategy.

Additionally, the measure provides a simple solution to study switches across the progress of decision tasks. Our present analysis shows that raters switched between IA modes even though task complexity was constant within treatments. Hence, we can't explain these switches as adaptation to the task environment. In our future research, we want to investigate different explanations for the occurring switching behavior, such as an accumulation of cognitive load. One explanation could be that raters deplete their cognitive resources while engaging in compensatory information integration and switch back to non-compensatory information integration to spare resources. Other potential explanations for the switches between IA modes could be changes in decision goals [33] or different levels of context complexity [42] through possible endogenous variations of similarity and conflicts between options on the screens. Further insights on the factors that evoke switches of IA behavior could then be used for the efficient design of IT facilitation. As an example, we could think of a responsive IT tool, which adapts the task environment (e.g. the number of presented options) to ensure optimal IA behavior during decision tasks by reacting to the identified factors.

References

1. Jensen, M.B., Hienerth, C., Lettl, C.: Forecasting the commercial attractiveness of user-generated designs using online data : an empirical Study within the LEGO user community. J. Prod. Innov. Manag. **31**, 75–93 (2014). https://doi.org/10.1111/jpim.12193
2. Beretta, M.: Idea selection in web-enabled ideation systems. J. Prod. Innov. Manag. **36**, 5–23 (2019). https://doi.org/10.1111/jpim.12439
3. Einhorn, H.J., Hogarth, R.M.: Behavioral decision theory: processes of judgment and choice. J. Account. Res. **19**, 1 (1981). https://doi.org/10.2307/2490959
4. Ferioli, M., Dekoninck, E., Culley, S., Roussel, B., Renaud, J.: Understanding the rapid evaluation of innovative ideas in the early stages of design. Int. J. Prod. Dev. **12**, 67 (2010). https://doi.org/10.1504/IJPD.2010.034313

5. Bullinger, A.C.B., Moeslein, K.: Innovation Contests – Where are we ? AMCIS 2010 Proceedings, pp. 1–9 (2010)
6. Hoornaert, S., Ballings, M., Malthouse, E.C., Van Den Poel, D.: Identifying new product ideas: waiting for the wisdom of the crowd or screening ideas in real time. J. Prod. Innov. Manag. **34**, 580–597 (2017). https://doi.org/10.1111/jpim.12396
7. Dawes, R.M.: The robust beauty of improper linear models in decision making. Am. Psychol. **34**, 571–582 (1979). https://doi.org/10.1037/0003-066X.34.7.571
8. Tversky, A.: Intransitivity of preferences. Psychol. Rev. **76**, 31–48 (1969). https://doi.org/10.1037/h0026750
9. Tversky, A.: Elimination by aspects: a theory of choice. Psychol. Rev. **79**, 281–299 (1972). https://doi.org/10.1037/h0032955
10. Gigerenzer, G., Goldstein, D.G.: Reasoning the fast and frugal way: models of bounded rationality. Psychol. Rev. **103**, 650–669 (1996). https://doi.org/10.1037/0033-295X.103.4.650
11. Patalano, A.L., Juhasz, B.J., Dicke, J.: The relationship between indecisiveness and eye movement patterns in a decision making informational search task. J. Behav. Decis. Mak. **23**, 353–368 (2010). https://doi.org/10.1002/bdm.661
12. Russo, J.E., Dosher, B.A.: Strategies for multiattribute binary choice. J. Exp. Psychol. Learn. Mem. Cogn. **9**, 676–696 (1983). https://doi.org/10.1037/0278-7393.9.4.676
13. Meißner, M., Oppewal, H., Huber, J.: Surprising adaptivity to set size changes in multi-attribute repeated choice tasks. J. Bus. Res. **111**, 163–175 (2020). https://doi.org/10.1016/j.jbusres.2019.01.008
14. Zuschke, N.: An analysis of process-tracing research on consumer decision-making. J. Bus. Res. **111**, 305–320 (2020). https://doi.org/10.1016/j.jbusres.2019.01.028
15. Payne, J.W., Bettman, J.R., Coupey, E., Johnson, E.J.: A constructive process view of decision making: multiple strategies in judgment and choice. Acta Psychol. (Amst) **80**, 107–141 (1992). https://doi.org/10.1016/0001-6918(92)90043-D
16. Johnson, E.J., Payne, J.W.: Effort and accuracy in choice. Manag. Sci. **31**, 395–414 (1985). https://doi.org/10.1287/mnsc.31.4.395
17. Payne, J.W., Bettman, J.R., Johnson, E.J.: Adaptive strategy selection in decision making. J. Exp. Psychol. Learn. Mem. Cogn. **14**, 534–552 (1988). https://doi.org/10.1037/0278-7393.14.3.534
18. Johnson, E.J., Shu, S.B., Dellaert, B.G.C., Fox, C., Goldstein, D.G., Häubl, G., Larrick, R.P., Payne, J.W., Peters, E., Schkade, D., Wansink, B., Weber, E.U.: Beyond nudges: tools of a choice architecture. Mark. Lett. **23**, 487–504 (2012). https://doi.org/10.1007/s11002-012-9186-1
19. Vom Brocke, J., Riedl, R., Léger, P.M.: Application strategies for neuroscience in information systems design science research. J. Comput. Inf. Syst. **53**, 1–13 (2013). https://doi.org/10.1080/08874417.2013.11645627
20. Lohse, G.L., Johnson, E.J.: A comparison of two process tracing methods for choice tasks. Organ. Behav. Hum. Decis. Process. **68**, 28–43 (1996). https://doi.org/10.1006/obhd.1996.0087
21. Horstmann, N., Ahlgrimm, A., Glöckner, A.: How distinct are intuition and deliberation? An eye-tracking analysis of instruction-induced decision modes. Judgm. Decis. Mak. **4**, 335–354 (2009). https://doi.org/10.2139/ssrn.1393729
22. Ryan, M., Krucien, N., Hermens, F.: The eyes have it: using eye tracking to inform information processing strategies in multi-attributes choices. Heal. Econ. (U. K.) **27**, 709–721 (2018). https://doi.org/10.1002/hec.3626
23. Glaholt, M.G., Reingold, E.M.: Eye movement monitoring as a process tracing methodology in decision making research. J. Neurosci. Psychol. Econ. **4**, 125–146 (2011). https://doi.org/10.1037/a0020692

24. Riedl, R., Brandstätter, E., Roithmayr, F.: Identifying decision strategies: a process- and outcome-based classification method. Behav. Res. Methods **40**, 795–807 (2008). https://doi.org/10.3758/BRM.40.3.795
25. Rosen, L.D., Rosenkoetter, P.: An eye fixation analysis of choice and judgment with multiattribute stimuli. Mem. Cogn. **4**, 747–752 (1976). https://doi.org/10.3758/BF03213243
26. Balcombe, K., Fraser, I., McSorley, E.: Visual attention and attribute attendance in multi-attribute choice experiments. J. Appl. Econ. **30**, 447–467 (2015). https://doi.org/10.1002/jae.2383
27. Payne, J.W.: Task complexity and contingent processing in decision making: an information search and protocol analysis. Organ. Behav. Hum. Perform. **16**, 366–387 (1976). https://doi.org/10.1016/0030-5073(76)90022-2
28. Glöckner, A., Herbold, A.-K.: An eye-tracking study on information processing in risky decisions: evidence for compensatory strategies based on automatic processes. J. Behav. Decis. Mak. **24**, 71–98 (2011). https://doi.org/10.1002/bdm.684
29. Velichkovsky, B.M., Rothert, A., Kopf, M., Dornhöfer, S.M., Joos, M.: Towards an express-diagnostics for level of processing and hazard perception. Transp. Res. Part F Traffic Psychol. Behav. **5**, 145–156 (2002). https://doi.org/10.1016/S1369-8478(02)00013-X
30. Ball, C.: A comparison of single-step and multiple-step transition analyses of multiattribute decision strategies. Organ. Behav. Hum. Decis. Process. **69**, 195–204 (1997). https://doi.org/10.1006/obhd.1997.2681
31. Payne, J.W., Braunstein, M.L.: Risky choice: an examination of information acquisition behavior. Mem. Cogn. **6**, 554–561 (1978). https://doi.org/10.3758/BF03198244
32. Wibmer, A., Wiedmann, F.M., Seeber, I., Maier, R.: Why less is more: an eye tracking study on idea presentation and attribute attendance in idea selection. In: 27th European Conference on Information Systems, pp. 1–14 (2019)
33. Shi, S.W., Wedel, M., Pieters, F.G.M.: (Rik): information acquisition during online decision making: a model-based exploration using eye-tracking data. Manag. Sci. **59**, 1009–1026 (2013). https://doi.org/10.1287/mnsc.1120.1625
34. Chernev, A.: Goal-attribute compatibility in consumer choice. J. Consum. Psychol. **14**, 141–150 (2004). https://doi.org/10.1207/s15327663jcp1401
35. Toubia, O., Netzer, O.: Idea generation, creativity, and prototypicality. Mark. Sci., 1–20 (2017). https://doi.org/10.1287/mksc.2016.0994
36. Blascheck, T., Kurzhals, K., Raschke, M., Burch, M., Weiskopf, D., Ertl, T.: State-of-the-art of visualization for eye tracking data. In: Borgo, R., Maciejewski, R., and Viola, I. (eds.) Proceedings of the Eurographics Conference on Visualization (EuroVis). The Eurographics Association (2014)
37. Rayner, K.: Eye movements and attention in reading, scene perception, and visual search. Q. J. Exp. Psychol. **62**, 1457–1506 (2009). https://doi.org/10.1080/17470210902816461
38. Sambandam, R.: Cluster analysis gets complicated. Mark. Res. **15**, 16–21 (2003)
39. Hair, J.F., Black, W.C., Babin, B.J., Anderson, R.E.: Multivariate Data Analysis. Pearson Education Limited, Harlow (2014). https://doi.org/10.1038/259433b0
40. Ward, J.H.: Hierarchical grouping to optimize an objective function. J. Am. Stat. Assoc. **58**, 236–244 (1963). https://doi.org/10.1080/01621459.1963.10500845
41. Caliński, T., Harabasz, J.: A dendrite method for cluster analysis. Commun. Stat. **3**, 1–27 (1974). https://doi.org/10.1080/03610927408827101
42. Pfeiffer, J., Duzevik, D., Rothlauf, F., Bonabeau, E., Yamamoto, K.: An optimized design of choice experiments: a new approach for studying decision behavior in choice task experiments. J. Behav. Decis. Mak. **28**, 262–280 (2015). https://doi.org/10.1002/bdm.1847

Relying on System 1 Thinking Leaves You Susceptible to the Peril of Misinformation

Spencer Early(✉), Seyedmohammadmahdi Mirhoseini, Nour El Shamy, and Khaled Hassanein

DeGroote School of Business, McMaster University, Hamilton, Canada
earlys@mcmaster.ca

Abstract. In the current era of unprecedented cultural and political tension, the growing problem of misinformation has exacerbated social unrest within the online space. Rectifying this issue requires a robust understanding of the underlying factors that lead social media users to believe and spread misinformation. We investigate a set of neurophysiological measures as they relate to users interacting with misinformation, delivered via social media. A rating task, requiring participants to assess the validity of news headlines, reveals a stark contrast between their performance when engaging analytical thinking processes versus automatic thinking processes. We utilize this observation to theorize intervention methods that encourage more analytical thinking processes.

Keywords: Misinformation · Neurophysiological measurement · EEG · Eye tracking

1 Introduction

The topic of fake news (a subset of the broader concepts of misinformation and disinformation henceforth referred to as misinformation) has become extremely prevalent in the realm of professional and social discourse. Mainstream media and popular social media platforms (e.g. Facebook, Twitter, etc.) have faced waves of criticism for their inability to effectively block the dissemination of misinformation [1]. This is partially the result of the problem's magnitude, whereby almost all social communication and media genres have been infected with various forms of misinformation. The political space was shook during the 2016 United States election where at least 25% of Americans opened a relevant misinformation webpage in the months leading to the election [1]. Additionally, we can see the cultural impact through the recent spread of extremely dangerous, and false, rhetoric regarding the dangers of vaccinations [2].

The technological aspects of misinformation research include (a) developing algorithmic approaches to distinguish misinformation and (b) designating information technology interventions to help users detect misinformation. In the past few years, the majority of research has been conducted on the first theme in which researchers utilize a combination of linguistic cues supplemented with machine learning network analysis,

F. D. Davis et al. (Eds.): NeuroIS 2020, LNISO 43, pp. 42–48, 2020.
https://doi.org/10.1007/978-3-030-60073-0_5

to analyze commonly spread online misinformation [3]. The literature under Theme (b) suggests that the current social media features such as fact-checkers and flagging systems are not fully effective in reducing users' belief in misinformation. One potential explanation is that these IT interventions are not addressing the real cause. For instance, flagging a piece of misinformation may not influence ideologically motivated users when it matches their pre-existing beliefs, a phenomenon referred to as confirmation bias [4].

People have a propensity to believe misinformation for numerous situational and dispositional factors, such as limited cognitive capacity and over-reliance on heuristics that makes individuals more susceptible to cognitive biases [5]. To effectively address the issue of misinformation we must first identify the key cognitive factors that lead to the belief in misinformation. This will enable the design of more effective interventions in order to reduce the incidence of social media users falling prey to misinformation.

The current study attempts to identify some of these key behavioral and cognitive factors that are correlated with a participant's likelihood to believe in misinformation. To this end, we utilize a set of physiological tools as potential predictors of participant performance in identifying misinformation. Building upon the dual-process theories of cognition, we use Electroencephalogram (EEG) to interpret System 1 (automatic, reflective, and effortless) and System 2 (deliberate, analytical, and effortful) thinking processes during decision making [6]. Further, eye-tracking pupillometry is used to assess fluctuations in pupil dilation, which indicates the intensity of users' cognitive load [7]. Identifying highly predictive factors amongst these measures will enable us to create a more robust experimental design, capable of conveying why people are susceptible to misinformation and how we can begin to rectify the underlying problem. Thus, the current study aims to answer one primary research question: *Is there any neurophysiological evidence to support the notion that System 1 thinking processes are associated with the belief in misinformation? And if so, how can we leverage this to reduce the belief and spread of misinformation?*

2 Literature Review

2.1 Dual Process Models of Cognition

The dual-process models point to the existence of two information processing modes in the human cognitive system that play a central role in evaluating arguments and forming impressions and beliefs [8]. The two modes have different processing principles and serve two distinct evolutionary purposes. The "associative processing mode" (i.e., System 1) relies on long-term memory to retrieve similarity-based information and form an impression about a stimulus [9]. This mode is a fast process that responds to an environmental cue in less than one second [10]. It searches for similar information to that cue in the long-term memory representation of one's experiences built up over the years [8]. In contrast, the "rule-based processing mode" (i.e., System 2) engages in an effortful and time-consuming process of searching for evidence and logic to make a judgment about a statement [8]. This processing mode is a slow and conscious process, which is under our deliberate control and has much fewer cognitive resources than the associative mode [11]. As cognitive misers, individuals are reluctant to engage in deliberative and effortful rule-based cognition [12].

The effect of cognitive biases on decision-making can be explained by the dual process model that recognizes the two contradictory impression formation modes: intuitive and analytical. Cognitive biases are more associated with intuitive and automatic processing of information in which users do not analytically evaluate an argument and trust their intuition or "gut-feeling". Intuitive processing is not only a main culprit of confirmation bias but also other biases that increase the likelihood that social media users will believe misinformation. (e.g., belief bias) [13].

Very few studies have been conducted to identify the neural correlates of System 1 and System 2. Pupillometry is considered as a measure of mental effort and cognitive states [7]. A study by Kahneman et al. [14] shows that utilizing system 2 resources is associated with an increase in pupil dilation. A recent EEG study was performed by Williams et al. [15] which replicated the Kahneman et al. experiment. They found that System 2 thinking is associated with frontal (Fz) theta rhythms (4–8 Hz) while System1 activities are correlated with increased parietal (Cpz) alpha rhysthms (10–12 Hz).

2.2 Research on Misinformation

Although research on misinformation existed for a long time, recently more researchers are showing interest in studying this phenomenon [16]. For example, in a recent study, participants were presented with articles containing misinformation, in the Facebook feed format and asked to indicate their accuracy. The results revealed via survey that most participants utilized personal judgement and familiarity with the articles source to distinguish misinformation [16]. A related study analyzed the effect of enabling co-annotations (i.e., the ability to make personal edits/additions) on social media news articles and how it impacts a user's interaction with misinformation [18]. The limited scale of the study demonstrated that the additional medium for discourse reduced the likelihood of perpetuating false information.

3 Methodology

An experiment was designed to study the cognitive mechanisms associated with belief in misinformation. Headlines were generated within two distinct categories. Control Headlines, in which the headline could be very easily identified as true or false (e.g., Ottawa is the capital of Canada), and polarizing headlines, in which the primary topic or figure mentioned is politically polarizing (e.g. Donald Trump, Barak Obama, Climate change, Abortion, etc.). A pilot study was designed to identify a list of politically polarizing terms. Fifty volunteers rated terms (e.g., Trump, Obama, climate change, planned parenthood, etc.) based on a 5-point scale, ranging from highly positive, to highly negative conveying personal viewpoints. The top 10 most divisive (strong view in either direction) terms were used to generate the politically polarizing headlines. The headlines were constructed based on popular controversial news articles published on social media and fact checked by Snopes website[1]. For each headline condition (Control/Polarizing), half the constructed headlines were false and the remaining half were

[1] https://www.snopes.com/fact-check/.

true. In the main experiment, the headlines were presented to the participants in a template, which was constructed to closely mimic the format used in many social media and news website platforms (Fig. 1). Participants were instructed to rate the accuracy of the presented headlines on a 4-point scale. Each participant evaluated 20 and 40 news headlines in the control and polarizing conditions, respectively.

Fig. 1. Headline presentation format, inlcuing 4-point rate scale.

Each participant session was audio and video recorded. We report the characteristics of the measurement tools according to the guidelines recommended by Müller-Putz, Riedl. Participants were fitted with the Cognionics (Cognionics Inc., CA) Quick-20 dry EEG headset with 20 electrodes located according to the 10–20 system, capable of sampling at 500 Hz. This device has a wireless amplifier with 24-bit AD resolution. An unobtrusive TOBII X2–60 (Tobii Technology AB) eye tracking module was attached to the participants' testing screen, capturing various metrics including gaze vectors, fixation points, and pupil dilation at 60 Hz. Thirteen participants, four female and nine male, participated in the preliminary study. Participant ages ranged between 18 and 55. Education level amongst participants ranged from a Bachelor's degree to Doctorate. Participant recruitment was conducted via email, as well as TV/newsletter advertisement. Full data collection is currently underway. Ethics approval was secured from the Ethics Research Board at the authors' university prior to any data collection.

3.1 Data Analysis

The EEG was preprocessed then fed into a decision tree classifier trained to predict the participant's belief in each presented headline (Fig. 2). The goal of applying a classifier to EEG data is to identify the EEG components associated with correct/wrong responses and investigate whether such components include the correlates of system1/system2. In the first step, the continuous EEG was epoched by segmenting the last 2 s of each trial. Then the preprocessing was performed by (1) applying an FIR filter (0.1 Hz–40 Hz) to remove the noise, (2) rejecting bad channels, (3) removing noisy epochs, (3) running independent component analysis (ICA) to identify artifacts such as eye blinks using

ADJUST plugin [19], and (4) rejecting bad components and reconstructing the EEG signal. We used Fast Fourier Transformation (FFT) to quantify the signal power based on four frequency ranges. Delta 0 to < 4 Hz, Theta 4 to < 8 Hz, Alpha 8 to 13 Hz, and Beta > 13 Hz. A total number of eighty features (20 Channels X four frequency powers) were fed into a decision tree classifier, which is a powerful binary classifier that maximizes the classification accuracy using information theory concepts such as entropy and mutual information [20]. 80% of the collected data was used for training and the remaining 20% for testing.

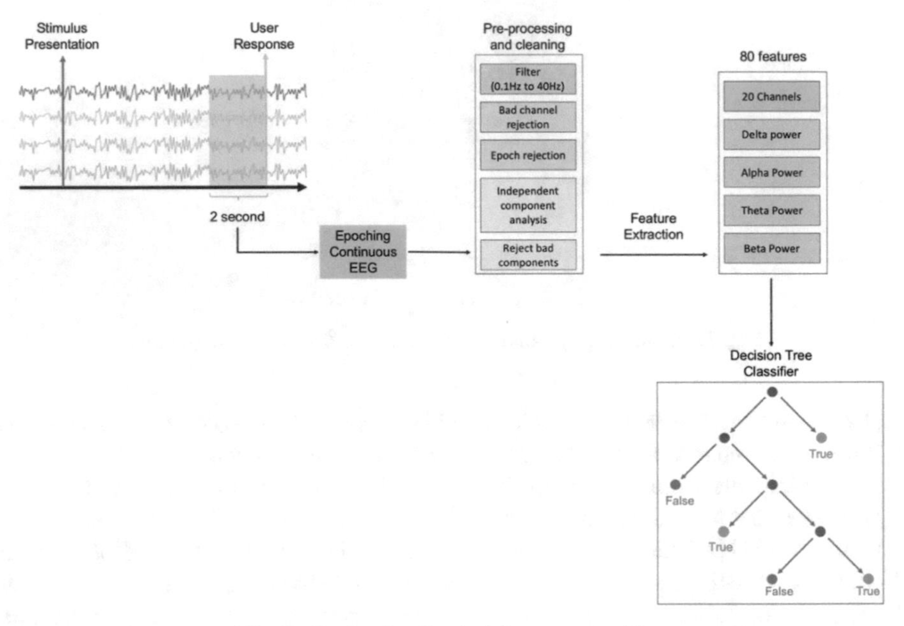

Fig. 2. Derivation for decision tree classifier.

Eye tracking data was prepossessed by removing blinks, distorted pupil recordings (e.g. participant looking away from the screen), as well as data points in which the tracker was unable to record one of the participant's eyes. A blank gray screen is presented for two seconds before each trial. This transition/reference screen is utilized as a baseline for each trial, allowing EEG and pupil values to stabilize from the previous trial. The baseline dilation for each trial is measured from the average, across both eyes, during this transition. The second measure is comprised of the average, across both eyes, dilation in the last 2 s of each trial (up to decision point). The pupil dilation difference (conveyed as a percentage), for each trial, is then calculated from these two measures.

4 Preliminary Results

Participants were able to correctly identify misinformation, for control headlines, at 96% accuracy. In contrast, performance for the politically polarizing headlines was significantly lower at 58% accuracy ($P < 0.05$). Similar performance patterns were observed in participants' response time where the average response time for control headlines was 6.7 s, while polarizing headlines trailed at 8.9 s ($P < 0.05$).

The two physiological measures, EEG and eye tracking, yielded varying results. The EEG decision tree classifier, constructed with 7 usable participant data sets, yielded a mean accuracy of 70%. EEG results also revealed that a majority incorrect responses were associated with parietal alpha activity (System 1 processes), while the majority of correct responses were associated with frontal theta activity (System 2 processes). Eye tracking pupil analysis yielded no distinct pattern or separation amongst control headlines, correct responses, and incorrect responses.

5 Discussion

EEG results suggest a distinction between System 1 (automatic) cognitive processes for incorrect responses and System 2 (analytical) for correct responses. This implies that when participants engage their analytical thinking they are much more likely to correctly identify misinformation. This result suggests that intervention methods that induce or encourage analytical thinking patterns are likely to reduce the acceptance and spread of misinformation. Having a user take an outsider's perspective, as well as critically approaching the topic from an opposing viewpoint, has shown to elicit System 2 processes [21]. Utilizing this notion of encouraging System 2 thinking processes as an intervention method provides a stark contrast to the ineffective methods attempted thus far (e.g., Fact checkers and flagging systems). Such methods merely declare whether a claim is likely fake with minimal effect on misinformation acceptance, as opposed to having the reader analytically critique their own perspective on the presented headlines as suggested by the preliminary results of this study.

5.1 Next Steps

The experimental design used in our study can be amended to apply and test the efficacy of intervention methods designed to induce System 2 activation. This can be accomplished in two varying methods. First, participants could be required to complete a training session using a software application prior to completing the experiment, or presented half way through. This software would train participants to think critically, providing the tools to breakdown each headline and approach it from multiple perspectives. The second method would incorporate a performance breakdown of how accurate the participant has been with regard to correctly identifying misinformation. This text box would include an analytical breakdown of each headline, sourced from agencies of varying political and social biases, as well as a percentage indicating general performance. These intervention methods can be assessed by having a select portion of future participants undergo this training, or be presented with the performance breakdown. We can then contrast their accuracy of identifying misinformation, as well as analyze their neurophysiological response, with participants who didn't undergo training or receive the performance breakdown to discern effectiveness of these interventions.

References

1. Guess, A.M., Nyhan, B., Reifler, J.: Exposure to untrustworthy websites in the 2016 U.S. election. 38.
2. Kata, A.: A postmodern pandora's box: anti-vaccination misinformation on the internet. Vaccine. **28**, 1709–1716 (2010). https://doi.org/10.1016/j.vaccine.2009.12.022
3. Conroy, N.J., Rubin, V.L., Chen, Y.: Automatic deception detection: Methods for finding fake news. Proc. Assoc. Inf. Sci. Tech. **52**, 1–4 (2015)
4. Moravec, P., Kim, A., Dennis, A.: Flagging Fake News: System 1 vs. System 2 (2018).
5. Arceneaux, K.: Cognitive biases and the strength of political arguments. Am. J. Political Sci. **56**, 271–285 (2012). https://doi.org/10.1111/j.1540-5907.2011.00573.x
6. Kahneman, D.: Thinking, Fast and Slow. Farrar Straus and Giroux, New York (2011)
7. Mathôt, S.: Pupillometry: psychology, physiology, and function. J. Cogn. **1**, 16 (2018)
8. Smith, E.R., DeCoster, J.: Dual-process models in social and cognitive psychology: conceptual integration and links to underlying memory systems. Pers. Soc. Psychol. Rev. (2016). https://doi.org/10.1207/S15327957PSPR0402_01
9. Wilson, T.D., Lindsey, S., Schooler, T.Y.: A model of dual attitudes. Psychol. Rev. **107**, 101–126 (2000). https://doi.org/10.1037/0033-295X.107.1.101
10. Bargh, J.A., Ferguson, M.J.: Beyond behaviorism: on the automaticity of higher mental processes. Psychol. Bull. **126**, 925–945 (2000). https://doi.org/10.1037/0033-2909.126.6.925
11. Evans, J.S.B.T., Stanovich, K.E.: Dual-process theories of higher cognition: advancing the debate. Perspect. Psychol. Sci. **8**, 223–241 (2013)
12. Fiske, S.T., Taylor, S.E.: Social Cognition. Mcgraw-Hill Book Company, New York (1991)
13. Torrens, D.: Individual differences and the belief bias effect: mental models, logical necessity, and abstract reasoning. Thinking Reasoning. **5**, 1–28 (1999). https://doi.org/10.1080/135467899394066
14. Kahneman, D., Peavler, W.S., Onuska, L.: Effects of verbalization and incentive on the pupil response to mental activity. Can. J. Psychol. Revue canadienne de psychologie. **22**, 186–196 (1968). https://doi.org/10.1037/h0082759
15. Williams, C., Kappen, M., Hassall, C.D., Wright, B., Krigolson, O.E.: Thinking theta and alpha: mechanisms of intuitive and analytical reasoning. NeuroImage **189**, 574–580 (2019)
16. Porter, E., Wood, T.J., Kirby, D.: Sex trafficking, Russian infiltration, birth certificates, and pedophilia: a survey experiment correcting fake news. J. Exp. Political Sci. **5**, 159–164 (2018)
17. Wood, G., Long, K.S., Feltwell, T., et al.: Rethinking engagement with online news through social and visual co-annotation. In: Proceedings of the 36th Annual ACM Conference on Human Factors in Computing Systems, CHI'18, pp. 1–12 (2018)
18. Flintham, M., Karner, C., Bachour, K., Creswick, H., Gupta, N., Moran, S.: Falling for fake news: investigating the consumption of news via social media. In: Proceedings of the 2018 CHI Conference on Human Factors in Computing Systems, pp. 376:1–376:10. ACM, New York (2018). https://doi.org/https://doi.org/10.1145/3173574.3173950.
19. Mognon, A., Jovicich, J., Bruzzone, L., Buiatti, M.: ADJUST: An automatic EEG artifact detector based on the joint use of spatial and temporal features. Psychophysiology **48**, 229–240 (2011)
20. Polat, K., Güneş, S.: Classification of epileptiform EEG using a hybrid system based on decision tree classifier and fast Fourier transform. Appl. Math. Comput. **187**, 1017–1026 (2007). https://doi.org/10.1016/j.amc.2006.09.022
21. Milkman, K.L., Chugh, D., Bazerman, M.H.: How can decision making be improved? Perspect. Psychol. Sci. **4**, 379–383 (2009). https://doi.org/10.1111/j.1745-6924.2009.01142.x

Neurophysiological Assessment of Ambivalence to Information

Akshat Lakhiwal[1](✉), Hillol Bala[1], and Pierre-Majorique Leger[2]

[1] Kelley School of Business, Indiana University, Bloomington, USA
{aklakh,hbala}@iu.edu
[2] HEC, Montreal, Canada
pierre-majorique.leger@hec.ca

Abstract. The proliferation of technologies has made information ubiquitously available to individuals who rely on it to make decisions or conduct transactions. We focus on how and why valence of information may elicit mixed reactions among individuals and potentially influence their decision-making process. Prior research in IS has primarily focused on positive and negative reactions to technology (in most cases separately). We examine the simultaneous presence of positive and negative dispositions—*ambivalence*. We theorize and show how ambivalence will elicit distinct behavioral responses and evoke attentional processes. We use electroencephalography (EEG) to conduct a within subject repeated measures laboratory experiment to illustrate these effects. Our results highlight that individuals experiencing ambivalence due to valence incongruent information exhibit a higher involvement of attentional processes than individuals who experience other types of information valence, i.e., positivity, negativity, and indifference. Individuals experiencing ambivalence also expressed different levels of behavioral intention to use a product from individuals who experienced positive, negative, and indifferent valence of information.

Keywords: Ambivalence · Information · Intention to use · Negativity bias · EEG

1 Introduction

The information age has not just triggered the digital revolution, but has also motivated a distinct shift to a knowledge-based society where the growth in technology-based innovation has led to an explosion in the world's capacity to store, communicate, and compute information [1–3]. It has enabled organizations to provide relevant content to individuals through personalization [1], and has also changed the way individuals fundamentally execute their decisions in their day-to-day life. Instead of relying simply upon passively available information from mass sources such as print or television, individuals can now actively seek information from search engines, mobile browsers, blogs, review platforms and brand websites accessible at their disposal [4]. This increasing ubiquity of information also implies that individuals have access to a wider array of content,

F. D. Davis et al. (Eds.): NeuroIS 2020, LNISO 43, pp. 49–57, 2020.
https://doi.org/10.1007/978-3-030-60073-0_6

carrying a stream of information about the attributes of the object of interest, where each attribute competes for attention, increasing its potential to influence decision making.

This increasing presence and accessibility of information and information sources also triangulate on the potential existence of incongruent information about same or different aspects of an entity, serving as the key motivation for our work. For instance, a mobile phone which has received a five-star critique rating for its performance on an online review platform, but a one-star rating from other users for its aesthetics would arouse mixed feelings and evaluative conflict for an individual interested in purchasing it. Such conflict due to the coexisting positive and negative dispositions is also experienced in a technology adoption environment, where the positive and negative appraisal of the attributes of a new technology have been shown to elicit perceptions of opportunity and threat [5] among individuals, respectively. Such individuals often experience a state of conflict and vacillation between their positive and negative dispositions [6], while using their coping and adaptation mechanisms [7, 8] to resolve it.

Thus, while most of the extant work in Information Systems (IS) research has considered the independent influence [9–13] of positive (e.g., opportunity, acceptance, perceived usefulness, or enjoyment) and negative (e.g. threat, resistance, perceived anxiety, avoidance, or technostress) attitudes on individual behavior, their coexistence and subsequent behavioral implications have received rather limited attention. This coexistence of positive and negative orientations toward the object of interest is known as *ambivalence* and it allows researchers to observe the bivariate nature of information and elicited attitudes, thereby allowing a deeper examination of behavioral effects elicited by positive and negative information [14]. In the contemporary environment, where individuals are exposed to a variety of pervasive technologies and are surrounded ubiquitous information sources, it is highly likely that they develop ambivalent dispositions.

In this study, we examine ambivalence and focus on how individuals experiencing it depict higher involvement of attentional processes in comparison to positivity, negativity or indifference (i.e., neutrality). Motivated by prior research in social psychology and recommendations by Bala et al. (2017) [15], we use electroencephalography (EEG) to neurophysiologically disassociate ambivalence from positivity, negativity and indifference. EEG allows us to capture and gain precise understanding of cognitive processes associated with ambivalence. We deploy *Event Related Potentials (ERP)* based measurement, which allows great temporal precision and enabled us to associate unique ERP components to ambivalence. By using ERPs, we seek answers to the following research questions:

RQ1: How are ambivalent reactions to information neurophysiologically different from positive, negative and indifferent reactions?
RQ2: How is the behavioral response due to ambivalence different from behavioral response due to positivity, negativity, and indifference?

2 Hypotheses

Our inquiry into ambivalence is motivated by theoretical arguments and anecdotal evidences of ubiquity of multi-valence information across a variety of contexts (including online platforms), where an individual might appraise positive and negative information in different sequences [16] and at different times, but can cumulatively experience evaluative conflict (in the form of ambivalence) during decision making. Consider the example of technology adoption in a healthcare setting, where an physicians might be cognizant of the efficiency benefits of a new electronic health records (EHR) system, but might still be threatened by the perceived loss of power due to its adoption [17]. As a psychological state, ambivalence is characterized by the person holding mixed feelings toward an object, and elicits cognitive, emotional and evaluative inconsistency [18–20]. Driven by this inconsistency, ambivalence has been shown to create a heightened state of arousal, marked by conflict, discomfort, and displeasure [21]. Individuals experiencing ambivalence are thus motivated to extensively involve their cognitive processes to assess the available information and achieve consistency in their intentions. While prior research in psychology has relied upon the cognitive conflict associated with ambivalence to observe activity in lateral prefrontal region and anterior cingulate cortex [22, 23] of the brain and thereby study its strength and associated evaluative difficulties, we focus on the information processing aspect of ambivalence to understand its effect on attentional processes.

Modulated by their state of heightened arousal, individuals experiencing ambivalence would exhibit high cognitive involvement and demand greater attention to process the conflicting pieces of information in order to achieve consistency in an evaluative context. Individuals would also use attention as a mechanism to dedicate their cognitive resources toward the opposing valences of information to resolve their decision-making difficulty [24, 25] as they are motivated to achieve evaluative consistency [19, 26]. For instance, research on balance theory and cognitive dissonance has shown that people are motivated to reduce their internal inconsistencies, and ambivalence is expected to have similar outcomes [18, 27]. This heightened cognitive involvement and subsequent attention would be characterized by the activation of the parietal cortex (near Pz electrode site) of the brain [28], which can be visualized by the P300 ERP component, a positive potential which peaks during the window of 300 to 500 ms (from the onset of the stimuli) [29, 30]. Thus, we posit,

H1: Ambivalence would lead to a higher activation of the P300 ERP component in comparison to positivity, or negativity, and indifference.

Research evaluating the strength of ambivalence has argued that ambivalence is a weaker form of attitude in comparison to univalent attitudes [23, 26], and can attenuate the relationship between evaluation and behavioral intentions, leading to delayed behavior [20, 31] or complete avoidance [32]. Univalent attitudes (positivity and negativity) on the other hand are stronger, and cognitively less expensive, thereby facilitating easier and faster decisions.

The comparison between ambivalence and indifference is even more interesting as both are characteristically weak attitudes and are often indifferentiable on the bipolar evaluation scale [33]. In an evaluative context, ambivalence has been shown to be associated with a reduced confidence [34], which can lead to heightened information processing and can increase its strength as an attitude [34]. In addition, prior research into ambivalence suggests that ambivalent attitudes are biased toward the negative affect, which is attitudinally stronger than positive affect in general, (*negativity bias* [35, 36]). Thus, we argue that because individuals experiencing ambivalence have a general biased inclination (toward negative affect) in comparison to those experiencing indifference, they would also exhibit a lower behavioral intention. However, being relatively weaker attitudes, ambivalence and indifference would still be less extreme (positive or negative) in comparison to people with univalent (more certain) evaluations.

H2a: Ambivalence and Indifference would lead to a higher intention to use in comparison to negativity
H2b: Ambivalence and Indifference would lead to a lower intention to use in comparison to positivity
H2c: Ambivalence would lead to a lower intention to use a product in comparison to indifference.

3 Methodology

We conducted a within subject controlled laboratory experiment, in line with previous studies which have examined ambivalence [15, 20, 37, 38]. Four (4) main conditions represented the 4 valence associations (positivity, negativity, indifference and ambivalence) of interest. A pretest was conducted to assess the validity of the stimuli with 13 participants (aged 22 to 25 years, 69% females) recruited in exchange for extra course credit from an introductory business course. For the main study, 22 participants were recruited from the university panel (Aged 18 to 30 years, 33% females) who received $30 for their participation. Data from 18 participants was used for the final analysis. Both the pretest and the EEG study were conducted after obtaining approval from the institutional review boards of our universities.

4 Treatment, Experimental Procedure and Measures

A focus group study was conducted with 5 undergraduates in their prefinal year of graduation to identify 180 items which were most relevant in day-to-day life of a college going student. The valence of each item was then manipulated by associating two attributes (adjectives) which were mined from Amazon product reviews using a logistic regression classifier. The attributes were associated in a way such that the overall evaluation of the item could be positivity (2 positive attributes), negativity (2 negative attributes), indifference (2 neutral attributes) or ambivalence (1 positive attribute and 1 negative attribute). The pretest was then conducted to assess the validity of the stimuli. 80 best items were selected for the main study based on the results of the pretest.

For the main study, each trial comprised of an item (with an associated brand indifferent image) presented for 2.5 s on a blank white screen and was followed sequentially by the two pre-assigned attributes (2.5 seconds per attribute) of the item (Fig. 1). Each presentation was followed by a fixation cross which was presented for 1.5 s. Participant reactions after the presentation of the second attribute represent our key region of interest in terms of the measured ERPs (as it represents the overall evaluation of the item based on the two attributes). Research in linguistics has shown that an average person takes about 200–300 ms to read and process a word [39, 40], which suggests that 2500 ms is arguably sufficient time for our investigation. After the second attribute, the participants responded to a battery of behavioral response questions. All trials were randomized and were presented blocks of 20.

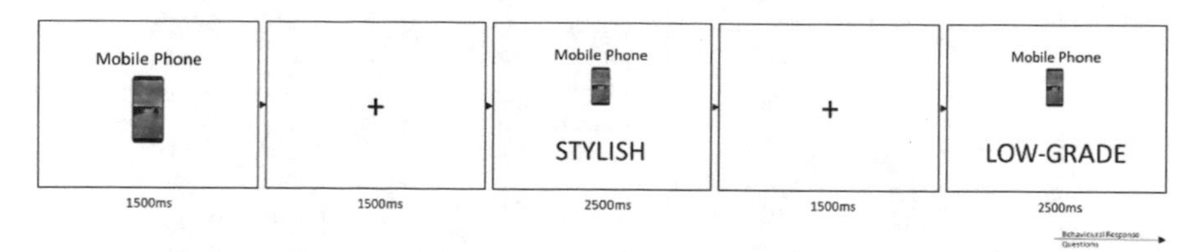

Fig. 1. Example of stimuli presentation sequence for each trial

The main study was conducted in the university's EEG laboratory. The EEG laboratory had two separate rooms. The first room, where the experiment was conducted was electrically shielded and soundproofed to avoid potential artefacts [41, 42]. The second room (control room) was where the experimenter could observe the data acquisition process from. The hardware (including the data collection computer) were in the control room. Once the participants arrived at the lab, they were greeted; informed consent was obtained. Participants were not allowed to wear glasses during the experiment and were required to put away their cellphones. Instructions were provided verbally as well as on their screens. During the experiment, participants were explicitly informed to minimize movements in order to reduce neurophysiological artefacts. We measured EEG using a 32-electrode EEG apparatus with a 10–20 system (arrangement). The vertex (electrode Cz) was used as the reference electrode and the impedance was kept below 15 Ω, with a sampling rate of 1000 Hz; LiveAmp amplifier[1] was used. ActiCap (See Footnote 1) cap tracker was used to localize electrodes with landmarks placed on the nasion and on the superior junction between the ear and the skull. The stimulus presentation as well as collection of behavioral responses was done using E-Prime software, while EEG signals were recorded using BrainVision (See Footnote 1) Recorder. Participants were shown a sample trial before the actual experiment began. Once the experiment was concluded, participants were required to complete the post-experiment questionnaire, were remunerated and dismissed.

The EEG data acquired through the 32 channel pre-amplified electrodes was analyzed using a combination of Brainvision analyzer and MATLAB. We followed the guidelines suggested in [42–44] while preprocessing the data. The sampling rate was set to 256 Hz and an infinite impulse response Butterworth filter on the EEG signal with a bandpass of 1–15 Hz was used [41, 42, 45]. Artefacts during trials were visually inspected and also cleaned using Artifact subspace reconstruction algorithm using MATLAB [46, 47]. The EEG data was segmented to epochs of 1700 ms (−200 ms to 1500 ms from the onset of the second attribute of the item) and were averaged for statistical analysis. Jackknife procedure with adjustment for sample size was further used to improve the signal to noise ratio of the waveforms before sampling for statistical analysis [30, 48–50]. The

[1] Brain Products GmbH, Gilching, Germany.

final data comprising of 18 jackknifed waveforms for each condition was sampled in the 300–500 ms time window (Grand averages in Fig. 2) to assess the peak amplitude of the P300 component at the Pz electrode [30, 42]

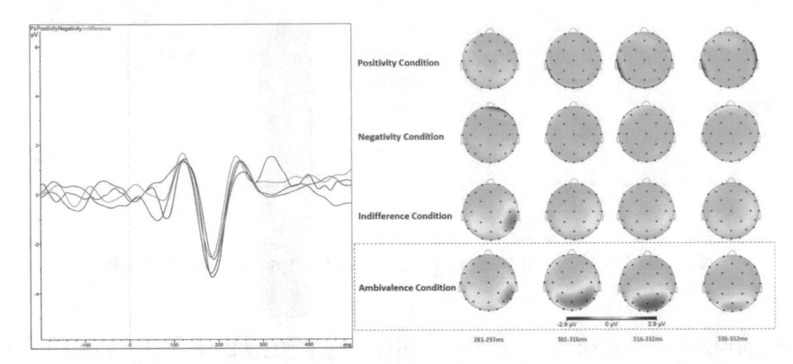

Fig. 2. *Left Panel*-P300 waveform at Pz electrode for ambivalence (black), indifference (red), positivity (blue) and negativity (green) condition. *Right Panel*-Brain Mappings for Positivity, Negativity, Indifference and Ambivalence condition at Pz electrode at around 300 ms from stimuli onset

Ambivalence was measured and calculated using two self-developed items based on the guidelines suggested by Thomson et al. [33]), on a five-point Likert scale. We also measured the participant's intention to use the product using a self-developed question ("Would you consider using this product?") on a seven-point Likert scale.

5 Analysis and Results

ANOVA results (F (3,2388) = 12.38, p < .01) of the pretest conducted using 180 items confirmed the validity of the stimuli, out of which 80 items were selected for the main study.

Results of ANOVA on EEG data support H1 as we find that the mean amplitude at the Pz electrode was significantly higher for the ambivalence group in comparison to the positivity $(F(1,34) = 4.42, p < .05)$, negativity $(F(1,34) = 4.65, p < .05)$ and indifference $(F(1,34) = 4.83, p < .05)$ group around 300 ms after onset of second attribute (Fig. 2). From our behavioral results, we also find support for H2a, H2b and H2c $(F(3,1436) = 432.77, p < .01)$ as the intention to use was third highest for the ambivalent group among the 4 treatment groups.

6 Discussion and Conclusion

Our research presents neurophysiological perspective on how individuals' ambivalent reactions to information elicit distinct attentional reactions and to the best of our knowledge is the first to associate an ERP potential with ambivalence. As a reaction defined by coexisting positive and negative emotions, ambivalence is highly prevalent in day-to-day lives, and has yet received limited attention from IS researchers. The study aims to draw attention toward such mixed reactions which influence individual behavior in every context.

We contribute by providing distinctly delineating neurophysiological and behavioral reactions to ambivalence in comparison to positivity, negativity and indifference. This alludes directly to online reviews and ratings where ambivalence causing information is often misrepresented as neutral. Our results also suggest that ambivalence leads to a lower intention to use and provide impetus for further introspection. While it is evident that ambivalence is characterized by an inclination toward negative affect, our results urge researchers to explore ambivalence as a transitionary state and examine factors which can increase behavioral intention for individuals who are in a state of flux and indecision due to ambivalence. We also highlight that individuals experiencing ambivalence potentially exhibit higher attention. Our examination thus provides great insights and opens avenues for IS researchers to explore the potential of cognitive processes, mechanisms and reactions associated with ambivalence.

References

1. Batra, R., Keller, K.L.: Integrating marketing communications: new findings, new lessons, and new ideas. J. Market. **80**(6), 122–145 (2016)
2. Hilbert, M., López, P.: The world's technological capacity to store, communicate, and compute information. Science **332**(6025), 60–65 (2011)
3. Pentland, A.: Social Physics: How Good Ideas Spread-the Lessons from a New Science. Penguin, London (2014)
4. Court, D., et al.: The consumer decision journey. McKinsey Q. **3**(3), 96–107 (2009)
5. Bala, H., Venkatesh, V.: Adaptation to information technology: a holistic nomological network from implementation to job outcomes. Manage. Sci. **62**(1), 156–179 (2016)
6. Stein, M.-K., et al.: Coping with information technology: mixed emotions, vacillation, and nonconforming use patterns. MIS Q. **39**, 367–392 (2015)
7. Beaudry, A., Pinsonneault, A.: Understanding user responses to information technology: A coping model of user adaptation. MIS Q. **29**, 493–524 (2005)
8. Beaudry, A., Pinsonneault, A.: The other side of acceptance: studying the direct and indirect effects of emotions on information technology use. MIS Q. **34**, 689–710 (2010)
9. Basuroy, S., Chatterjee, S., Ravid, S.A.: How critical are critical reviews? The box office effects of film critics, star power, and budgets. J. Market. **67**(4), 103–117 (2003)
10. Chen, Z., Lurie, N.H.: Temporal contiguity and negativity bias in the impact of online word of mouth. J. Mark. Res. **50**(4), 463–476 (2013)
11. Mauri, A.G., Minazzi, R.: Web reviews influence on expectations and purchasing intentions of hotel potential customers. Int. J. Hospital. Manage. **34**, 99–107 (2013)
12. Reinstein, D.A., Snyder, C.M.: The influence of expert reviews on consumer demand for experience goods: a case study of movie critics. J. Ind. Econ. **53**(1), 27–51 (2005)
13. Wang, Z., et al.: Saliency effects of online reviews embedded in the description on sales: moderating role of reputation. Decis. Support Syst. **87**, 50–58 (2016)
14. Walden, E.A., Browne, G.J., Larsen, J.T.: Ambivalence and the bivariate nature of attitudes in information systems research. In: Proceedings of the 38th Annual Hawaii International Conference on System Sciences. IEEE (2005)
15. Bala, H., Labonté-LeMoyne, E., Léger, P.-M.: Neural correlates of technological ambivalence: a research proposal. In: Davis, F., Riedl, R., vom Brocke, J., Léger, P.M., Randolph, A. (eds.) Information Systems and Neuroscience, pp. 83–89. Springer, Cham (2017)

16. Hogarth, R.M., Einhorn, H.J.: Order effects in belief updating: the belief-adjustment model. Cogn. Psychol. **24**(1), 1–55 (1992)
17. Lapointe, L., Rivard, S.: A multilevel model of resistance to information technology implementation. MIS Q. **29**, 461–491 (2005)
18. Heider, F.: Attitudes and cognitive organization. J. Psychol. **21**(1), 107–112 (1946)
19. van Harreveld, F., Nohlen, H.U., Schneider, I.K.: The ABC of ambivalence: Affective, behavioral, and cognitive consequences of attitudinal conflict. In: Advances in Experimental Social Psychology, pp. 285–324. Elsevier (2015)
20. Van Harreveld, F., Van der Pligt, J., de Liver, Y.N.: The agony of ambivalence and ways to resolve it: introducing the MAID model. Pers. Soc. Psychol. Rev. **13**(1), 45–61 (2009)
21. Nordgren, L.F., Van Harreveld, F., Van Der Pligt, J.: Ambivalence, discomfort, and motivated information processing. J. Exp. Soc. Psychol. **42**(2), 252–258 (2006)
22. Cunningham, W.A., Raye, C.L., Johnson, M.K.: Implicit and explicit evaluation: fMRI correlates of valence, emotional intensity, and control in the processing of attitudes. J. Cogn. Neurosci. **16**(10), 1717–1729 (2004)
23. Luttrell, A., et al.: Neural dissociations in attitude strength: distinct regions of cingulate cortex track ambivalence and certainty. J. Exp. Psychol. Gen. **145**(4), 419 (2016)
24. Potter, R.E., Balthazard, P.: The role of individual memory and attention processes during electronic brainstorming. MIS Q. **28**, 621–643 (2004)
25. Shen, W., Hu, Y., Ulmer, J.R.: Competing for attention: an empirical study of online reviewers' strategic behavior. MIS Q. **39**(3), 683–696 (2015)
26. Maio, G.R., Bell, D.W., Esses, V.M.: Ambivalence and persuasion: the processing of messages about immigrant groups. J. Exp. Soc. Psychol. **32**(6), 513–536 (1996)
27. Festinger, L.: Conflict, Decision, and Dissonance (1964)
28. Coull, J.T.: Neural correlates of attention and arousal: insights from electrophysiology, functional neuroimaging and psychopharmacology. Prog. Neurobiol. **55**(4), 343–361 (1998)
29. Sutton, S., et al.: Evoked-potential correlates of stimulus uncertainty. Science **150**(3700), 1187–1188 (1965)
30. Luck, S.J.: An Introduction to the Event-Related Potential Technique. MIT press, Cambridge (2014)
31. Monteith, M.J.: Self-regulation of prejudiced responses: implications for progress in prejudice-reduction efforts. J. Pers. Soc. Psychol. **65**(3), 469 (1993)
32. Luce, M.F., Bettman, J.R., Payne, J.W.: Choice processing in emotionally difficult decisions. J. Exp. Psychol. Learn. Mem. Cogn. **23**(2), 384 (1997)
33. Thompson, M.M., Zanna, M.P., Griffin, D.W.: Let's not be indifferent about (attitudinal) ambivalence. Attitude Strength Antecedents Consequences **4**, 361–386 (1995)
34. Jonas, K., Broemer, P., Diehl, M.: Attitudinal ambivalence. Eur. Rev. Soc. Psychol. **11**(1), 35–74 (2000)
35. Broemer, P.: Relative effectiveness of differently framed health messages: the influence of ambivalence. Eur. J. Soc. Psychol. **32**(5), 685–703 (2002)
36. Baumeister, R.F., et al.: Bad is stronger than good. Rev. Gen. Psychol. **5**(4), 323–370 (2001)
37. Nohlen, H.U., van Harreveld, F., Cunningham, W.A.: Social evaluations under conflict: negative judgments of conflicting information are easier than positive judgments. Soc. Cogn. Affect. Neurosci. **14**(7), 709–718 (2019)
38. Nohlen, H.U., et al.: Evaluating ambivalence: social-cognitive and affective brain regions associated with ambivalent decision-making. Soc. Cogn. Affect. Neurosci. **9**(7), 924–931 (2014)
39. Osterhout, L., Holcomb, P.J.: Event-related brain potentials elicited by syntactic anomaly. J. Mem. Lang. **31**(6), 785–806 (1992)
40. Darnell, K.: Discriminating Between Syntactic and Semantic Processing: Evidence from Event-related Potentials (1995)

41. Duncan, C.C., et al.: Event-related potentials in clinical research: guidelines for eliciting, recording, and quantifying mismatch negativity, P300, and N400. Clin. Neurophysiol. **120**(11), 1883–1908 (2009)
42. Müller-Putz, G.R., Riedl, R., Wriessnegger, S.C.: Electroencephalography (EEG) as a research tool in the information systems discipline: foundations, measurement, and applications. CAIS **37**, 46 (2015)
43. Tivadar, R.I., Murray, M.M.: A primer on electroencephalography and event-related potentials for organizational neuroscience. Organ. Res. Methods **22**(1), 69–94 (2019)
44. Woodman, G.F.: A brief introduction to the use of event-related potentials in studies of perception and attention. Atten. Percept. Psychophys. **72**(8), 2031–2046 (2010)
45. Léger, P.-M., et al.: Precision is in the eye of the beholder: Application of eye fixation-related potentials to information systems research. Association for Information Systems (2014)
46. Chang, C.-Y., et al.: Evaluation of artifact subspace reconstruction for automatic EEG artifact removal. In: 40th Annual International Conference of the IEEE Engineering in Medicine and Biology Society (EMBC). IEEE (2018)
47. Mullen, T., et al.: Real-time modeling and 3D visualization of source dynamics and connectivity using wearable EEG. In: 35th annual international conference of the IEEE engineering in medicine and biology society (EMBC). IEEE (2013)
48. Knott, V., et al.: EEG power, frequency, asymmetry and coherence in male depression. Psychiatry Res. Neuroimaging **106**(2), 123–140 (2001)
49. Gevins, A., et al.: Monitoring working memory load during computer-based tasks with EEG pattern recognition methods. Hum. Factors **40**(1), 79–91 (1998)
50. Kiesel, A., et al.: Measurement of ERP latency differences: a comparison of single-participant and jackknife-based scoring methods. Psychophysiology **45**(2), 250–274 (2008)

Behavior Regulation in Social Media: A Preliminary Analysis of Pupil Size Change

Yu-feng Huang(✉) and Feng-yang Kuo

National Sun Yat-Sen University, Kaohsiung, Taiwan
evanhuang@mis.nsysu.edu.tw, bkuo@mail.nsysu.edu.tw

Abstract. Social media users may regulate their behaviors to follow norms of their online communities. This regulation process, however, might be too transient to be captured using self-reports and therefore is suitable for a NeuroIS investigation. Previously, in an event-related potential (ERP) experiment designed to study this regulation process, Huang, Kuo, and Lin [1] found that this regulation process could be reflected in an ERN-like ERP, and the ERP's magnitude is correlated with people's internet privacy concern. In this work-in-progress we seek to use eye-tracking to replicate their findings. Here we report our current results of pupil size anslyses, which so far are consistent with the previous ERP findings.

Keywords: Pupil size · Eye-tracking · Behavior regulation · Social media

1 Introduction

Behavior regulation is a critical cognitive process that calibrates people's behaviors with their intentions [2]. Failure of regulation can decrease behavior-intention alignment that occurs during the introduction of an information system [3]. Evidence has shown that self-regulation is a limited cognitive resource [4, 5], and this limited resource can be further impaired by factors such as mental work load [6]. Therefore, whether people can appropriately distribute their resource into tasks is critical for their regulation success.

Social media users regulate their online behaviors. Previous research in social media has shown that users may be motivated to conform to their online community' social norms, because violating such norms can lead to various types of social sanctions, including isolation, public humiliation, and/or cyberbullying [7]. To avoid these sanctions, social media users need to activate a regulative process to monitor their online behaviors and issue warnings (e.g., the feeling of "Oops") when behavior errors are committed. However, it remains unclear what factors affect users' distribution of resources for behavior regulation when they are using social media.

The regulation process can be transient or non-verbalizable, and therefore might be difficult to capture using self-reports. Huang, Kuo, and Lin [1] (the HKL study) used electroencephalogram (EEG) and designed an event-related potential (ERP) experiment to look for the evidence of this regulation process. In the experiment they asked participants to decide whether to press "like" toward a series of social-sensitive (with sex

F. D. Davis et al. (Eds.): NeuroIS 2020, LNISO 43, pp. 58–63, 2020.
https://doi.org/10.1007/978-3-030-60073-0_7

information) and social-insensitive (without sex information) app logos and compare the ERPs associated with the two types of logos. Pressing like toward a social-sensitive logo can be considered a regulation failure because this behavior is inappropriate in most social media usage scenarios. Their findings indicate that, compared with social-insensitive logos, when participants press like toward social-sensitive logos there is an ERP similar to the classical error-related negativity (ERN), which has been strongly associated with behavior regulation. This ERN-like pattern provides the first evidence of the existence of a self-regulation process in social media usage. Besides, the authors also showed that the magnitude of the ERN-like pattern is correlated with users' internet privacy concern (IPC) [8], suggesting that users with strong privacy concern might have devoted more cognitive resources for regulation.

In this work-in-progress, we seek to replicate HKL [1] to find evidence of the regulation. Moreover, instead of using EEG or ERP, in this research we use pupil size change to detect the regulation process for the following reasons: (1) pupil size change is useful to indicate people's devotion of cognitive resources [9–11], and therefore can help researchers to track whether and when the regulation process occurs during social media usage; (2) a cross-tool replication can enhance the scientific robustness of the previous findings [12]; and (3) a comparison of the results from EEG and eye-tracking increases our understanding toward these two tools, and therefore helps future researches to choose their tool to investigate NeuroIS topics.

So far we have collected 48 participants. In this work-in-progress we report our current pupil analyses, and provide a first-step discussion regarding the nature of behavior regulation and a comparison between EEG and eye-tracking findings.

2 Methods

2.1 Task and Stimulus

The HKL's paradigm, including the stimulus set, is used in this study. Specifically, the experiment presented an app logo in the center of the screen for 150 ms, followed by a blank screen (randomized between 500 and 1000 ms). By pressing the corresponding buttons, participants were required to decide whether to "like" the logo during the blank screen (i.e., before the next logo appeared). This short decision time can minimize the involvement of deliberation and increase the frequency of slip in regulation. There were 20 sensitive and 60 non-sensitive logos that were presented in random order. This procedure was repeated 12 times, resulting in 960 trials for each participant. Note that participants performed the task both in the private and the public condition (480 trials for both conditions). In the private condition, participants imagined that their task is performed in private (no one else will see). In the public condition, participants imagined that the apps were posted on their Facebook page. Their evaluations were visible to their friends. In this work-in-progress we aggregate across both conditions for subsequent analysis.

Eye movement data were collected using the EyeLink 1000 system at 250 Hz sampling rate (SR Research, Mississauga, Canada). To increase eye movement data quality, we perform the following steps: a nine-point calibration was executed before the experiment, a drift correction was performed before each block, and asked participants to be seated with their chins put on a chin rest. All stimuli were displayed on a 19″ LCD screen at a levelled distance of 60 cm to participants' eyes.

3 Participants and Procedure

Currently we have successfully collected eye movement data from 48 healthy adult volunteers (18 males, mean age = 20.94 years old). All participants reported to have normal or corrected-to-normal vision. After the experimenter finished the eye-tracking calibration and read the instruction, participants finished the task alone in the eye-tracking room. After the experiment they also completed the IPC questionnaire before they received their compensation and debrief.

4 Analysis and Result

4.1 The Quantification of Pupil Size Change

The current analysis focuses on pupil size change [13]. We extracted pupil data from the raw samples using Data Viewer software (SR Research, Mississauga, Canada). Pupil size extracted from this software is in arbitrary units that is linear with true diameter, and is normalized following the procedure by [13]. Our analysis time span started at the onset of a logo and ended right before the onset of the next one. Baseline was set to the duration of the presence of the stimuli (i.e., 150 ms). After the normalization, a positive percentage of pupil size change indicates pupil dilation, and a negative percentage indicates pupil constriction.

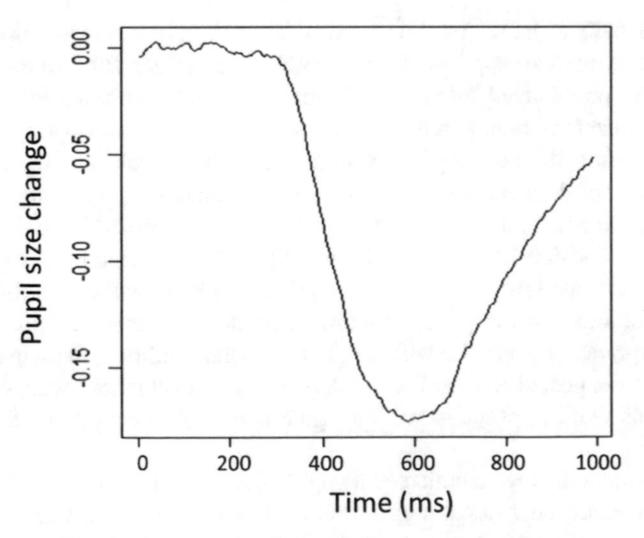

Fig. 1. An example of a normalized pupil size change

We found that in this study the pupil size first started to constrict hundreds of milliseconds after the onset of a stimulus (time 0) and then started to regress to baseline (Fig. 1). Therefore, for each trail the point with the most constriction was extracted for subsequent analysis. This value was sorted and averaged according to a participant's response (like and dislike) and logo type (sensitive and non-sensitive). Statistical analysis on pupil size change was performed on participants' averages. Non-response trials (9.19%) were excluded from analysis.

5 Results

Results indicate that participants avoid liking a sensitive logo, showing the regulative behavior. The paired t-test indicates that when a sensitive logo is presented, the frequency that the participants like the logo is 32.51%, significantly lower than the frequency when a non-sensitive logo is presented (62.02%, $p < .01$). More importantly, the frequency that participants like a sensitive logo is correlated with his/her IPC score ($r = -.32$, $p = .03$), indicating that the higher the IPC the less likely that a participant would like a sensitive logo. This correlation cannot be observed in the non-sensitive logos ($r = .07$, $p = .65$). This result shows that participants' privacy concern is related with their regulative behavior.

We further examine the pupil size change associated with participants' decision whether to like a sensitive logo. The paired t-test shows that when a sensitive logo is presented, participants' pupil is more constricted (−12.73%) when they like the logo than when they dislike it (−12.11%, $p = .04$). More importantly, a participant's pupil size difference between like and dislike is marginally correlated with his/her IPC scores ($r = -.26$, $p = .08$), providing preliminary evidence that the higher the IPC the stronger the pupil constriction when people like a sensitive logo. This correlation is not observed in the non-sensitive logos ($p = .69$). The findings indicate that people might devote more cognitive resources, reflected in the measurement of pupil size change, when they like a sensitive logo than like a non-sensitive logo and users' privacy concern is related with the cognitive resource devoted for regulation.

6 Discussion

This ongoing study provides behavioral evidence that people regulate their behavior when they use social media. They avoid liking a sensitive logo, which in most cases can violate norms in online communities. Furthermore, people with stronger privacy concern [8] are even less likely to like these sensitive logos, suggesting people's trait can relate to their motivation of regulation. This behavioral regulation pattern is also associated with pupil size change. People's pupils constrict more when they like a sensitive logo than when they dislike it, and people with stronger privacy concern has stronger constriction. The results of this study are consistent with HKL's study, which also shows that the ERN-like brain wave pattern is more evident when people like a sensitive logo than a non-sensitive one, and that the magnitude of this brain wave pattern is also correlated with the same internet privacy concern scale [8]. In conclusion, in this study we interpret the pupil size change as devotion of certain cognitive recourses [9], and therefore our finding indicates that people's devotion of cognitive resources for regulation increases with their privacy concern.

The significance of this study is therefore threefold. First, to our knowledge, this study is probably the first to use pupil size change to study regulation in social media. We show that pupil size change can be a valuable measure to detect some cognitive activities that might be transient or non-verbalizable in the online environment. Second, we use eye-tracking to replicate the findings previously reported using EEG and ERP, enhance robustness of the previous findings [1]. Third, we show that pupil size change, like EEG and ERP, has good temporal resolution and is therefore suitable for finding transient differences in cognitive activities. However, we want to note that compared with EEG and ERP, experiment results from pupil size change are more difficult to interpret because, unlike EEG and ERP data, these results are less specifically associated with certain cognitive functions. For example, using experiment designs and tracing its signal source in the brain, researchers generally agree that the functional significance of ERN can be narrowed down to self-regulation [14]. However, this link between a pupil size change and a specific functional significance is less clear. While the interpretation that pupil size change indicates devotion of cognitive resources is plausible and the most likely [11], so far we cannot completely rule out alternative interpretations, including the relationship between pupil response and problem difficulty or people's interest in a task [15, 16]. Therefore, pupil data are more suitable for studies that only seek to detect whether there is a difference between manipulations or those that have applied experimental paradigms with limited alternative interpretations. Future work of this study includes the comparison of pupil size change between the private and the public condition. Computer-mediated communication studies have shown that the strength of social influence can enhance people's regulation behavior [17, 18] and therefore predict that the pupil size change differ in the two conditions. Future work should provide analysis to test this prediction.

References

1. Huang, Y.-f., Kuo, F.-Y.B., Lin, C.S.: Behavior regulation in social media: a neuroscientific investigation. In: ICIS 2017 Proceedings. Seoul, South Korea (2017)
2. Hofmann, W., Schmeichel, B.J., Baddeley, A.D.: Executive functions and self-regulation. Trends Cogn. Sci. **16**(3), 174–180 (2012)
3. Polites, G.L., Karahanna, E.: The embeddedness of information systems habits in organizational and individual level routines: Development and disruption. MIS Q. **37**(1), 221–246 (2013)
4. Carver, C.S., Scheier, M.F.: Attention and Self-Regulation. Cambridge University Press, New York (1981)
5. Baumeister, R.F., Heatherton, T.F., Tice, D.M.: Losing Control: How and Why People Fail at Self-Regulation. Academic, San Diego (1994)
6. Ward, A., Mann, T.: Don't mind if i do: disinhibited eating under cognitive load. J. Pers. Soc. Psychol. **78**(4), 753–763 (2000)
7. Willard, N.E.: Cyberbullying and cyberthreats: responding to the challenge of online social aggression, threats, and distress. Research Press, Champaign Illinois (2007)
8. Dinev, T., Hart, P.: Internet privacy concerns and their antecedents-measurement validity and a regression model. Behav. Inf. Tech. **23**(6), 413–422 (2004)

9. Shojaeizadeh, M., et al.: Detecting task demand via an eye tracking machine learning system. Decis. Support Syst. **116**, 91–101 (2019)
10. Kahneman, D., Beatty, J.: Pupil diameter and load on memory. Science **154**(3756), 1583–1585 (1966)
11. Fehrenbacher, D.D., Djamasbi, S.: Information systems and task demand: an exploratory pupillometry study of computerized decision making. Decis. Support Syst. **97**, 1–11 (2017)
12. Conrad, C., Bailey, L.: What can neurois learn from the replication crisis in psychological science? In: Davis, F.D. (ed.) Information Systems and Neuroscience, pp. 129–135. Springer, Switzerland (2019)
13. Einhäuser, W., et al.: Pupil dilation reflects perceptual selection and predicts subsequent stability in perceptual rivalry. Proc. Natl. Acad. Sci. **105**(5), 1704–1709 (2008)
14. Luu, P., Tucker, D.M., Makeig, S.: Frontal midline theta and the error-related negativity: neurophysiological mechanisms of action regulation. Clin. Neurophysiol. **115**(8), 1821–1835 (2004)
15. Hess, E.H., Polt, J.M.: Pupil size in relation to mental activity during simple problem-solving. Science **143**(3611), 1190–1192 (1964)
16. Hess, E.H., Polt, J.M.: Pupil size as related to interest value of visual stimuli. Science **132**(3423), 349–350 (1960)
17. Kiesler, S., Siegel, J., McGuire, T.W.: Social psychological aspects of computer-mediated communication. Am. Psychol. **39**(10), 1123–1134 (1984)
18. Spears, R., Lea, M.: Social influence and the influence of the 'social' in computer-mediated communication. In: Lea, M. (ed.). Harvester Wheatsheaf (1992)

A NeuroIS Investigation of the Effects of a Digital Dark Nudge

Francis Joseph Costello, Jin Ho Yun, and Kun Chang Lee(✉)

Sungkyunkwan University, Seoul, South Korea
joe.costehello@gmail.com, jin.ho.yun90@gmail.com,
kunchanglee@gmail.com

Abstract. Based on the behavioral economic theory, the recent uptake in the use of Nudge Theory, has been seen in many companies aiding consumers' decision-making processes. However, recently online channels have started to use this technique for profiteering purposes, coined in this paper as a digital dark nudge. While cases of good nudge have been extensively studied, research on their dark sides and the effects these may have are absent and unclear. We demonstrate in a pilot study on 92 participants proof of concept and thus conceptualize the next steps in analyzing this phenomenon through the lens of NeuroIs. Through this process we aim to identify whether digital dark nudges will be detrimental to a consumers' buying intentions and whether they may have a long-term negative impact on those engaged in the use of them.

Keywords: Digital dark nudging · Online commerce · Nudge theory · fNIRS · NeuroIS

1 Introduction

For over a decade now, most companies have been engaging in electronic channels (i.e. e-commerce and m-commerce), forcing incumbent organizations to replace physical proximity with online convenience. This has led to a wider engagement from consumers in digital platforms for satisfying their product and service needs [1]. This new economic reality is transforming the way companies do business and has meant the need to explore new types of strategies for engaging consumers onto their online channels [2]. One avenue some companies have taken is the use of behavioral economics (BE) within decision support systems [3], influencing the way users engage with and view the designs presented on these platforms. One BE theory widely implemented is that of Nudge theory or known online as a digital nudge [4]. Digital nudging is based on the well-known theory: *Nudge Theory*. Winning Richard Thaler the Nobel Memorial Prize in Economic Sciences in 2017, the rise and use of this type of BE is gradually gaining popularity and has helped to nudge people into improving their economic and health decisions, known as *libertarian paternalism* [5]. The use of online digital nudges is premised on three principles that help guide their use, namely: all nudges should be transparent and

F. D. Davis et al. (Eds.): NeuroIS 2020, LNISO 43, pp. 64–70, 2020.
https://doi.org/10.1007/978-3-030-60073-0_8

never misleading; and it should be easy to opt out, preferably within one click; and, there should be a good reason to believe that the behavior being encouraged will be in that person's interest to engage [5].

Recently, however, the use of nudge-like tactics has been found to be negatively influencing design patterns and the overall customer experience on online channels. In 2015 for the New York Times, Thaler highlighted private sector cases in which Nudge Theory was being utilized, however, while violating his three key principles from nudge theory, claiming "The offer was misleading, opting out was cumbersome, and the entire package did not seem to be in the best interest of a potential subscriber" [6]. Despite this warning, similar patterns have emerged in many online platforms [7] where profiteering on the inherent economic behavioral traits that humans have has been seen [3].

Despite this type of online behavior being first explored within the user interface field, known as a dark pattern [8], this research has been limited to dark pattern taxonomies and research on how prominent they are in website design [7, 8]. Our research attempts to move past this and look at the individual level effects of these dark nudges (including their use of dark patterns). For these reasons we have incorporated nudge theory as the underlying motivating principle of this study [6]. Furthermore, nudge theory is an exploration of individual decision-making through a BE lens and thus can be directly measured through NeuroIS tools [9]. With the digital nudging field starting to blossom, attempting to understand ethical concerns early on with empirical analysis seems a timely and valid study [10]. Ample evidence exists of good nudges [11–15], but, till now, no study has attempted to empirically test the phenomenon of dark nudging in the context of e-commerce (hereby coined as a digital dark nudge (DDN)). Secondly, no study has coupled this investigation with an exploration of the neurophysiological underpinnings that occur during a decision-making task involving a DDN. This paper therefore looks to rectify this through the lens of a NeuroIS investigation by using functional near-infrared spectroscopy (fNIRS) in order to explore this decision-making process.

The research objectives of this study are to: (1) define what a dark nudge is and how it is presented in modern e-commerce. After a dark nudge is operationalized, we will (2) perform a behavioral pilot study to acquire proof of concept. We also aim to identify if it has a negative effect on consumers behavioral responses upon the e-commerce shopping experience. Through this process, we will (3) identify ways to design a future fNIRS study in which we will compare the neural underpinnings when a consumer is exposed to distinct digital dark versus good nudge conditions.

2 Planned Research and Expected Outcomes

2.1 A Typology Digital Dark Nudges

In an attempt to directly operationalize a dark nudge, we can view it as any advertisement or website design that does not follow all or at least acouple of Thaler's three principles of transparency, in the best interest of the consumer, and easy to opt out [5, 6]. Many examples of the use of dark nudges have been seen in dark pattern website design [16], however, we attempt to operationalize them using BE and their individual level effects on human decision-making. In an example, one common dark pattern known as a hidden cost is a common strategy whereby the initial price of an item (usually shown in a discount type format), hikes up in price as the user moves through the purchase page. In contrast, a good nudge would be upfront about these types of costs within the purchase process. Through the lens of a dark nudge and BE, this is known as a shrouded attribute [17], the use of hidden adds-on and surcharges in a purchasing scenario. Profiteering on human behavioral instincts and ignoring the promotion of people's own interests has shown that libertarian paternalism has been potentially hijacked in favor of profits [18]. Thus, the ethical concerns of this need to be empirically explored [10].

As consumers use online channels for their consumption, they are constantly having to make BE decisions. In order to capture the underlying processing of consumers' decision-making in the DDN context, we will employ a neurocognitive examination using an fNIRS methodology. We will investigate how young consumers process decision-making when exposed to shrouded attributes (i.e. hidden costs). The application of fNIRS within cognition research is highly reliable for measuring the values of oxygenation in hemoglobin in the prefrontal cortex [19]. In our context, as young consumers have extensive experiences with digital platforms including online search, purchases, and social networking, we have little anticipation to identify emotional aspects toward a DDN. Therefore, our study aligns with an analysis on the cognitive aspects. Specifically, we expect to observe the dorsolateral prefrontal cortex (DLPFC) activity when young consumers are confronted with a DDN. The DLPFC has previously been explored for identifying higher-order functions that include cognition and problem solving [9]. Prior literature also points to the role of the DLPFC activation seen in the decisions to accept or reject unfair offers as part of a wider decision-making function involving other regions of the brain. This suggests that the DLPFC is potentially part of a network that modulates the relative impact of fairness motives and self-interest goals within decision making [20].

2.2 Hypotheses

As identified, there is a rise in the use of DDNs within online channels [7] and therefore understanding the effects these have on consumers is needed. In the short-term companies have all to gain from implementing DDNs, however, it in the long run we hypothesize this can only lead to negative consequences whereby the consumer becomes aware and smart to such schemes [18]. To borrow from the BE literature, myopic consumers will ultimately become sophisticated to such tactics played by willing companies [17]. Ultimately consumers will opt for more transparent processes whereby DDNs are not employed. Thus, the following hypotheses are proposed based on the literature thus far:

H1: For the Digital Dark Nudge Condition, Consumers Are More Likely to Recognize Hidden Costs, and Thus Will Decrease the Likelihood of a Future Purchase Decision.

H2: For the Digital Dark Nudge Condition, Consumers Are More Likely to Recognize Hidden Costs, and Thus Will Decrease the Likelihood of a Future Recommendations.

Additionally, from the viewpoint of a neurophysiological response, when consumers are presented with such DDN stimulus they have to make a decision on whether to buy or not. However, before they make this decision, consumers are expected to stop to make judgements, such as value, leading to an increased cognitive process [20]. As previously explored, there are multiple potential brain regions that play a role in this cognitive decision-making [9, 19, 20] and thus, we hypothesize:

H3 [1]: *Oxygenation Hemoglobin Values in the DLPFC Will Be Greater in Response to a Digital Dark Nudge Compared to a Good Nudge Condition.*

3 Behavioral Pilot Study

3.1 Experimental Design

We recruited 92 college-aged undergraduate participants (female = 61) for the pilot test. Additionally, we opted to employ a one-way (dark vs. Good) between-subjects design whereby participants were randomly assigned to each of the condition based on the software randomization technique. Participants were presented a story that told them they had a specific budget for a product/service and that they had spent one hour searching the internet before stumbling on the presented product/service. We prepared two price categories in order to find out if any price differences could be found, with $100 and $300 being chosen as acceptable prices for the given product/service scnereos.

3.2 Manipulation of Conditions

In the case of the good nudge scenario (see Fig. 1) a screen showed the main product, original price (including extra costs) and then the final price in a larger font with a bold style to make the final price clear. Next, participants were asked about purchase intention before being presented with a final payment screen. Next valence and arousal from a SAM test [21] was presented in order to understand the emotion of the participant after knowing the fact. Next two questions were asked based on future repeat purchase intention and likelihood of future recommendations to friends and family.

For the DDN condition, the design of the stimuli was slightly different. For the same product, we included a discount price page before the main purchasing page and kept this price the focus of the advertisement (i.e. $80 was the discount price and this was advertised in greater font size and bold on the final page, despite the true price being $100 which was seen in small font and not bold). The same layout was used: i.e. SAM test, future purchase intention, and recommendation in order to statistically compare the participants' opinions on the two conditions.

3.3 Variables and Measures

Four dependent variables existed within the pilot test. First, we wanted to test the emotion of the participants after they had seen either one of the two conditions. This was done with SAM on a 9-point Likert scale [21]. Secondly, participants were asked about their

[1] This hypothesis is specific to the planned fNIRS pilot study that will be conducted in the near future.

future repeat purchase intention after seeing the stimulus. The next dependent variable was the likelihood of a future recommendation to friends and family. Both questions were measured with a 7-Likert scale item in order to allow for a test between each condition. Further, we employed three control variables: age, gender, and prior experience with purchasing a product/service online based on a scale of never, 1–2 times, more than 3 times (Fig. 2).

Fig. 1. First phase for the good nudge.

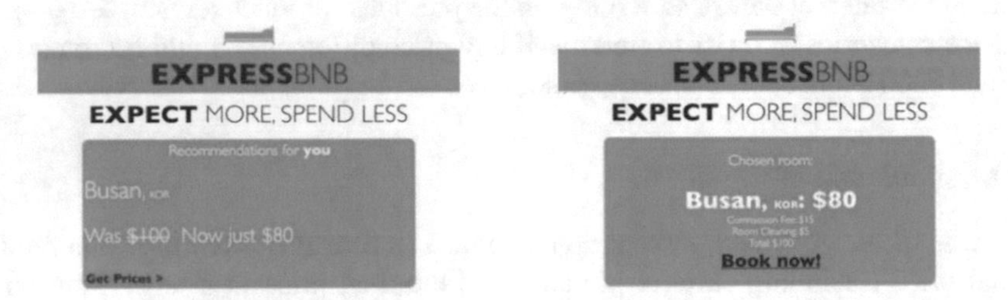

Fig. 2. First and second phase for the dark nudge

3.4 Pilot Test Results

Based on our initial expectations, the pilot test results were consistent with what was originally hypothesized (H1, H2). Behavioral results from two repeated e-commerce conditions (i.e. AirBnb and SkyScanner) were averaged for each dark and good nudge, respectively. To begin with, initial purchase decision showed no significant differences between the two conditions ($p > .10$). Additionally, SAM, measured by valence and arousal, showed the same reaction ($p > .10$). In regard to good and dark nudges, these results reflect the fact that most people will continue with a purchase and feel indifference in their emotion. This is where DDNs have been seen to be effective in that they can attract customers into a purchase decision without alerting the consumers to the fact that it may not be honest.

However, the results from the future purchase intention and recommendation revealed significant differences between the two conditions. One-way analysis of covariance (ANCOVA) results revealed that participants showed higher future purchase intention for the good nudge ($M = 4.16$) than the one for the DDN ($M = 3.38$, $F(1, 88) = 7.494$, $p = .007$, $\eta^2 = .078$) whilst controlling for gender, age and online purchasing experience. They also showed a higher likelihood of future recommendation for the good nudge ($M = 4.20$) than for the DDN condition ($M = 3.40$, $F(1, 88) = 7.898$, $p = .006$, $\eta^2 = .082$) whilst controlling for gender, age and online purchasing experience.

3.5 Planned Data Acquisition and Analysis

We further plan to perform the fNIRS to study the dissociable neurophysiological reaction to good and dark nudges. The within-subject design will be employed as the stimuli will be presented sequentially within each good and DDN condition. 12 trials will be repeated for each condition, presenting a diverse range of scenarios from companies engaged in online channels.

For the fNIRS analysis, the OBELAB and NIRS-SPM toolbox will used to measure both oxygenation and deoxygenation hemoglobin values. The preprocessed data for each subject will be converted to the General Linear Model (GLM). Then, the group-level analyses will be tested to compare between the DDN and good nudge condition.

4 Discussion and Contribution

Based on the results of a pilot study we have found that the potential impact of a digital dark nudge could have an effect on consumers decisions to engage economically with an online provider of services and goods. Ultimately, this could lead to a reduced likelihood of future purchase intention as well as recommendations for that given product/service and thus needs further attention. Within this working paper we have provided early evidence that should now be further investigated thorough the use of NeuroIS tools in order to understand the underlying neurophysiological effects at play coupled with the behavioral data obtained in this study.

References

1. Manu, A.: Transforming Organizations for the Subscription Economy. Routledge, London (2017)
2. Choudhury, V., Karahanna, E.: The relative advantage of electronic channels: a multidimensional view. MIS Q. **32**, 179–200 (2008)
3. Arnott, D., Gao, S.: Behavioral economics for decision support systems researchers. Decis. Support Syst. **122**, 113063 (2019)
4. Weinmann, M., Schneider, C., vom Brocke, J.: Digital nudging. Bus. Inf. Syst. Eng. **58**, 433–436 (2016)
5. Thaler, R.H.: The power of nudges, for good and bad. The New York Times **26**, 1–5 (2015)
6. Thaler, R.H., Sunstein, C.R.: Nudge: Improving Decisions About Health, Wealth, and Happiness, 1st edn. Penguin, Michigan (2009)
7. Mathur, A., Acar, G., Friedman, M.J., et al.: Dark patterns at scale: findings from a crawl of 11K shopping websites. In: Proceedings of the ACM on Human-Computer Interaction, vol. 3(CSCW), pp. 1–32 (2019)
8. Gray, C.M., Kou, Y., Battles, B., et al.: The dark (patterns) side of UX design. In: Conference on Human Factors in Computing Systems - Proceedings 2018, pp. 1–14, April 2018

9. Dimoka, D., Gupta, A., et al.: On the use of neurophysiological tools in IS research: developing a research agenda for NeuroIS. MIS Q. **36**, 679 (2012)
10. Lembcke, T.-B., Engelbrecht, N., Brendel, A.B., Kolbe, L.: To nudge or not to nudge: ethical considerations of digital nudging based on its behavioral economics roots. In: ECIS 2019 Proceedings, pp 1–17 (2019)
11. Benartzi, S., Beshears, J., Milkman, K.L., et al.: Should governments invest more in nudging? Psychol. Sci. **28**, 1041–1055 (2017)
12. Capraro, V., Jagfeld, G., Klein, R., et al.: Increasing altruistic and cooperative behaviour with simple moral nudges. Sci. Rep. **9**, 1–11 (2019)
13. Laran, J., Janiszewski, C., Salerno, A.: Nonconscious nudges: encouraging sustained goal pursuit. J. Consum. Res. **46**, 307–329 (2019)
14. Romero, M., Biswas, D.: Healthy-left, unhealthy-right: can displaying healthy items to the left (versus right) of unhealthy items nudge healthier choices? J. Consum. Res. **43**, 103–112 (2016)
15. Spence, C.: Gastrophysics: nudging consumers toward eating more leafy (salad) greens. Food Qual. Prefer. **80**, 103800 (2020)
16. Brignull H, Darlo A (2019) WHAT ARE DARK PATTERNS? https://www.darkpatterns.org/. Accessed 12 Mar 2020
17. Gabaix, X., Laibson, D.: Shrouded attributes, consumer myopia, and information suppression in competitive markets. Q. J. Econ. **121**, 505–540 (2006)
18. White, M.D.: The Manipulation of Choice: Ethics and Luibertarian Paternalism. Springer, New York (2013)
19. Doi, H., Nishitani, S., Shinohara, K.: NIRS as a tool for assaying emotional function in the prefrontal cortex. Front. Hum. Neurosci. **7**, 1–6 (2013)
20. Knoch, D., Pascual-Leone, A., Meyer, K., et al.: Diminishing reciprocal fairness by disrupting the right prefrontal cortex. Science **314**, 829–832 (2006)
21. Bradley, M.M., Lang, P.J.: Measuring emotion: the self-assessment manikin and the semantic differential. J. Behav. Ther. Exp. Psychiatry **25**, 49–59 (1994)

Technostress Measurement in the Field: A Case Report

Thomas Fischer[1(✉)] and René Riedl[1,2]

[1] University of Applied Sciences Upper Austria, Steyr, Austria
thomas.fischer@jku.at, rene.riedl@fh-steyr.at
[2] Johannes Kepler University Linz, Linz, Austria

Abstract. Contemporary technostress research is mainly based on studies in laboratory settings and online surveys. To foster technostress research in the field, we compared four data collection methods, a blend of self-reports and physiological measurements, in the context of a case study in one organization. Over three non-consecutive workweeks, 16 participants filled out online surveys, wrote an online diary, wore a chest strap to measure their heart rate, and measured their blood pressure using a wrist-worn device. All four methods were assumed to imply a low level of intrusiveness as it enabled self-measurement by the participants without the need for continuous researcher intervention. The four data collection methods are compared based on six major criteria to determine measurement quality (i.e., reliability, validity, sensitivity, diagnosticity, objectivity, and intrusiveness). We find that each data collection method has its strengths and weaknesses. What follows is the need for mixed methods designs in technostress field studies.

Keywords: Blood pressure · Case report · Case study · Field study · Heart rate · Survey · Technostress

1 Introduction

Technostress (hereafter TS) is a phenomenon that arises from "[d]irect human interaction with ICT, as well as perceptions, emotions, and thoughts regarding the implementation of ICT in organizations and its pervasiveness in society in general…" ([1], p. 18). Research has demonstrated that TS can have detrimental effects such as substantial activations of physiological stress mechanisms (e.g., elevated release of stress hormones, [2]) or reduced individual performance (e.g., [3]). To date, most of the research on TS has been conducted in laboratory settings and/or has relied heavily on self-report measures [4, 5]. Although surveys and laboratory studies can also be useful to create practical insights on the emergence of technostress in a large number of organizations (e.g., [6, 7]), we argue that creating insights that are more specific to a certain organization and can then be translated into organization-specific interventions (e.g., through tailored stress management trainings, [8]), requires research within organizations (i.e., research *in situ*, [9]).

F. D. Davis et al. (Eds.): NeuroIS 2020, LNISO 43, pp. 71–78, 2020.
https://doi.org/10.1007/978-3-030-60073-0_9

Investigations of this kind are often accompanied by substantial effort for both researchers (e.g., time to identify an appropriate case organization, effort to coordinate data collection and to actually collect data through personal interviews or observations) and study participants (e.g., work interruptions to collect data, feeling of being monitored). Hence, it is crucial to deliberately select data collection methods that minimize effort for all involved stakeholders, yet provide the necessary scientific rigor, and also allow for insights that are relevant to researchers and practitioners. In the present case report, we defined a set of measures that fulfill these criteria and tested them in a longitudinal field study. The measures we defined constitute a blend of self-reports and physiological measurements. As the human stress response in general, and hence also the TS response, are based on two mechanisms, physiological (e.g., [1]) *and* psychological (refer to work by Lazarus, as cited in [5]), such a blend is critical to establish a maximum of insight in TS research.

2 Case Study Setting and Data Collection Methods

From December 2017 to June 2018 we investigated TS within a case organization in Austria. The company is a publisher that employs approximately 25 individuals at its headquarters. 16 of these individuals participated in a longitudinal study, which involved the application of a set of data collection methods over a period of three non-consecutive workweeks (one in March, one in May, and one in June).

A set of four data collection methods was chosen for this case study. We decided to select two self-report methods (i.e., online surveys and an online diary) and two neurophysiological methods (i.e., heart rate measurement and blood pressure measurement) as these types of methods have been most frequently employed in previous TS research [4]. It has to be noted though that there is a wide variety of additional neurophysiological measures that have previously been applied in technostress studies [10]. For example, Tams et al. [3] collected salivary alpha-amylase as a physiological indicator of stress in addition to self-reports. In the context of this study though, the management of the case organization expressed their concerns regarding more intrusive measures such as markers of endocrine system activity (e.g., hormones collected through blood, urine, or saliva samples, [1]). Therefore, measures of autonomic nervous system activity were favored, as related data can be gathered through devices that many individuals are already familiar with (e.g., devices to track sportive activity, which also gather heart rate data).

Online surveys were administered at the end of each workweek through an online platform (i.e., SoSciSurvey[1]). During the first workweek it took participants on average 45 min to fill out the survey, while in the two other weeks completion took participants around 20 min. The difference was mainly caused by questions related to individual characteristics such as personality or attitudes that were only included in the first week. The *online diary* was based on the "Day Reconstruction Method" proposed by Kahneman et al. [11] and involved participants using an online tool implemented using Microsoft Sharepoint® to indicate their activities and used technologies throughout the day. Participants were instructed to report in 30 min blocks and they were able to select

[1] https://www.soscisurvey.de/ [last access on 02/25/2020].

from previously generated keywords (e.g., a list of activities relevant to their job). *Heart rate* (HR) measurement was implemented using a consumer-grade chest strap used commonly for sportive activities (i.e., Polar H7, [12]) in combination with a smartphone app to gather the collected data (see also [13] and [14] for more details on this setup). Participants put on the chest strap when they came into the office and started measurement, which led to continuous data collection throughout the workday. *Blood pressure* (BP) measurement was conducted by participants themselves using a wrist-worn device (i.e., Omron RS8, [15]). In line with protocols for longitudinal studies (see, for example, [16] and [17]), participants were instructed to measure twice a day, shortly after they arrived at the office and again shortly before they left the office. At each measurement point (upon arrival and when they left), BP was measured two times and then averaged. Please refer to the online material of the article by Fischer and Riedl [18] for further details on the applied data collection procedures (in German).

3 Methods Evaluation

To compare the four data collection methods, the quality criteria for measurement methods presented by Riedl et al. [19] are used (see Table 1). In addition to these criteria, the subjective experiences of the researchers are also involved in the evaluation of the suitability of the data collection methods for field research.

Table 1. Quality criteria for measurement methods based on Riedl et al. [19], p. xxix

Criterion	Criterion descriptions
Reliability	*Reliability.* The extent to which a measurement instrument is free of measurement error, and therefore yields the same results on repeated measurement of the same construct
Validity	*Validity.* The extent to which a measurement instrument measures the construct that it purports to measure
Sensitivity	*Sensitivity.* A property of a measure that describes how well it differentiates values along the continuum inherent in a construct
Diagnosticity	*Diagnosticity.* A property of a measure that describes how precisely it captures a target construct as opposed to other constructs
Objectivity	*Objectivity.* The extent to which research results are independent from the investigator and reported in a way so that replication is possible
Intrusiveness	*Intrusiveness.* The extent to which a measurement instrument interferes with an ongoing task, thereby distorting the investigated construct

Reliability. *Online surveys* were mainly based on existing instruments, which have been applied repeatedly and demonstrated their reliability (e.g., [20]). The *online diary* included a pre-configured list of activities, though these activities were in many cases abstracted so that they would be relevant to all participants. Hence, this level of abstraction can be a potential cause of reduced reliability as it leaves room for interpretation.

According to Riedl et al. [19], *HR* measures are reliable generally, though our used data collection setup was a source of reduced reliability. For example, the Bluetooth connection between chest strap and smartphone was not always given, which led to rare instances of data loss or skewed data. For *BP* measurement a variety of interfering factors exists (e.g., arm position, room temperature, noise level, [16]). Therefore, in the context of this study, even with substantial participant training, doubtful data was not uncommon (e.g., unexpectedly high or low values).

Validity. Specific *survey* instruments for the measurement of aspects of TS (e.g., stressors, [20]) exist. Yet, their content validity has been called into question before (e.g., [21]). The *online diary* also included the option to enter text (e.g., to describe instances of computer hassles); yet, it is less specific as compared to surveys. For neurophysiological measurement methods in general, Riedl et al. [19] highlight that individual physiological indicators should not be interpreted in isolation as they can, in many cases, represent different constructs. This need for triangulation with other sources of data was particularly visible for BP in our study, which can be influenced by many interfering variables and at the same time was only measured at discrete points in time (as compared to HR, which was measured continuously and therefore, based on the amount of data available, allows for more fine-grained analyses).

Sensitivity. *Survey* items were usually measured using a scale from 1 (lowest level of agreement) to 7, and each construct usually included several items to increase sensitivity (e.g., as opposed to more global statements such as "Did you feel stressed?"). The *diary* had the potential to provide us with more fine-grained data (e.g., specific stressful events), yet we found that individuals usually only reported uncommon or uniquely stressful events (e.g., program shutdowns), which skewed the data available from this source. With one data point per second, *HR* provided us with the most fine-grained data, with a potential for high sensitivity (e.g., responses to stressful stimuli in seconds, [14]). *BP* provided us with data that was only more fine-grained than survey data (measurement twice per day as compared to once per week), which limits its sensitivity significantly (e.g., as compared to HR).

Diagnosticity. Items included in *surveys* are often general statements that can be potentially related to many constructs. Hence, repeated applications of the same instruments are needed to prove their diagnosticity (e.g., by demonstrating discriminant validity in comparison with potentially related constructs). High diagnosticity can therefore only be ensured for established survey instruments. In the case of the diary, instances of TS could be reported, though the reported events were often abstracted (i.e., a whole situation rather than a specific stressor) and therefore potentially mixed with other forms of stress (e.g., usability problems with a system and work-related stress due to time pressure at the same time). For HR and BP, diagnosticity cannot be evaluated in isolation as there is no specific TS marker or some comparable metric that would be specific for this phenomenon. Hence, when it comes to specifically measuring stress due to ICT, self-report instruments are superior as compared to neurophysiological measures. The rationale is that self-report measures can be tailored to measure the specific construct of interest on a fine-grained level, whereas neurophysiological measures allow for more general insights into stress physiology that do not necessarily allow for a straight-forward distinction between stress responses based on the type of stressor (e.g., whether a spike in HR was caused by an ICT-related problem or some other acute stressor such as task overload or social conflict is, in field settings, usually not totally clear based on HR data alone).

Objectivity. Several biases have been found to potentially influence the responses to *surveys* (e.g., common method bias, [22]; social desirability bias, [23]). Although anonymity was guaranteed in our study, wrong answers made with deliberation cannot be completely ruled out. Guaranteeing anonymity was an even bigger problem for the diary, as entries can, in most cases, be directly traced back to specific individuals (e.g., due to specific combinations of activities). HR and BP are not as open for manipulation as self-reports, with HR being measured continuously on a level of granularity that makes specific manipulations difficult (e.g., deliberately lowering HR in our study context is hardly possible), though this possibility would have existed for BP (e.g., by elevating the arm slightly during measurement to lower BP). It has to be noted though that there are also systematic confounders that can affect the data collected through neurophysiological measures (e.g., drinking coffee or smoking in the case of BP [16]). Hence, in order to ensure objectivity, the circumstances of measurement have to be documented carefully.

Intrusiveness. At this point, we use the response rates as a potential indicator for the participants' effort related to each method, and this effort, in turn, was used to assess a method's potential intrusiveness in the work routine. Completing the *survey* led to 20 to 30 min of effort for participants per workweek, and the response rates diminished over the study period (15 completed surveys in week 1, 10 in week 2, 8 in week 3). Filling out the *diary* resulted in 20 to 30 min of effort per workday. In the first study week, we had 57 days with at least one entry by a study participant (i.e., 71% of a max. Value of 80, where 80 = 16 individuals × 5 workdays). This value diminished to 45 in week 2 (56%) and 37 in week 3 (46%). In the optimal case, study participants only had to invest around five minutes of their workday into the collection of *HR* data (i.e., to put on the chest strap and start data collection and then stop data collection and take down the chest strap again). Yet, due to unexpected technical problems (e.g., lack of a Bluetooth connection between chest strap and smartphone), this was not always the case. In the first study week, we received 47 usable HR data sets (59% of 80), 36 in week 2 (45%) and 44 in week 3 (55%). *BP* measurement required about 10 min per participant per workday. In the first study week, we received 76 usable datasets (48% of 160, where 160 = 16 individuals × 5 workdays × 2 measurements upon arrival and just before they left the company), 59 in week 2 (37%) and 40 in week 3 (25%).

4 Ranking of Data Collection Methods and Conclusion

Based on the discussion in the previous section and our own experiences from the reported case study, we created a comparative ranking of the four data collection methods using the six quality criteria. The results of this ranking are portrayed in Table 2, with 1 indicating the best suitability of a method with respect to a certain criterion. It should be noted that this ranking is relative and based on the subjective impressions of the researchers involved in this specific case study. Yet, the experiences that we made in our field study constitute a viable basis for future studies. Future studies could, for example, explore in which types of research settings this ranking may change or which types of additional research methods (e.g., experience sampling as a mix of a survey and a diary, [24]) could be even better suited for a comparable research setting.

Table 2. Comparative ranking of the data collection methods (rank 1 = best method, …)

	Surveys	Diary	HR	BP
Reliability	1	3	2	4
Validity	1	2	3	4
Sensitivity	3	4	1	2
Diagnosticity	1	2	3	4
Objectivity	2	4	1	3
Intrusiveness	2	3	1	4

First, this study showed that consumer-grade devices can be used to collect data in the field without substantial interventions by involved researchers (see, for example, [25] or [26] on this "lifelogging" approach to data collection in the context of organizational stress research). Perhaps more importantly though, this study showed that each data collection method has strengths and weaknesses, and therefore several methods should be used in combination to create a more holistic picture of the TS process.

For example, *survey* instruments are crucial to assess results of individual appraisal and they constitute a low-effort approach to the initial assessment of perceptions (e.g., related to general topics such as the organizational climate). The *diary* can then be used to collect perceptual data on a more fine-grained level as a means to enhance survey data (e.g., specific events that occur on a regular basis and could be sources of TS). In addition to self-reports and its potential limitations (e.g., social desirability), physiological measures can then be used as a means to gather data unobtrusively and in a continuous fashion (i.e., in the case of *HR*). This continuous stream of data can then be triangulated with self-report data (e.g., to identify and prioritize the most straining instances of TS), which then creates the basis for interventions (e.g., stress reduction training). The effective combination of these four data collection methods during a longitudinal study such as the one that was presented in this article is dependent on the purpose of a study though. For example, if there is a need to identify specific stressors, then more emphasis should be put on the frequent application of self-report measures to assess distress appraisal, while the assessment of physiological well-being would demand more intense use of biological measures (e.g., including devices that can capture data on a clinical level of quality). In addition, the timing of the use of certain types of measures (e.g., more quantitative measures such as surveys and physiological measures or more qualitative measures such as diaries) also depends on the purpose of the study (see, for example, introductory books into mixed methods research such as [27, 28] for the rationale behind certain orders in which methods can be applied).

Acknowledgement. This research was funded by the Austrian Science Fund (FWF) as part of the project "Technostress in organizations" (project number: P 30865) at the University of Applied Sciences Upper Austria.

References

1. Riedl, R.: On the biology of technostress: literature review and research agenda. DATA BASE Adv. Inf. Syst. **44**, 18–55 (2013)
2. Riedl, R., Kindermann, H., Auinger, A., Javor, A.: Technostress from a neurobiological perspective - system breakdown increases the stress hormone cortisol in computer users. Bus. Inf. Syst. Eng. **4**, 61–69 (2012)
3. Tams, S., Hill, K., Ortiz de Guinea, A., Thatcher, J., Grover, V.: NeuroIS - alternative or complement to existing methods? Illustrating the holistic effects of neuroscience and self-reported data in the context of technostress research. J. Assoc. Inf. Syst. **15**, 723–753 (2014)
4. Fischer, T., Riedl, R.: Technostress research: a nurturing ground for measurement pluralism? Commun. Assoc. Inf. Syst. **40**, 375–401 (2017)
5. Fischer, T., Riedl, R.: Theorizing technostress in organizations: a cybernetic approach. In: Thomas, O., Teuteberg, F. (eds.) Proceedings of the 12th International Conference on Wirtschaftsinformatik, pp. 1453–1467 (2015)
6. Tams, S., Ahuja, M., Thatcher, J., Grover, V.: Worker stress in the age of mobile technology. The combined effects of perceived interruption overload and worker control. J. Strateg. Inf. Syst. **29**, 101595 (2020)
7. Tams, S., Thatcher, J.B., Grover, V.: Concentration, competence, confidence, and capture. An experimental study of age, interruption-based technostress, and task performance. J. Assoc. Inf. Syst. **19**, 857–908 (2018)
8. Arnetz, B.B.: Techno-stress: a prospective psychophysiological study of the impact of a controlled stress-reduction program in advanced telecommunication systems design work. J. Occup. Environ. Med. **38**, 53–65 (1996)
9. Fischer, T., Riedl, R.: Neurois in situ: on the need for neurois research in the field to study organizational phenomena. In: Liang, T.-P., Yen, N.-S. (eds.) Workshop on Information and Neural Decision Sciences, pp. 20–21 (2014)
10. Fischer, T., Riedl, R.: The status quo of neurophysiology in organizational technostress research: a review of studies published from 1978 to 2015. In: Davis, F.D., Riedl, R., Vom Brocke, J., Léger, P.M., Randolph, A.B. (eds.) Information Systems and Neuroscience, 10, pp. 9–17. Springer, Cham (2015)
11. Kahneman, D., Krueger, A.B., Schkade, D.A., Schwarz, N., Stone, A.A.: A survey method for characterizing daily life experience: the day reconstruction method. Science **306**, 1776–1780 (2004)
12. Plews, D.J., Scott, B., Altini, M., Wood, M., Kilding, A.E., Laursen, P.B.: Comparison of heart-rate-variability recording with smartphone photoplethysmography, polar H7 chest strap, and electrocardiography. Int. J. Sports Physiol. Perform. **12**, 1324–1328 (2017)
13. Baumgartner, D., Fischer, T., Riedl, R., Dreiseitl, S.: Analysis of heart rate variability (HRV) feature robustness for measuring technostress. In: Davis, F.D., Riedl, R., Vom Brocke, J., Léger, P.-M., Randolph, A. (eds.) Information Systems and Neuroscience: NeuroIS Retreat 2018, pp. 221–228. Springer, Heidelberg (2018)

14. Kalischko, T., Fischer, T., Riedl, R.: Techno-unreliability: a pilot study in the field. In: Davis, F.D., Riedl, R., Vom Brocke, J., Léger, P.-M., Randolph, A. (eds.) Information Systems and Neuroscience: NeuroIS Retreat 2019, pp. 137–145. Springer, Heidelberg (2019)
15. Takahashi, H., Yoshika, M., Yokoi, T.: Validation of Omron RS8, RS6, and RS3 home blood pressure monitoring devices, in accordance with the European Society of Hypertension International Protocol revision 2010. Vasc. Health Risk Manage. **9**, 265–272 (2013)
16. Fischer, T., Halmerbauer, G., Meyr, E., Riedl, R.: Blood pressure measurement: a classic of stress measurement and its role in technostress research. In: Davis, F.D., Riedl, R., Vom Brocke, J., Léger, P.-M., Randolph, A.B. (eds.) Information Systems and Neuroscience. Gmunden Retreat on NeuroIS 2017, pp. 25–35. Springer, Cham (2017)
17. Ogedegbe, G., Pickering, T.: Principles and techniques of blood pressure measurement. Cardiol. Clin. **28**, 571–586 (2010)
18. Fischer, T., Riedl, R.: Messung von digitalem Stress im organisationalen Umfeld. Erfahrungen aus einer Fallstudie. HMD - Praxis der Wirtschaftsinformatik **57**, 218–229 (2020)
19. Riedl, R., Davis, F.D., Hevner, A.R.: Towards a NeuroIS research methodology: intensifying the discussion on methods, tools, and measurement. J. Assoc. Inf. Syst. **15**, i–xxxv (2014)
20. Ragu-Nathan, T.S., Tarafdar, M., Ragu-Nathan, B.S., Tu, Q.: The consequences of technostress for end users in organizations: conceptual development and empirical validation. Inf. Syst. Res. **19**, 417–433 (2008)
21. Fischer, T., Pehböck, A., Riedl, R.: Is the technostress creators inventory still an up-to-date measurement instrument? Results of a large-scale interview study. In: Ludwig, T., Pipek, V. (eds.) Proceedings of the 14th International Conference on Wirtschaftsinformatik, pp. 1834–1845 (2019)
22. Podsakoff, P.M.: Self-reports in organizational research: problems and prospects. J. Manage. **12**, 531–544 (1986)
23. Hebert, J.R., Clemow, L., Pbert, L., Ockene, I.S., Ockene, J.K.: Social desirability bias in dietary self-report may compromise the validity of dietary intake measures. Int. J. Epidemiol. **24**, 389–398 (1995)
24. Atz, U.: Evaluating experience sampling of stress in a single-subject research design. Pers. Ubiquit. Comput. **17**, 639–652 (2013)
25. Fischer, T., Riedl, R.: Lifelogging for Organizational Stress Measurement. Theory and Applications . Springer, Cham (2018)
26. Fischer, T., Riedl, R.: Lifelogging as a viable data source for NeuroIS researchers: a review of neurophysiological data types collected in the lifelogging literature. In: Davis, F.D., Riedl, R., Vom Brocke, J., Léger, P.-M., Randolph, A. (eds.) Information Systems and Neuroscience: Gmunden Retreat on Neurois 2016, pp. 165–174. Springer, Heidelberg (2016)
27. Tashakkori, A., Teddlie, C. (eds.): Handbook of Mixed Methods in Social & Behavioral Research. SAGE Publications, Thousand Oaks (2003)
28. Creswell, J.W.: Research Design: Qualitative, Quantitative, and Mixed Methods Approaches. SAGE Publications, Thousand Oaks (2003)

Do Users Respond to Challenging and Hindering Techno-Stressors Differently? A Laboratory Experiment

Christoph Weinert(✉), Katharina Pflügner, and Christian Maier

Information Systems and Services, University of Bamberg, Bamberg, Germany
{christoph.weinert,katharina.pfluegner,
christian.maier}@uni-bamberg.de

Abstract. Techno–stressors are typically hindering for users. These then cause adverse user responses, such as techno-exhaustion, which in turn result in reduced task performance. Latest technostress research adds two types of stressors: hindrance techno-stressors (HTS) and challenge techno–stressors (CTS). Using that knowledge, this research-in-progress paper develops a research model assuming that both types of techno-stressors lead to different user responses (e.g., motivation, techno-exhaustion, arousal) and, in turn, have a different impact on task performance. To validate that empirically, we propose a mixed-experimental research design following a pre-post approach with three different treatments (e.g., HTS, CTS, control) using among other different biomarkers (e.g., SC, sAA, cortisol) to measure arousal. The expected contributions and future steps are discussed.

Keywords: Challenge/Hindrance stressors · Technostress · Exhaustion · Arousal · Task performance · Skin conductance · Alpha-amylase · Cortisol

1 Introduction

The use of Information Systems (IS), mainly if they do not work appropriately, challenges or hinders users. This has the potential to create technostress [1], which is stress caused by using an IS [2]. Most qualitative and quantitative examinations focus on the hindering side of technostress [2–4]. Here techno-stressors, which are technology-related conditions potentially stimulating technostress [1], lead to adverse user responses. For example, users are psychologically exhausted [4, 5] and show physiological body reactions such as sweaty hands [6, 7] and the release of cortisol [8]. All these responses then potentially reduce the task performance of users [3, 4, 9].

Psychological research shows that both a good and a bad side of stress exist by differentiating between challenge-stressors and hindrance-stressors [10, 11]. Challenge-stressors (CS) are evaluated as an opportunity because users can learn and gain personal growth. On the contrary, hindrance-stressors (HC) are evaluated as hindering such that they obstruct the users from the achievement of their goals, such as they can work faster and more time-efficient [12].

F. D. Davis et al. (Eds.): NeuroIS 2020, LNISO 43, pp. 79–89, 2020.
https://doi.org/10.1007/978-3-030-60073-0_10

However, although these stressors – challenging techno-stressors (CTS) and hindrance techno-stressors (HTS) are introduced to the technostress context [12], it is currently not known how users respond to CTS and HTS and how these responses influence task performance. The user in terms of his or her psychological and physiological responses might differ when encountering CTS and HTS, and hence, have a different effect on task performance. Therefore, the research question is:

How do user responses differ when perceiving a CTS or HTS? How do CTS and HTS influence task performance?

The paper is structured as follows. Next, a short overview of the challenge and hindrance framework is given to introduce challenge and hindrance stressors. Afterward, the transactional process of technostress is explained. The combination of both represents the aim of the present research. Subsequently, the research model is developed, and the design of the laboratory experiment is outlined. Lastly, the paper discusses the expected contributions of the research-in-progress paper.

2 Theoretical Background

2.1 Technostress

Technostress is defined as the perception of stress caused by the use of IS [1]. Technostress is based on the transactional theory of stress [13, 14] and should be understood as a transactional process comprising techno-stressors, user responses, and outcomes [2].

Techno-stressors are technology-related stimuli [1], which lead to psychological and physiological user responses and, in turn, to behavioral outcomes [15, 16].

Psychological user responses reflect psychological reactions to a techno-stressor [16]. Different reactions have been examined in previous literature, such as job and user satisfaction [2], burnout [16, 17], or addiction [18]. However, most investigations interpret psychological user responses as techno-exhaustion in prior technostress literature [4, 5, 9, 19], which is based on emotional exhaustion from burnout research [20, 21] and understood as the feeling of tension and depletion of emotional resources [20].

Physiological user responses include bodily responses to techno-stressors such as cardiovascular, biochemical and gastrointestinal symptoms [15] and has been the focus of several investigations of users' biological responses to techno-stressors [8, 22, 23] (see [22] for an extensive description of physiological user responses). It is mostly understood as physiological arousal (hereafter arousal) which activates the autonomic nervous system and is manifested by emotional sweating [7, 24, 25],, blood pressure [26], pupil dilation [27], faster heartbeat [28] as well as an increase of the cortisol level [23].

Behavioral user responses include behavioral responses to the techno-stressors, such as lower task performance [3, 9, 29].

2.2 Challenge and Hindrance Framework

Based on the transactional stress theory [13, 14], the challenge and hindrance (CH) framework consider *"stress as an individual's psychological response to a situation in which there is something at stake for the individual"* [10]. The framework differentiates, following the primary appraisal process within stress theory [13], between challenge and hindrance stressor. *"The situation is appraised with respect to whether it is potentially challenging– beneficial or threatening– harmful"* [10]. The framework suggests that challenge stressors might be positively and hindrance stressors negatively related to performance. The framework states that two mechanisms are responsible for the effect of why these stressors influence performance. First, challenge stressors increase the motivation to learn, and hindrance stressors reduce this motivation, which leads to a different effect on performance. Second, both stressors lead to exhaustion, which reduces performance (see Fig. 1 [10]).

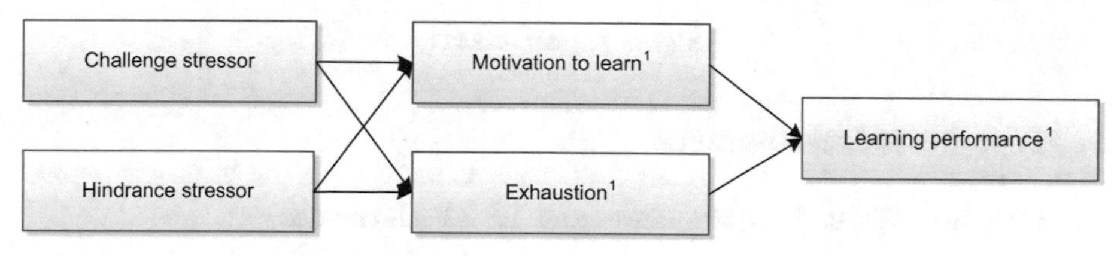

Fig. 1. Challenge and hindrance framework [10]

2.3 The Present Research

Extent technostress literature mostly considers bad stress and neglects the differentiation between HTS which are defined as *"stressors that hinder progress toward personal goal attainment and accomplishments"* [12] and CTS which is understood as *"stressors that present opportunities for individual learning and personal growth"* [12]. The present research combines the CH framework with technostress literature by concentrating on CTS and HTS and, more importantly, focuses on their different effect on task performance. Thereby, psychological literature indicates that the effect of HTS and CTS on task performance is mediated by motivation and exhaustion [10]. NeuroIS literature on technostress demonstrates that besides the psychological user responses (e.g., motivation, exhaustion) also physiological user responses in terms of arousal acts as a mediator [22, 26, 30] [1]. As shown in Fig. 2, the effect of motivation has been adapted to the technostress context. Techno-exhaustion is an essential part of both research streams, and arousal has been derived from extent technostress literature, indicating that techno-stressors lead to psychological (techno-exhaustion) as well as physiological (arousal) user responses, which are responsible for the effect of CTS and HTS on task performance.

[1] As the paper focuses on the different user responses and their effect on task performance the relationship between the mediators will only be considered as covariables but a theoretical relation should be the focus of future research.

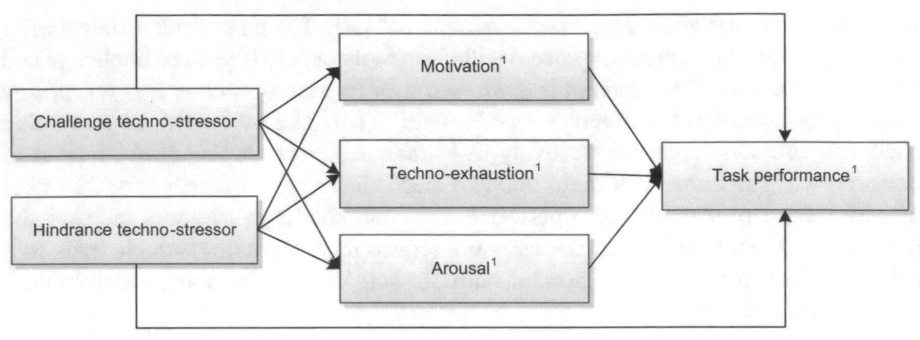

Fig. 2. Present research

3 Hypotheses Development

3.1 CTS and HTS on User Responses and Task Performance

The CH framework demonstrates that challenge stressors positively influence motivation, whereas hindrance stressors reduce it [10]. Empirical evidence also validates this relationship [10]. The reason might be that the encountering of CTS is characterized by opportunity so that users aim to learn and gain personal growth [12]. Besides, CTS are related to higher approach orientation, such as learning and openness [31]. Hence, the willingness to learn and approach orientation implies a high motivation. On the contrary, HTS are characterized by hindering the achievement [12] and avoidance orientation, such as withdrawal, which indicates lower motivation [31]. Therefore, it is assumed that:

H1a: Users who encounter a CTS have a higher motivation than users who encounter a HTS.

The negative influence of HTS on techno-exhaustion has been empirically validated [3–5, 9]. The underlying rationale subsumes that the hindrance, by using the IS, leads to the tension and depletion of one's emotional resources and so increases techno-exhaustion. The negative effect of CS has also been shown in psychological research [10]. One explanation is that the processing of information associated with the desire to learn and grow personally in the challenge-situation is also consuming emotional resources, which results in higher techno-exhaustion. Hence, it is assumed that CTS and HTS have both a negative effect on techno-exhaustion so that we assume that:

H1b: Users who encounter a CTS have the same high techno-exhaustion level than users who encounter a HTS.

The effect of HTS on arousal has been empirically shown in several investigations [8, 24]. One explanation might be that in a HTS situation, the body needs more oxygen, but the arteries remain narrowed, resulting in individuals being more aroused because the body has to work harder to get oxygen [31]. On the contrary, initial evidence demonstrates that the body also responds to CTS [27]. The need for more oxygen also characterizes the arousal response towards a CTS. However, the arteries dilate, and more oxygen can be transported such that the body is less aroused [31]. Also, related literature indicates that individuals show a more efficient arousal pattern in challenging situations [32]. The difference in the arousal level can also be reasoned by psychological literature, indicating that CS leads only to an activation of the sympathetic-adrenal-medullary (SAM). In contrast, the encounter of HS leads to the activation of SAM as well as hypothalamic-pituitary-adrenal (HPA) axis [31], indicating a higher arousal level. Therefore, we assume that:

H1c: Users who encounter a CTS are less aroused than users who encounter a HTS.

As mentioned above, the encounter of CTS is characterized by opportunity. For example, users who are forced to process a vast amount of tasks quickly might think that they can gain acknowledgment from co-workers and supervisors by trying to work as fast as they can, which results in high task performance. Contrary, HTS are characterized by hindrance [12], and several investigations demonstrate that HTS reduces task performance [16, 33]. For instance, users who are hindered by interruptions think that they cannot manage the task, resulting in lower task performance. Therefore, it is assumed that:

H1d: Users who encounter a CTS have a better task performance than users who encounter a HTS.

3.2 User Responses on Task Performance

Previous literature indicates that motivation [10, 34] increases task performance. Motivation refers to the willingness to accomplish the task. Fundamental behavioral theories demonstrate that the willingness to perform a behavior increases the actual behavior [35]. Hence, the higher the motivation to accomplish the task, the higher the task performance. As a CTS and HTS differ, it is assumed that high motivation leads to high task performance when encountering a CTS, and on the contrary, that low motivation leads to low task performance when encountering a HTS. Therefore, it is assumed that:

H2a: Motivation is positively associated with task performance when encountering a CTS or HTS.

The negative effect of techno-exhaustion on task performance has been shown previously [3, 9]. The encounter with a techno-stressor leads to a feeling of tension and depletion of one's emotional resources [5] and, in turn, the energy to manage the situation is limited. Thus, task performance is at a lower level [3]. As both - CTS and HTS – consume emotional resources, it is assumed that techno-exhaustion reduces task performance in both situations. Therefore, it is assumed:

H2b: Techno-exhaustion is negatively associated with task performance when encountering a CTS or HTS.

Technostress literature shows that arousal influences task performance [9]. The relationship between arousal and task performance follows a u-shaped curve indicating that an increase in arousal results in a rise in task performance up to a given arousal level where the effect turns, and task performance reduces [36]. Assuming that CTS and HTS lead to different arousal patterns-one more efficient than the other [31, 32], the arousal level in a challenge situation is lower and has not reached the turning point such that it increases task performance, whereas the arousal level in a hindrance situation has passed the turning point and is too high so that it harms task performance [32]. Therefore, it is assumed that:

H2c: Arousal increases task performance when encountering a CTS and decreases task performance when encountering a HTS.

4 Methodology

To validate the research model, the paper follows an experimental approach. The design, procedure, manipulation, tasks and technology, and the measurement techniques used are described next.

Design. The laboratory experiment follows a uni-factorial pre-post mixed design concentrating on the factor techno-stressor with three factorial-levels: CTS, HTS, which represents the experimental groups and control group.

Manipulation. To intentionally manipulate a CTS and HTS, the paper focused on specific facets from each, which have been previously manipulated. From the ten facets of CTS and HTS [12] the paper focuses on time urgency as a CTS and hassle or obstacles as a HTS. By manipulating time urgency, the paper is based on an experimental manipulation that has been previously conducted [37]. Time urgency or pressure is defined as the situation where individuals think that they have not enough time to accomplish their tasks in the given time [37]. Subjects had three minutes to accomplish the task, which was not possible at that time duration. Additionally, the subjects were every 30 s reminded how much time they had left [37]. Moreover, the paper derives the manipulation of *hassle or obstacles* from prior work [8, 38] by aiming to simulate a computer breakdown in the form of an error message. The subjects encounter for a total of three minutes each 30 s a new error message.

Tasks and Technology. The experiment contains three tasks (see Fig. 3). The first (pre) and third (post) tasks are used to capture the change in task performance when encountering a CTS and HTS. Therefore, we adapted a memory test that has already been used in previous technostress experiments [3, 9]. The second task is part of the manipulation; subjects work on a work-related task by using MS SharePoint to simulate a real work environment.

Measurement. In the experiment, subjective and objective data are measured [30]. Valid measurement scales are used to capture techno-exhaustion [3] and motivation [34]. Arousal is objectively measured in line with past literature by using several biomarkers [39] because the encounter of techno-stressors activates the autonomic nervous system

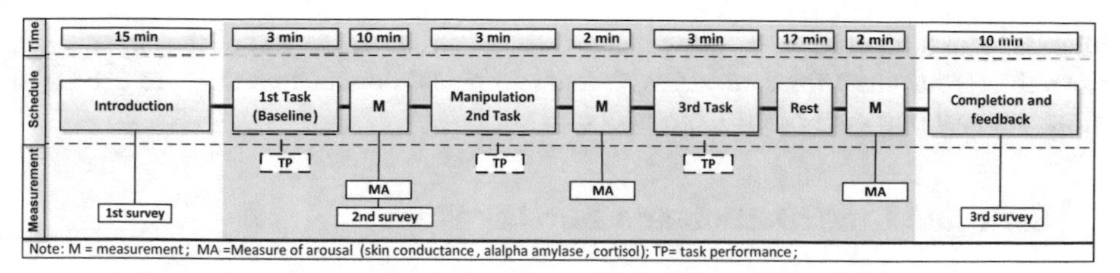

Fig. 3. Experimental procedure

(ANS) as well as the HPA axis [22]. The ANS encompassing a sympathetic and a parasympathetic arm is responsible for immediate response to stressor encounters. The activation of the sympathetic arm in specific the SAM leads to an increase in sweat gland activity measured as skin conductance (SC) [7, 24, 25] and in salivary alpha-amylase (sAA) [23, 40]. SC measures the activity of the ANS, capturing the conductivity of the skin. Three different pathways exist which go from the nervous system to the sweat glands [41, 42]. Theses pathways are taken by encountering stimuli, which lead to an activation of the eccrine sweat glands causing a sweat secretion. An endosomatic method will be applied, which does not use an external current for the measure by using a MentalBioScreen K3 device. sAA is an enzyme which reflects SAM activity and can be measured by salivary analysis [22, 40, 43]. This biomarker has been used previously in the technostress context [23]. The activation of the HPA axis is especially assumed when encountering HTS [31]. Cortisol is a stress hormone in humans that indicates the activation of the HPA axis, which can also be measured by salivary analysis. The activation of the HPA axis is slower and therefore, the release of cortisol takes 20 to 25 min after the stressor has been encountered [8].

Task performance is measured by the number of matching pairs uncovered in the allocated time [3]. To control the manipulation validity, the challenge and hindrance scale by Benlian [12] is used. Also, the main control variables within the context of the technostress will be captured.

Procedure. Subjects are randomly assigned by using a stratified randomization strategy (strata: gender) [44] to allocate men and women equally because of the cortisol measurement. For example, the cortisol level is affected by the female menstrual cycle such that an equal assignment of men and women is needed as well as the control of such individual gender effects.

The procedure of the experiment is shown in Fig. 3. During the introduction, the subjects are fitted with the SC equipment and fill out the first survey. After that, the first task begins where the subjects work on the memory task, which is used to measure task performance. Afterward, the arousal level is measured, and the subjects have to fill out the second survey. The manipulation phase starts by intentionally triggering the CTS and HTS (see manipulation). Subsequently, arousal is again measured. The third task follows where the subjects again work on the memory task to capture task performance. A resting period follows during which the subjects can relax until the last measurement phase starts, where the arousal level (especially the measurement of cortisol because it takes 20 min until cortisol is fully released) is measured. At the end of the experiment, the subjects fill out a third survey with demographical details, and they are debriefed about the real purposes of the experiment.

5 Expected Contribution and Further Steps

The paper aims to investigate the effect of CTS and HTS on user responses such as motivation, techno-exhaustion, and arousal and, in turn, on task performance. To do so, we focus on the challenge and hindrance framework and discuss it in relation to existing technostress research. We expect that the results of the planned experiment contribute to technostress literature [1, 12] by offering further insights into how CTS and HTS influence the user and his or her behavior. More importantly, the paper extends previous literature concentrating on techno-exhaustion [4, 5, 9] and arousal [8, 22, 23] by revealing whether these user responses differ when encountering CTS or HTS. Among others, while there are first indications that both influence the user, it is unclear whether there are different paths or processes through which either CTS or HTS have an influence. Moreover, we further strengthen insights on challenge and hindrance technostress [12] by providing theory in how these influence physiological user responses. The present paper extends previous examinations by investigating whether arousal differs when perceiving a CTS or HTS. In addition, investigating the effect of motivation, techno–exhaustion and arousal on task performance might extend previous technostress literature [3, 9], which focuses primarily on task performance. In this context, the role of arousal might be significant as CTS might lead to a more efficient arousal pattern by pupping more oxygen through the body, such that the subject has a task performance, whereas HTS might result in a too high arousal level because the arteries constrict and less oxygen is transmitted, which decreases task performance. In sum, we aim to provide a more coherent perspective on how CTS and HTS influence a user's life and thereby respect that techno-stressors typically influence the user physiologically, psychologically and behaviorally.

The experiment will be conducted in the second quarter of 2020 with flowing data analyses, so that the research can be finished in the third quarter of 2020.

6 Conclusion

Resent technostress literature indicates that not only HTS exist but also CTS. Hence, the research-in-progress paper aims to investigate the different user responses towards both–CTS and HTS, and in turn, their influences on task performance. A research model considering the three user responses motivation, techno-exhaustion, and arousal, as well as their effect on task performance, is developed. To validate the research model, an experimental design is presented, which includes different biomarkers (i.e., SC, sAA, cortisol) to measure arousal, among others. Expected contributions and further steps are discussed.

References

1. Tarafdar, M., Cooper, C.L., Stich, J.-F.: The technostress trifecta - techno eustress, techno distress and design. Theoretical directions and an agenda for research. Inf. Syst. J. **29**, 6–42 (2019)
2. Ragu-Nathan, T.S., Tarafdar, M., Ragu-Nathan, B.S., Tu, Q.: The consequences of technostress for end users in organizations: conceptual development and empirical validation. Inf. Syst. Res. **19**, 417–433 (2008)
3. Tams, S., Thatcher, J.B., Grover, V.: Concentration, competence, confidence, and capture. An experimental study of age, interruption-based technostress, and task performance. J. Assoc. Inf. Syst. **19**, 857–908 (2018)
4. Maier, C., Laumer, S., Weinert, C., Weitzel, T.: The effects of technostress and switching stress on discontinued use of social networking services. A study of Facebook use. Inf. Syst. J. **25**, 275–308 (2015)
5. Ayyagari, R., Grover, V., Russell, P.: Technostress: technological antecedents and implications. MIS Q. **35**, 831–858 (2011)
6. Weinert, C., Maier, C., Laumer, S., Weitzel, T.: Technostress mitigation: an experimental study of social support during a computer freeze. J. Bus. Econ. (Forthcoming)
7. Weinert, C., Maier, C., Laumer, S.: What does the skin tell us about information systems usage? A literature-based analysis of the utilization of electrodermal measurement for IS research. In: Davis, Fred D., Riedl, R., vom Brocke, J., Léger, P.-M., Randolph, Adriane B. (eds.) Information Systems and Neuroscience. LNISO, vol. 10, pp. 65–75. Springer, Cham (2015). https://doi.org/10.1007/978-3-319-18702-0_9
8. Riedl, R., Kindermann, H., Auinger, A., Javor, A.: Technostress from a neurobiological perspective. On the biology of technostress: literature review. Bus. Inf. Syst. Eng. **4**, 61–69 (2012)
9. Tams, S., Hill, K., Ortiz de Guinea, A., Thatcher, J., Grover, V.: NeuroIS—alternative or complement to existing methods? Illustrating the holistic effects of neuroscience and self-reported data in the context of technostress research. J. Assoc. Inf. Syst. **15**, 723–752 (2014)
10. LePine, J.A., LePine, M.A., Jackson, C.L.: Challenge and hindrance stress. Relationships with exhaustion, motivation to learn, and learning performance. J. Appl. Psychol. **89**, 883–891 (2004)
11. LePine, M.A., Zhang, Y., Crawford, E.R., Rich, B.L.: Turning their pain to gain. Charismatic leader influence on follower stress appraisal and job performance. Acad. Manag. J. **59**, 1036–1059 (2016)
12. Benlian, A.: A daily field investigation of technology-driven stress spillovers from work to home. MIS Q. (2019, forthcoming)
13. Lazarus, R.S., Folkman, S.: Stress, Appraisal, and Coping. Springer, New York (1984). https://doi.org/10.1007/978-1-4419-1005-9
14. Lazarus, R.S.: Psychological Stress and the Coping Process. McGraw-Hill, New York (1966)
15. Cooper, C.L., Dewe, P., O'Driscoll, M.P.: Organizational Stress. A Review and Critique of Theory, Research, and Applications. Sage Publications, Thousand Oaks (2001)
16. Maier, C., Laumer, S., Wirth, J., Weitzel, T.: Technostress and the hierarchical levels of personality. A two-wave study with multiple data samples. Eur. J. Inf. Syst. **62**, 1–27 (2019)
17. Srivastava, S.C., Chandra, S., Shirish, A.: Technostress creators and job outcomes: theorising the moderating influence of personality traits. Inf. Syst. J. **25**, 355–401 (2015)
18. Tarafdar, M., Maier, C., Laumer, S., Weitzel, T.: Explaining the link between technostress and technology addiction for social networking sites. A study of distraction as a coping behavior. Inf. Syst. J. **62**, 51 (2019)

19. Maier, C., Laumer, S., Eckhardt, A., Weitzel, T.: Giving too much social support. Social overload on social networking sites. Eur. J. Inf. Syst. **24**, 447–464 (2014)
20. Maslach, C., Schaufeli, W.B., Leiter, M.P.: Job Burnout. Ann. Rev. Psychol. **52**, 397–422 (2001)
21. Moore, J.E.: One road to turnover: an examination of work exhaustion in technology professionals. MIS Q. **24**, 141–168 (2000)
22. Riedl, R.: On the biology of technostress. Literature review and research agenda. DATA BASE Adv. Inf. Syst. **44**, 18–55 (2013)
23. Galluch, P.S., Grover, V., Thatcher, J.B.: Interrupting the workplace: examining stressors in an information technology context. J. Assoc. Inf. Syst. **16**, 1–47 (2015)
24. Riedl, R., Kindermann, H., Auinger, A., Javor, A.: Computer breakdown as a stress factor during task completion under time pressure: identifying gender differences based on skin conductance. Adv. Hum.-Comput. Interact. **2013**, 1–7 (2013)
25. Teubner, T., Adam, M., Riordan, R.: The impact of computerized agents on immediate emotions, overall arousal and bidding behavior in electronic auctions. J. Assoc. Inf. Syst. **16**, 838–879 (2015)
26. Fischer, T., Halmerbauer, G., Meyr, E., Riedl, R.: Blood pressure measurement: a classic of stress measurement and its role in technostress research. In: Davis, F.D., Riedl, R., vom Brocke, J., Léger, P.-M., Randolph, A.B. (eds.) Information Systems and Neuroscience. LNISO, vol. 25, pp. 25–35. Springer, Cham (2018). https://doi.org/10.1007/978-3-319-67431-5_4
27. Eckhardt, A., Maier, C., Buettner, R.: The influence of pressure to perform and experience on changing perceptions and user performance: a multi-method experimental analysis. In: Proceedings of the 33rd International Conference on Information Systems - ICIS 2012 (2012)
28. Baumgartner, D., Fischer, T., Riedl, R., Dreiseitl, S.: Analysis of heart rate variability (HRV) feature robustness for measuring technostress. In: Davis, Fred D., Riedl, R., vom Brocke, J., Léger, P.-M., Randolph, Adriane B. (eds.) Information Systems and Neuroscience. LNISO, vol. 29, pp. 221–228. Springer, Cham (2019). https://doi.org/10.1007/978-3-030-01087-4_27
29. Tarafdar, M., Tu, Q., Ragu-Nathan, T.S.: Impact of technostress on end-user satisfaction and performance. J. Manage. Inf. Syst. **27**, 303–334 (2010)
30. Fischer, T., Riedl, R.: Technostress research. A nurturing ground for measurement pluralism? Commun. Assoc. Inf. Syst. **40**, 375–401 (2017)
31. Akinola, M., Kapadia, C., Lu, J.G., Mason, M.F.: Incorporating physiology into creativity research and practice. The effects of bodily stress responses on creativity in organizations. AMP **33**, 163–184 (2019)
32. Uphill, M.A., Rossato, C.J.L., Swain, J., O'Driscoll, J.: Challenge and threat. A critical review of the literature and an alternative conceptualization. Fron. Psychol. **10**, 1255 (2019)
33. Tarafdar, M., Pullins, E.B., Ragu-Nathan, T.S.: Technostress. Negative effect on performance and possible mitigations. Inf. Syst. J. **25**, 103–132 (2015)
34. Salehan, M., Kim, D., Kim, C.: Use of online social networking services from a theoretical perspective of the motivation-participation-performance framework. J. Assoc. Inf. Syst. **18**, 141–172 (2017)
35. Fishbein, M., Ajzen, I.: Belief, Attitude, Intention and Behavior. An Introduction to Theory and Research. Addison-Wesley, Reading (1975)
36. Yerkes, R.M., Dodson, J.D.: The relation of strength of stimulus to rapidity of habit-formation. J. Comp. Neurol. Psychol. **18**, 459–482 (1908)
37. Pearsall, M.J., Ellis, A.P.J., Stein, J.H.: Coping with challenge and hindrance stressors in teams. Behavioral, cognitive, and affective outcomes. Organ. Behav. Hum. Decis. Processes **109**, 18–28 (2009)

38. Riedl, R., Fischer, T.: System response time as a stressor in a digital world: literature review and theoretical model. In: Nah, F.F.-H., Xiao, B.S. (eds.) HCIBGO 2018. LNCS, vol. 10923, pp. 175–186. Springer, Cham (2018). https://doi.org/10.1007/978-3-319-91716-0_14
39. Vogel, J., Auinger, A., Riedl, R.: Cardiovascular, neurophysiological, and biochemical stress indicators: a short review for information systems researchers. In: Davis, Fred D., Riedl, R., vom Brocke, J., Léger, P.-M., Randolph, Adriane B. (eds.) Information Systems and Neuroscience. LNISO, vol. 29, pp. 259–273. Springer, Cham (2019). https://doi.org/10.1007/978-3-030-01087-4_31
40. Petrakova, L., Boy, K., Mittmann, L., Möller, L., Engler, H., Schedlowski, M.: Salivary alpha-amylase and noradrenaline responses to corticotropin-releasing hormone administration in humans. Biol. Psychol. **127**, 34–39 (2017)
41. Dawson, M.E., Schell, A.M., Courtney, C.G.: The skin conductance response, anticipation, and decision-making. J. Neurosci. Psychol. Econ. **4**, 111–116 (2011)
42. Boucsein, W.: Electrodermal Activity. Springer, Boston (2012). https://doi.org/10.1007/978-1-4614-1126-0
43. Harmon, A.G., Towe-Goodman, N.R., Fortunato, C.K., Granger, D.A.: Differences in saliva collection location and disparities in baseline and diurnal rhythms of alpha-amylase. A preliminary note of caution. Horm. Behav. **54**, 592–596 (2008)
44. Kang, M., Ragan, B.G., Park, J.-H.: Issues in outcomes research. An overview of randomization techniques for clinical trials. J. Athletic Training **43**, 215–221 (2008)

Advancing NeuroIS from a Dialectical Perspective

Lars Taxén(✉)

Department of Computer and Information Science, Linköping University, Linköping, Sweden
lars.taxen@gmail.com

Abstract. In spite of decades of research, the essence of central IS phenomena such as information, the IT artifact, and the Information System, remain unsettled. This aggravates the identification and definition of their neural correlates, which in turn may stymie the future potential of NeuroIS. To this end, the purpose of this paper is to define an intellectual core for the NeuroIS field, which is based on the dialectics between the individual and her social environment. Such a core enables a reconceptualization of information, the IT artifact, and the Information System as distinct ontological phenomena, however dialectically related. Accordingly, their neural correlates can be more distinctly defined, which opens up for alternative research questions. An example of a NeuroIS study addressing such questions is provided. Thus, the paper contributes to the formulation of a foundation for the future advancement of NeuroIS.

Keywords: Dialectics · Information · IT-artifact · Information system · Neural correlates · Action · Theory of functional systems · NeuroIS

1 Introduction

NeuroIS is an Information System (IS) subfield that "examines topics lying at the intersection of IS research and neurophysiology and the brain sciences" [1]. Accordingly, NeuroIS draws on and contributes to findings in both the IS and neuroscience disciplines.

The NeuroIS field has made substantial progress over the last decade [2, 3]. However, the nature of central IS phenomena such as information, the IT artifact, and the IS has not been satisfactorily answered in spite of decades of research. The essence of information "has plagued research on information systems since the very beginning" [4, p. 363]. The "vastly inconsistent definitions of the term 'the IT artifact' … demonstrate why it no longer means anything in particular" [5, p. 47]. Likewise, mutually incompatible definitions of ISs have been suggested [6]. Lee amply summarizes the state of play in the IS discipline: "Virtually all the extant IS literature fails to explicitly specify meaning for the very label that identifies it. This is a vital omission, because without defining what we are talking about, we can hardly know it" [7, p. 338]. This discussion is so far by and large unheeded in the NeuroIS community.[1]

[1] For a case in point, see e.g. [2], where "IT artefact" and "IS artefact" are interchangeably referred to.

F. D. Davis et al. (Eds.): NeuroIS 2020, LNISO 43, pp. 90–99, 2020.
https://doi.org/10.1007/978-3-030-60073-0_11

However, the nature of any discipline is to ask research questions "that are not being asked by other disciplines, or questions that other disciplines are incapable of asking" [26, p. 1]. The precarious state of the IS relatum of NeuroIS raises concerns about what NeuroIS research contributes to the *IS discipline.* As vom Brocke and Liang [27, p. 218] underline, NeuroIS "studies that originate from applying neuroscience strategies of inquiry, rather than from contributing to IS theory, are not beneficial to the field." Without a clear understanding of what constituted IS phenomena, NeuroIS faces the double risk of researching topics that are either non IS-relevant or being addressed satisfactorily in other disciplines [26].

To address this issue, IS phenomena need to be further theorized in such a way that reliable methods and tools can be developed in search for their neural correlates [25]. Stated differently, how can central IS phenomena such as information, the IT artifact, and the IS be articulated so that their neural correlates can be defined more distinctly?

To answer this question, an alternative formulation of the fundamental assumptions underlying the IS field is proposed. A fecund point of departure is the fact that "living systems are units of interactions; they exist in an ambience. From a purely biological point of view they cannot be understood independently of that part of the ambience with which they interact: the niche; nor can the niche be defined independently of the living system that specifies it" [30, p. 2]. Consequently, the individual and the social realms *mutually constitute* each other – a *dialectical* perspective (DP for short).

In previous NeuroIS contributions [8, 33, 34, 35, 36], I have addressed various aspects of such a dialectical perspective. The purpose of this paper is to assemble these fragments into a coherent intellectual *core* [10] of the IS discipline, which can be used to articulate neural correlates of IS phenomena.

The line of argument proceeds as follows. After a short outline of dialectics, a mental model – the Theory of Functional Systems (TFS) [16] – for the dialectics between the individual and her environment is described. The IS phenomena information, the IT artifact, and the IS are then related to various stages in TFS. This model is further articulated in terms of neurobiological prerequisites for *acting*, which of course is central to NeuroIS – brains have developed for acting relevantly in the world to the benefit of our species survival.

The dialectical foundation so conceived is used to reconceptualize information, the IT artifact, and the IS as separate ontological phenomena, however dialectically related. Accordingly, their neural correlates can be more distinctly defined, which in turn opens up for stating alternative research questions. An example of a prospective study NeuroIS study for investigating the user interface of an ERP-system is outlined. In conclusion, the paper contributes to the formulation of a theoretical foundation for the further advancement of NeuroIS.

2 Outlining the Dialectical Perspective

Dialectics has a long philosophical tradition from Aristoteles, Hegel, Marx and others [11]. The individual – social dialectics was originally formulated by Marx in his first thesis on Feuerbach:

> The main defect of all hitherto-existing materialism—that of Feuerbach included—is that the Object [*der Gegenstand*], actuality, sensuousness, are conceived only in the form of the object [*Objekts*], or of contemplation [*Anschauung*], but not as human sensuous activity, practice [*Praxis*], not subjectively [12].

The dialectical synthesis Marx proposes is Praxis, in which the object is seen simultaneously as an independently existing, recalcitrant material reality, and a goal or purpose or idea that we have in mind. This stance is referred to as *der Gegenstand* where "*gegen* means against, towards, contrary to, signaling a reality that offers resistance to our efforts and desires, and *der Stand* means category or state of affairs" [13, p. 404, original emphasis].

Thus, the social reality produced and the producers mutually constitute each other: "The very nature or character of a man is determined by what he does or his *praxis*, and his products are concrete embodiments of this activity" [14, p. 44, original emphasis]. The essence of this understanding is that the meaning of phenomena we attend as relevant is "dependent on the orientation and action of people toward them" [15, p. 68]. For NeuroIS, this means that IS phenomena cannot be profoundly understood as detached from the human element.

2.1 The Individual – Social Dialectics

The praxis stance implies that we need to find a conceptualization of action that comprises both the individual and social realms. Such a model is provided by Anokhin's Theory of Functional System (TFS; see Fig. 1), which outlines the main neurobiological functions involved in action [16, 17, 18].

Two groups of functions are identified depending on which kind of nerves are actuated: *afferent* ones going from the periphery of the body to the brain, and *efferent* ones going from the brain to effectors such as muscles or glands.

According to TFS, action proceeds in the following stages. In *Afferent synthesis*, sensations from the external world (*situational afferentation*), previous experiences retained from *memory*, and *motivation* are integrated into a coherent Gestalt of the situation – the perception of a pattern or structure as a whole. Based on this Gestalt, a decision of *what* to do, *how* to do, and *when* to do is taken. *Decision making* involves two functions: anticipation of the expected result (*acceptor of result*) and the formation of an *action program*: "if I act in this way, I assume this will result". *Releasing afferentation* sets off *Efferent excitation*, in which the action is performed. The result is evaluated against the anticipated result via *Back-afferentation*. Depending on the outcome of the evaluation, the sequence is repeated or stopped. The entire episode is then retained in memory for acting relevantly in future, similar situations.

The profound implication of the TFS model is that our neurobiological system enables us to confer signhood onto *any* perceived elements in our environment that we find meaningful and relevant. Something attended in the environment evokes a neural response leading to action. In order to articulate this thinking further, the mental functions in Fig. 1 are called *biomechanical factors*. The corresponding environmental elements are called *communal factors*.

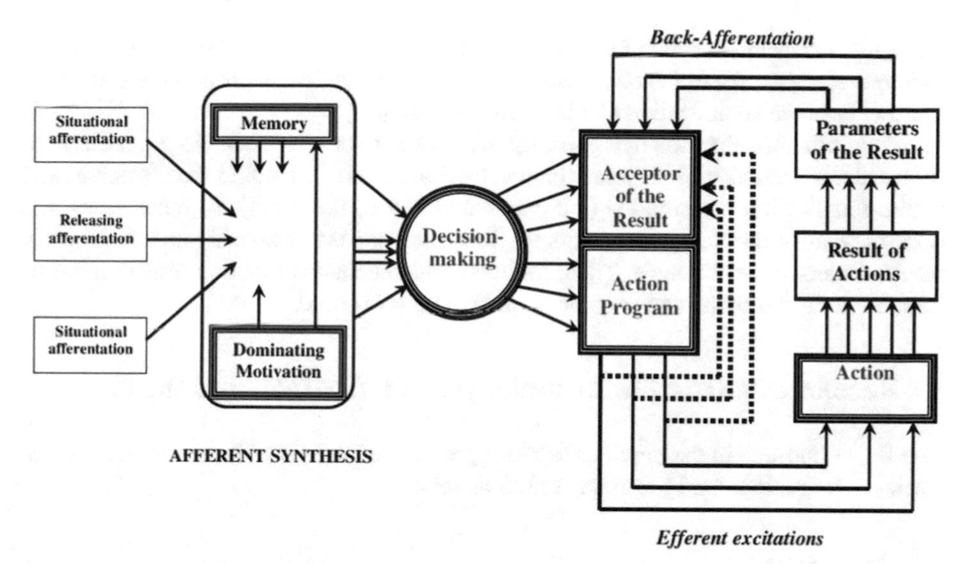

Fig. 1. General architecture of an individual functional system. Thick lines mark internal functions, while thinner lines indicate functions that depend on interaction with the environment. Adapted from [17, p. 115]. With permission.

2.2 Communalization

The TFS describes requisite stages for individual action. However, such actions are usually performed as social acts in a joint, collaborative activity. Joint action is "the larger collective form of action constituted by the fitting together of the lines of behavior of the separate participants…. Joint actions range from a simple collaboration of two individuals to a complex alignment of the acts of huge organizations or institutions" [15, p. 70]. To understand joint action, it is necessary to distinguish between two types of individual actions. When, for example, a pianist gives a recital, she performs an *autonomous* act [19, p. 19]. There is no other musician involved. When the same pianist plays in a piano trio, she also preforms an individual act, but now together with other musicians. Such individual acts, "performed only as parts of joint actions", are called *participatory* ones [19]. Communal factors are requisite for both types of actions. However, in joint actions, these factors become *common identifiers* [15, p. 71] around which participatory actions gravitate.

In order for communal factors to become common identifiers, these have to go through a process of *institutionalization*, i.e. making them into established elements of a community; more or less taken for granted. An example of a common identifier in a musical community is the score, which has evolved over a long period into its present form [28]. Correspondingly, each prospective musician needs to through a process of *socialization* by which she learns to behave in a way that is acceptable to the community. This amounts to learning how to read scores, and how to play according to norms signified by marks on the score.

In the literature, the relationships between socialization and institutionalization are usually investigated from either a socialization or an institutional perspective. However, these two perspectives are in fact related – neither can exist without the other. In order to emphasize this, the term *communalization* is introduced to foreground the dialectical relation between socialization and institutionalization.

Thus, communalization of a musician involves both processes, however on different timescales – socialization in years (biomechanical factors internal to the musician) and institutionalization in centuries (the communal factor; the score). In other cases, the opposite may be the case as, for example when a new app is introduced into a community used to manage similar apps. Then, individual biomechanical factors remain more or less the same, while the app is communalized in a short time.

3 Reconceptualizing Information, the IT Artifact, and the iS

The DP as outlined in the previous sections provides a theoretical basis for reconceptualizing information, the IT artifact, and IS as follows.

3.1 Information

Information is conceived as the Gestalt resulting from the Afferent synthesis stage in TFS. Thereby, the individual *informs* herself about the situation at hand. A profound implication is that "information is constituted – not just interpreted – or symbolically represented and exchanged – but actually constituted *as* information" [20, p. 13, original emphasis]. Thus, information is an inherently individual phenomenon, however requisite on the external world to be constituted.

3.2 The IT Artifact

An IT artifact is a physical artifact comprised of software running on hardware. Such an artifact is intentionally designed to be informative. It should be instrumental in informing the individual about the state of things in the world. Thus, the IT artifact contributes to the constitution of individual information in the Afferent synthesis stage in Fig. 1. However, the IT artifact may also be involved in the other stages. In the Efferent excitation stage, the artifact may be used in action; the result of which is back-propagated in the Back-afferentation stage for evaluation in the acceptor of results. Accordingly, the IT artifact – as in fact any artifact – has both an afferent and efferent side; it must be recognized as relevant, and that something can be done with it.

For the individual, the artifact may proceed from more or less unintelligible to a useful artifact, fluently employed in order to achieve the task at hand. This process can be seen as the gradual formation of an autonomous, *individual* "information system", comprised of the individual's neurobiological structure and the IT artifact.

However, the computerized IT artifact is not the only artifact contributing to the constitution of individual information systems. Hevner et al. suggest that "*constructs* (vocabulary and symbols), *models* (abstractions and representations), *methods* (algorithms and practices), and *instantiations* (implemented and prototype systems)" [39, p. 77, italics in original] also should be considered as IT artifacts.

Such artifacts are sometimes regarded as "nonmaterial" (see e.g. [40]), but in fact, these are just as material as everything else we can perceive and act upon. For example, a model needs to materialized somewhere – on paper, on a computer screen, as a mock-up, talked about, etc. The model may be intangible, as formulated in our minds, but it cannot be *nonmaterial*.[2] Our neural and biological constitution is surely material. If there were such things as "nonmaterial" elements in the world, we could not even talk about them; much less act with them. They would simply be inaccessible to human experience. Accordingly, the discussion of materiality versus nonmateriality is an impasse.

The point here is that *any* perceivable impression from the outside world may in principle contribute to the constitution of information. What is important is whether such impressions are *relevant* or not for the situation at hand.

3.3 The IS

To become useful in a community, an IT artifact has to be communalized into that community. A new ERP-system is just an artifact among others until users begin forming idiosyncratic information systems, clustering, as it were, around the artifact. This process renders the artifact into a communal factor – it becomes an institutionalized Information System comprised of the IT artifact and the ensemble of user's individual information systems in various stages of formation.

During communalization, the IT artifact will be adapted to the specifics of the community, but it will *remain an artifact*. Accordingly, we "do not need to put humans inside the boundary of the IT artifact in order to make these artifacts social" [29, pp. 93–94].

The gist of this conceptualization is that information, the IT artifact, and the IS are ontologically separate phenomena, however dialectically related to each other. One cannot sensibly be understood without the others. Consequently, studying information, the IT artifact, and the IS as isolated, unrelated phenomena will at best provide incomplete results, and at worst fallacious.

4 Articulating Action

The TFS as outlined in Sect. 2 is in essence a theory of action. Brains "evolved to control the activities of bodies in the world... the mental is inextricably interwoven with body, world and action [9, p. 527]. Consequently, a prime concern for NeuroIS is to identify neural correlates of action, and to investigate how these correlates relate to IS phenomena conceptualized as in Sect. 3.

I have proposed that communal factors signifying *objects*, *contexts*, *spaces*, *times*, *norms*, and *transitions* between communities are requisite phenomena characterizing

[2] However, the neural effects of attending and acting with a model may of course be documented by NeuroIS tools and methods.

action.[3] The neural correlates of these factors are conceived as the biomechanical factors *objectivating*, *contextualizing*, *spatializing*, *temporalizing*, *habitualizing*, and *refocusing*.[4] In mainstream neuroscience, the neural underpinnings of such factors are well researched. However, the phenomena that these are neural correlates of, are seldom attended.[5]

In the DP, these six dimensions of action are regarded as a totality that I call the *activity modalities* [e.g. 8, 37]. All modalities are requisite and mutually constituting each other. If anyone modality cannot be actuated due to some neurobiological impairment, action is hampered or downright impossible. Accordingly, the *dialectical relation between the individual and social realms is seen as comprised of the activity modalities*. Importantly, the modalities are necessary but certainly not sufficient for acting. Other mental functions, such as intentions, emotions, trust, etc., also need to be considered. However, the importance of the modalities is that they bridge the neural and social realms. Hence, they direct attention to the *relation* between disciplines.

5 Exemplifying the Dialectical Approach

So far, I have outlined the general features of a dialectical foundation for NeuroIS. To exemplify how this foundation may be employed in NeuroIS studies, I will use the SAP user interface in Fig. 2 [from 38]:

Numbers indicate areas which can be interpreted as the communal factors *object* (the Primary Sync Object "Purchase Order" [1]), *context* (the activity "Build Sync" [2]) *space* (the various encircled entities related to "Purchase Order" [3]), *time* (the sequence of activities from left to right [4]), *norm* (the Sync Number "D47–220-1277" [5]), and *transition* (the "Change Document" indicating a transition to another Sync Object [6]). In the DP interpretation, these communal factors are neural correlates of the biomechanical factors *objectivating, contextualizing, spatializing, temporalizing, habitualizing,* and *refocusing.*

This gist of this argumentation is that any user of the SAP system scans the screen, looking for visual cues signifying the six types of communal factors in order to inform themselves how to act. This suggests a prospective NeuroIS study for investigating the efficiency of the user interface as follows. Independent variables may be manipulations of the interface, such as repositioning of the factors on the screen (e.g. placing the focal object towards the center rather than as now, towards the periphery), coloring the factors differently, devising different icons signifying the factors, removing one or more of the factors, etc.

Dependent variables may be coupled to the various stages in the TFS model in Fig. 1. For example, the time from first glance of the interface to doing something with it, indicates that the Afferent synthesis stage in competed, i.e., the user has informed

[3] These particular factors were identified from long-term observations of and reflections over projects developing complex telecom systems in industry [31,32].

[4] Such mental functions need to be seen as Complex Functional Systems [23] since "no specific function is ever connected with the activity of one single brain center. It is always the product of the integral activity of strictly differentiated, hierarchically interconnected centers" [24, p. 171].

[5] Common characterizations are "world" [e.g. 21], "environment" [e.g. 22] and the like.

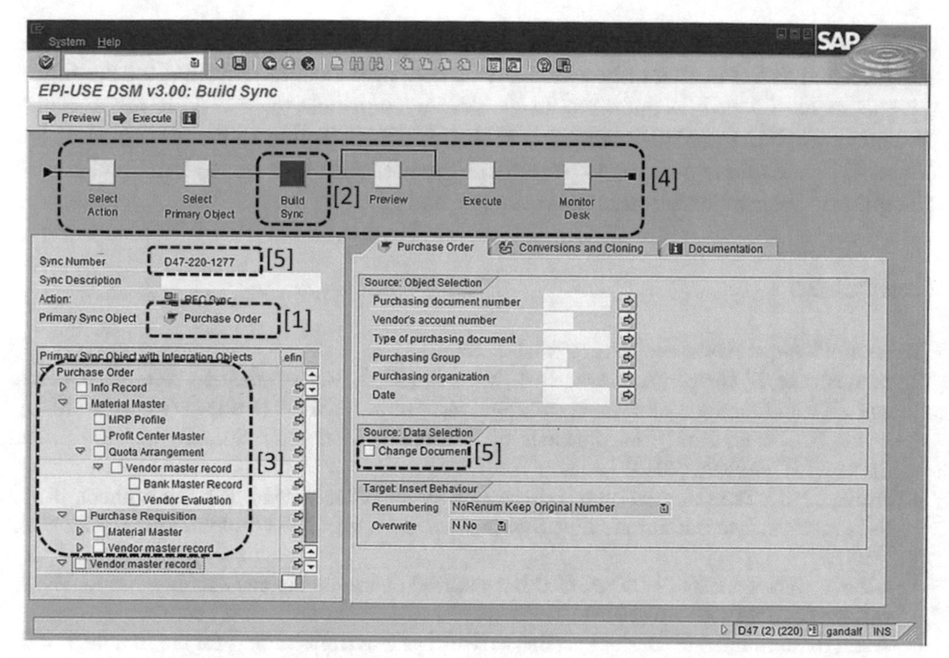

Fig. 2: Example of a SAP Screenshot (Source: Original SAP Screenshot from https://softkat.ueu.org/software/mysap.html) (Data Sync Manager (DSM) from EPI-USE Labs is a solution for copying of data from production to non-production SAP systems (https://www.epiuse.com/products/dsm-product-suite/overview)).

herself about the task and taken a decision to act. Eye-tracking tools may be used to analyze how fast the factors are identified on the screen. This may be augmented with a technostress analysis. Information-seeking stopping behavior indicating the efficiency of Back-afferentation stage might be analyzed by EEG methods. And so on. The result of such studies would be guidelines for designing user interfaces in an optimal way with respect to the totality of the activity modalities.

Another area for NeuroIS studies to investigate, is the communalization of the SAP system into a relevant IS in the organization. How should it be designed to support the clustering of autonomous individual "information system" as efficiently as possible? How easy is it to adapt the SAP artifact to specific community needs? Which other artifacts, such as models, can be employed to speed up the communalization process? Such questions involve both the constitution of individual information, as well as the architecture and properties of the SAP system. Measures needs to be developed indicating the progress of communalization, and how such measures can be analyzed with NeuroIS tools and methods.

6 Conclusion

In a recent paper [2], vom Brocke et al. suggest four key research areas for the future of the NeuroIS field: IS design, IS use, emotion research, and neuro-adaptive systems. This paper proposes an additional key area: elaborating a dialectical core underpinning the NeuroIS and IS fields. Not only would such a core provide a firm ground for the future advancement of NeuroIS, but it would in addition contribute to specifying the meaning of central IS phenomena; something that that the IS discipline so far has failed to do. NeuroIS is in a unique position to pursue this avenue, since it strives to transit otherwise disciplinary-focused research in neuroscience and IS.

References

1. NeuroIS. https://www.neurois.org/what-is-neurois/
2. vom Brocke, J., Hevner, A., Léger, P.M., Walla, P., Riedl, R.: Advancing a NeuroIS research agenda with four areas of societal contributions. Eur. J. Inf. Syst. (EJIS) **29**(1), 9–24 (2020)
3. Riedl, R., Léger, P.M.: Fundamentals of NeuroIS - Information Systems and the Brain. Springer, Heidelberg (2016)
4. Boland, R.J.: The information of information systems. In: Boland, R.J., Hirschheim, R.A. (eds.) Critical Issues in Information Systems Research, pp. 363–379. John Wiley, New York (1987)
5. Alter, S.: The concept of 'IT artifact' has outlived its usefulness and should be retired now. Inf. Syst. J. **25**(1), 47–60 (2015)
6. Alter, S.: Defining information systems as work systems: implications for the IS field. Eur. J. Inf. Syst. **17**(5), 448–469 (2008)
7. Lee, A.S.: Retrospect and prospect: information systems research in the last and next 25 years. J. Inf. Technol. **25**(4), 336–348 (2010). https://doi.org/10.1057/jit.2010.24
8. Taxén, L.: Towards reconceptualizing the core of the IS field from a neurobiological perspective. In: Davis, F.D., Riedl, R., vom Brocke, J., Léger, P.-M., Randolph, A.B. (eds.) Information Systems and Neuroscience - Gmunden Retreat on NeuroIS 2017, pp. 201–210. Springer Cham (2018). DOI: https://doi.org/https://doi.org/10.1007/978-3-319-67431-5_23
9. Love, N.: Cognition and the language myth. Lang. Sci. **26**(6), 525–544 (2004)
10. Sidorova, A., Evangelopoulos, N., Valacich, J.S., Ramakrishnan, T.: Uncovering the intellectual core of the information systems discipline. MIS Q. **32**(3), 467–482 (2008)
11. Wong, W.: Understanding dialectical thinking from a cultural-historical perspective. Philos. Psychol. **19**(2), 239–260 (2006). https://doi.org/10.1080/09515080500462420
12. Marx, K.: The German Ideology: Including Theses on Feuerbach and Introduction to the Critique of Political Economy. Prometheus Books, Amherst (1998) https://www.marxists.org/archive/marx/works/1845/theses/index.htm. Originally published in 1845
13. Adler, P.: The evolving object of software development. Organization **12**(3), 401–435 (2005)
14. Bernstein, R.: Praxis and Action. University of Pennsylvania Press, Philadelphia (1999)
15. Blumer, H.: Symbolic Interactionism: Perspective and Method. Prentice-Hall, Englewood Cliffs (1969)
16. Red'ko, V.G., Prokhorov, D.V., Burtsev, M.B.: Theory of functional systems, adaptive critics and neural networks. In: Proceedings of International Joint Conference on Neural Networks, Budapest, pp. 1787–1792 (2004)
17. Toomela, A.: Biological roots of foresight and mental time travel. Integr. Psychol. Behav. Sci. **44**, 97–125 (2010)
18. Toomela, A.: Towards understanding biological, psychological and cultural mechanisms of anticipation. In: Nadin, M. (ed.) Anticipation: Learning from the Past. The Russian/Soviet Contributions to the Science of Anticipation, pp. 431–456. Springer, Cham (2015)
19. Clark, H.H.: Using Language. Cambridge University Press, Cambridge (1996)

20. Garfinkel, H.: Toward a Sociological Theory of Information. Paradigm Publishers, Boulder (2008)
21. Knudsen, E.I.: Fundamental components of attention. Annu. Rev. Neurosci. **30**, 57–78 (2007)
22. Gratch, J., Marcella, S.: The architectural role of emotions in cognitive systems. In: Gray, W.D. (ed.) Integrated Models of Cognitive Systems, pp. 230–242. Oxford University Press, Oxford (2007)
23. Luria, A.R.: LS Vygotsky and the problem of localization of functions. Neuropsychologia **3**, 387–392 (1965)
24. Akhutina, T.V.: LS Vygotsky and AR Luria: foundations of neuropsychology. J. Russ. East Eur. Psychol. **41**(3/4), 159–190 (2003)
25. Abend, G.: What are neural correlates neural correlates of? BioSocieties **12**, 415–438 (2017)
26. Hassan, N.R.: Constructing the right disciplinary IS questions. In: Twenty-third Americas Conference on Information Systems, Boston, 2017, pp. 1–10 (2017)
27. vom Brocke, J., Liang, T.P.: Guidelines for neuroscience studies in information systems research. J. Manage. Inf. Syst. **30**(4), 211–234 (2014)
28. Hoskin, K.: Spacing, timing and the invention of management. Organization **11**(6), 743–757 (2004)
29. Goldkuhl, G.: The IT artefact: an ensemble of the social and the technical? – a rejoinder. Syst. Signs Actions **7**(1), 90–99 (2013)
30. Maturana, H.R.: Biology of Cognition. BCL Report # 9.0. University of Illinois, Urbana (1970)
31. Taxén, L.: A Framework for the Coordination of Complex Systems' Development. Dissertation No. 800. Linköping University, Dep. of Computer & Information Science (2003). https://liu.diva-portal.org/smash/record.jsf?searchId=1&pid=diva2:20897
32. Taxén, L.: Using Activity Domain Theory for Managing Complex Systems. Information Science Reference. Information Science Reference (IGI Global) Hershey PA. ISBN 978–1–60566–192–6 (2009)
33. Taxén, L.: Information systems and activity modalities – a coordination perspective on the interaction between the neural and the social. In: Gmunden Retreat on NeuroIS 2011, Gmunden, Austria, 26–28 June 2011
34. Taxén, L.: An investigation of model quality from the activity modality perspective. In: Gmunden Retreat on NeuroIS 2012, Gmunden, Austria, 3–6 June 2012
35. Taxén, L.: Reconceptualizing information systems from the activity modality perspective. In: Gmunden Retreat on NeuroIS 2013, Gmunden, Austria, 1–4 June 2013
36. Taxén, L.: An investigation of the nature of information systems from a neurobiological perspective. In: Davis, F., Riedl, R., vom Brocke, J., Léger, P.-M., Randolph, A. Information Systems and Neuroscience, Gmunden Retreat on NeuroIS 2015, pp. 27–34. Springer, Dordrecht (2015). https://doi.org/10.1007/978-3-319-18702-0
37. Taxén, L: Reviving the individual in sociotechnical systems thinking. In: Kowalski, S., Bednar, P., Nolte, A., Bider, I. (eds.) 5th International Workshop on Socio-Technical Perspective in IS development (STPIS 2019), Stockholm, pp. 1–8, 11 June 2019. https://ceur-ws.org/Vol-2398/
38. Taxén, L., and Riedl, R.: Understanding coordination in the information systems domain. J. Inf. Technol. Theory Appl. (JITTA), **17**(1), Article 2, 5–40 (2016). https://aisel.aisnet.org/jitta/vol17/iss1/2
39. Hevner, A., March, S.T., Park, J., Ram, S.: Design science in information systems research. MIS Q. **28**(1), 75–105 (2004)
40. Faulkner, P., Runde, J.: Technological objects, social positions, and the transformational model of social activity. MIS Q. **37**(3), 803–818 (2013)

An Eye-Tracking Study of Differences in Reading Between Automated and Human-Written News

Chenyan Jia[1(✉)] and Jacek Gwizdka[2]

[1] School of Journalism, University of Texas at Austin, Austin, USA
chenyanjia@utexas.edu

[2] School of Information, University of Texas at Austin, Austin, USA
neurois2020@gwizdka.com

Abstract. An eye-tracking experiment ($N = 24$) was conducted to study differences in reading between automated and human-written news. This work adopted expectation-confirmation theory to examine readers' prior expectations and actual perceptions of both human-written news and automated news. Results revealed that nine eye-tracking variables were significantly different when people read automated news vs. human-written news. Findings also showed promising classification results of 31 eye-tracking-derived features. Self-reported results showed that the readability of human-written news was perceived as significantly higher than that of automated news.

Keywords: Automated journalism · Human-written news · Eye-tracking · Expectation-confirmation theory · Readability

1 Introduction and Background

As increasing number of news organizations have adopted automation technology to generate news [1], auto-generated content is spreading faster than anticipated. Automated journalism refers to the news generated by algorithms with little human intervention [2, 3]. While automation was designed to free up human journalists from repetitive work, the readability of automated journalism remains underexplored. Readability is the ease with which one can read a text, which depends largely on the writer's capacity to simplify words, grammar, and sentences [4]. It can be considered as a linguistic measure, particularly for news stories [5]. The readability of automated journalism has been mostly discussed in the journalistic realms. Given that developers are unlikely to fully disclose their algorithms [6], some scholars in the communication field have measured the readability of automated journalism by conducting experimental study to ask readers rate their perceptions of news [e.g. 7–9]. Others focused on the texts generated by algorithms by conducting qualitative textual analysis and content analysis [10]. Previous researchers found that people rate human-written news as more readable than automated news [7, 8, 11]. For instance, Clerwall [7] adapted the readability measure that was used to evaluate

F. D. Davis et al. (Eds.): NeuroIS 2020, LNISO 43, pp. 100–110, 2020.
https://doi.org/10.1007/978-3-030-60073-0_12

print news stories [12] into the field of automated journalism, namely credibility, liking, quality, and representativeness. Graefe and colleagues [13] conducted an exploratory factor analysis and found that news readability included four dimensions, respectively coherence, conciseness, comprehensiveness, and descriptiveness. These previous studies showed that readers favoured human-written news instead of automated news in terms of readability [9, 13].

Since news stories generally contain longer words and sentences than other writings [14, 15], some researchers quantified the concept of readability by using Flesch-Kincaid reading ease score which is based on the average sentence length and average syllables per word [16, 17]. A higher score indicates an easier-to-read text. An alternative way to examine the reading process is to use eye-tracking tools. Eye movements can reflect the reading process [18] and reveal whether readers encounter difficulties in text comprehension [19, 20]. Previous studies found that the level of domain knowledge, types of task, cognitive abilities and individual differences had effects on participant's reading behaviour [21–24]. Eye-tracking researchers have also combined the Flesch-Kincaid measures to evaluate the comparability of stimuli and the ease of reading. Some previous eye-tracking studies also used this score to evaluate whether the stimuli can be easily read by participants [e.g. 25, 26]. Very few previous studies, however, have used eye-tracking tools to examine the difference between how people read automated news and human-written news. Therefore, this study used eye-tracking measures to detect the differences in reading between automated news and human-written news. We conducted an exploratory analysis of eye-tracking data and self-reported data in order to compare reading processes of automated news and human-written news.

2 Theoretical Framework

Expectation-confirmation theory (ECT) was adopted as the theoretical framework. ECT is a cognitive theory which relates consumer satisfaction, post-purchase behaviour, and the adoption of new technologies [27, 28]. It has been originally used to compare pre-consumption expectations and product performance [29]. ECT assumes that satisfied consumers are more likely to repurchase the product than dissatisfied ones [29, 30]. Prior expectation refers to consumers' anticipation toward the product before they use the product [31]. Actual perception refers to consumers' evaluation of the product gained from their actual use experience.

ECT has been transferred to the context of automated news by Haim and Graefe [8]. They expected that if people were positively surprised by a news' quality, they would assign higher ratings; otherwise, they would assign lower ratings [8]. Based on results of previous studies, we hypothesized that people's prior expectations (**H1**) and actual perceptions (**H2**) of the readability of human-written news would be higher than that of automated news; Readers' actual perceptions of the readability would be lower than their prior expectations for (**H3a**) human-written news and higher for (**H3b**) automated news. We also investigated on which eye-tracking measures would differ significantly when people read automated vs. human-written news (**RQ**)? .

To better understand people's perception of automated journalism, we further explored the extent of confirmation, satisfaction, and continuous intention by using ECT

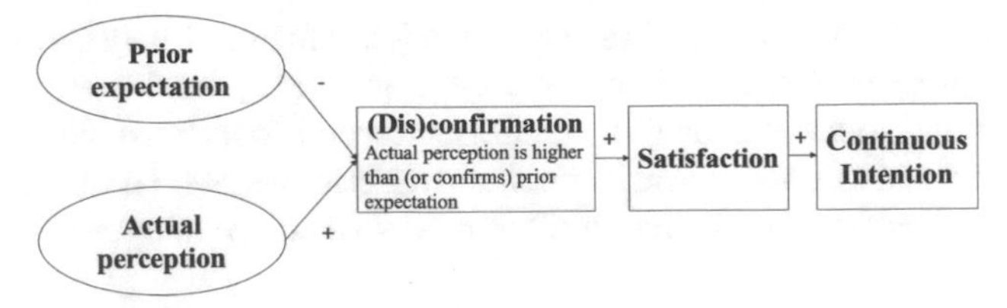

Fig. 1. Expectation-confirmation theory (ECT) model.

model. We hypothesized that readers' extent of confirmation would be positively associated with their actual perceptions of the automated journalism **(H4a)** and negatively associated with their prior expectations of automated journalism **(H4b)**; Readers' extent of confirmation would be positively associated with their satisfaction with automated journalism **(H5);** Reader's extent of continuous intention would be positively associated with their satisfaction with automated journalism **(H6)** (Fig. 1)**.**

3 Method

We conducted a two (article source: human-written or automated news) by three (topic: financial, earthquake and sports news) within-subject experiment in 2019. Participants ($N = 24$, $M_{\text{age}} = 29.4$, 54% female) were pre-screened for their native or near-native level of English, and normal to corrected-to-normal vision. Participants were recruited from among undergraduate and graduate students at the University of Texas at Austin in the United States. They received 19.04 years of education on average ($SD = 2.33$, $Median = 18$). Before the experiment, 62.5% of participants had read automated journalism.

Participants were asked to fill out a pre-test questionnaire, in which they needed to rate their prior expectations for human-written and automated news by using a readability measurement adapted from [7] on a 7-point scale. Then participants were asked to read three sets of news stories, respectively finance news, earthquake news and sports news. These three sets were selected because they were among the most developed automated journalism areas. Each set contained one automated story and one human-written news story with similar Flesch-Kincaid reading ease and level score, similar number of words, and the same topic (see Table 1). Mann-Whitney U test showed no significant difference in words count ($U = 2, p = .275$), Flesch-Kincaid Ease ($U = 3, p = .513$), and Flesch-Kincaid Level ($U = 2$, $p = .275$) between two types of stories. Therefore, we were overall confident that our selected stimuli were comparable and would not affect the validity of the experiment.

The order of articles was randomized, and the article type was unknown to participants. In order to avoid the situation where participant can speculate the sources of articles based on the other article within one set, we added one confounding stimuli (civic human-written news). After reading each article, participants rated their actual perceptions of each article. At the end of the experiment, researchers showed participants the answers with the correct sources of articles. Participants completed a post-test questionnaire in which they rated their confirmation, satisfaction, and continuous intention

Table 1. Stimuli features.

No.	Topic	Author (A/H)	Source	Words count	Flesch-Kincaid ease	Flesch-Kincaid level
1	Finance	A	AP	106	38.7	10.9
2	Finance	H	Business Insider	153	31.4	13.9
3	Earthquake	A	LA Times	96	44.5	11.8
4	Earthquake	H	Salt Lake Tribune	104	34.8	13.1
5	Sports	A	AP	114	55	8.6
6	Sports	H	AP	240	58.7	9.8
7	Civic	H	LA Times	275	49.3	12.3

Note: A is automation news; H is human-written news; Article 7 is a confounding stimulus

after reading both human-written and automated news. Eye movement during reading articles was recorded using the Tobii TX-300 eye-tracker (Tobii AB, www.tobii.com) with 1920 × 1080 screen by iMotions software (iMotions A/S, www.imotions.com). The research was approved by the university's IRB. Each session lasted about 25 min.

In the eye-tracking data analysis, the independent variable was article type (automated news or human-written news). Dependent variables were obtained from eye-tracking data exported by using the I-VT fixation detection algorithm [32]. We cleaned the data by removing task data with low quality eye-tracking (quality $< 70\%$, which corresponded to about 7% of all tasks). We used the following types of variables: fixation count, regression count, fixation duration, saccade duration, length (in pixels), velocity, and angle (four categories). In order to rule out the potentially confounding factor of different article length, we normalized selected dependent variables by dividing them by the article length in words. Saccade angle was categorized to indicate (approximately) eye movement 1) forward ($-10°$; $10°$), 2) backward ($170°$; $180°$) in the same text line. Angles outside these ranges were categorized as either 3) forward, or 4) backward above/below the text line.

Measures of prior expectations and actual perceptions measures were adapted from [7]. Participants were asked to rate 13 items readability measurement through a 7-point scale, respectively "the story is well-written", "the story is concise", "the story is comprehensive", "the story is coherent", "the story is clear", "the story is pleasing to read", "the story is lively", "the story is interesting", "the story is enjoyable", "the story is exiting", "the story is objective", "the story is fair", and "the story is unbiased". Two items were reverse coded ("the story is boring"; "the story is biased"). The reliability of this measurement was high, *Cronbach's* $\alpha = .91$, $M = 4.18$, $SD = 1.06$. The measure of confirmation was adapted from [27]. Participants were asked to rate two items on a 7-point Likert scale (from 1 - very low to 7 - very high). "My reading experience with automated news was better than what I expected." "Overall, most of my expectations of automated news were confirmed." The reliability of this measurement was *Cronbach's*

$\alpha = .50$, $M = 8.71$, $SD = 2.08$. The reliability was relatively low. Therefore, we solely used the second item in the data analysis. The measure of satisfaction was adapted from Spreng and Olshavsky [33]. Four items were used to measure satisfaction on a 7-point scale. "My reading experience with automated news makes me feel satisfied." "My reading experience with automated news makes me feel pleased." "My reading experience with automated news makes me feel contented." "My reading experience with automated news makes me feel delighted." The reliability of this measurement is *Cronbach's* $\alpha = .95$, $M = 12.96$, $SD = 5.64$. The measure of continuous intention was adapted from [29] This measure had one item: "I intend to read automated news again." We designed two attention check questions. Check 1 was embedded in one measurement. Two items in the journalistic quality measurement were reverse coded. Check 2 was a question about the content of one earthquake article "Where did the earthquake take place?". 95.8% participants passed both checks.

4 Results

Due to not-normal distribution of variables and the lack or homogeneity of variance, we used non-parametric Mann-Whitney U test (M-W U) to analyse eye-tracking data. Given the exploratory nature of our research, we conducted individual M-W U tests for each variable (Table 2). Results showed that nine eye tracking measures were significantly different between periods of automated news and human-written news.

When people read automated news, their maximum saccade length and variability of saccade length (standard deviation), regression counts, maximum saccade duration, maximum saccade speed were all significantly lower than for human-written news. Four saccade angle counts were also all significantly smaller when reading automated news than human-written news. This result indicated that there were fewer forward and fewer backward saccades when reading automated news as compared to reading human-written news.

We ran classifications using Weka 3.8 [34] on 31 eye-tracking features including variables listed in Table 2. The number of data points for the two classes was balanced (67 data points in each class). However, due to the relatively small sample size, we generated 100% more data samples using two methods a) random generation with replacement; b) synthetic sample generation SMOTE [35]. We presented best classification results obtained by applying random forest classifier with 10-fold cross-validation [36] (Table 4) and the best features (Table 3) for the sampling method which yielded the best results (random sampling with replacement). The best features overlapped with the statistically significant results shown in Table 2. For classification results, we report accuracy, precision, recall, false positive (FP), false negative (FN), area under ROC curve (ROC AUC) and F-measure (F1). FP (also referred to as type I error) shows the percentage of incorrect predictions for human-written news, while FN (type II error) for automated news. All classifications were either good or very good. The best (F-measure 99.3%) result was obtained for random sampling with replacement, but even the worst result obtained for intact sampling (F-measure 72.1%) was relatively good.

Analyses of Variance (ANOVAs) were used to test **H1**, **H2** and **H3**. **H1** predicted readers' expectations of readability of human-written news would be higher than those

Table 2. Descriptive statistics and Mann-Whitney U tests for eye-tracking variables.

Variable	Automation Mean (SD)	Human Mean (SD)	M-W U statistics
Dwell time on article (per word)	330.9 (158.76)	325.59 (111.53)	$U = 2100; p = .52$
Fixation duration total/word	232.94 (126.73)	228.42 (96.06)	$U = 2167; p = .73$
Mean fixation duration	204.99 (22.28)	206.95 (19.13)	$U = 2064; p = .42$
Fixation count (per word)	1.12 (0.55)	1.09 (0.41)	$U = 2209; p = .87$
Max saccade length	975.93 (102.79)	1271.31 (273.94)	$U = 855; p < .001$***
Mean saccade length	197.12 (32.39)	201.11 (31.34)	$U = 2023; p = .324$
Max. saccade duration	911.49 (1340.65)	1012.96 (937.31)	$U = 1700.5; p = .015$*
Mean saccade duration	96.7 (65.26)	96.49 (52.79)	$U = 2175; p = .757$
Min. saccade velocity	0.19 (0.25)	0.11 (0.12)	$U = 1778.5; p = .038$*
Max. saccade velocity	11.5 (1.43)	12.8 (2.06)	$U = 1405; p < .001$***
Regression count	21.12 (13.43)	28.42 (16.86)	$U = 1536.5; p = .002$**
Saccade angle fwd count	66.76 (34.49)	103.07 (52.93)	$U = 1128; p < .001$***
Saccade angle back count	27.93 (18.75)	39.3 (24.54)	$U = 1430.5; p < .001$***
Saccade angle fwd up/down	11.73 (9.99)	19.28 (15.51)	$U = 1466.5; p < .001$***
Saccade angle back up/down	11.1 (6.17)	15.82 (7.46)	$U = 1304; p < .001$***

Durations in milliseconds; length in screen pixels; *$p < .05$ **$p < .01$ *** $p < .001$
Sample Size (Automation/Human): 67/67

Table 3. Classification results – best features shown for best classification only.

Data sampling	Best features (in the order of weights from Information Gain Ranking Filter)
Random with replacement	Max Saccade Length, Mean Fixation Duration, w_fixCount_S, w_fixCount_R, Min. Saccade Velocity, w_fixDur_R_tot, Max. Saccade Velocity, pRRR, Saccade angle fwd Count, std_saccade_dur, Mean Saccade Length, Dwell time on article (per word)

of automated news, which was fully supported. Readers' expectations of human-written news ($M = 4.75, SD = .87$) were significantly higher than those of automated journalism ($M = 3.59$, $SD = .91$), $F(1, 47) = 20.06$, $p < .001$ (as shown in Table 5).

H2 predicted that readers' actual perceived readability of human-written news would be higher than those of automated news. **H2** was supported on every topic, $F(1, 143) =$

Table 4. Classification results (Random Forest with 10-fold cross validation).

Data sampling	Samples A/H	Accuracy [%]	Precision [%]	Recall [%]	FP/FN [%]	ROC AUC [%]	F-measure [%]
Intact (no sampling)	6767	74.6	75.4	74.6	16.4/34.3	76.6	72.1
Random with replacement	**134/134**	**99.3**	**99.3**	**99.3**	**1/1**	**1**	**99.3**
SMOTE (synthetic)	134/134	86.2	87.1	86.2	1/2.2	94.8	86.1

Table 5. Differences between prior expectations and actual perceptions (mean (SD)).

	Automation	Human	ANOVA	*p*
Prior expectation	3.59 (0.91)	4.75 (0.87)	F(1, 47) = 20.06	.00***
Actual perception	4.05 (0.78)	4.45 (0.92)	F(1, 143) = 7.89	.006**
ANOVA	F(1,94) = 5.61	F(1,95) = 2.01	N/A	N/A
p	.02*	.16	N/A	N/A

7.89, $p = .006$, as shown in Table 6. A two-way ANOVAs also showed that story topic produced a significant main effect on the perceived readability, $F(2, 140) = 5.77$, $p = .004$, $\eta^2 = .08$. Specifically, the perception of earthquake news ($M = 4.58$, $SD = .76$) was significantly higher than of both financial news ($M = 4.05$, $SD = .80$), $p = .07$, and sports news ($M = 4.11$, $SD = .97$), $p = .02$. The perceived readability of financial news was lower than sports news, but not significant, $p = .975$.

Table 6. Two-way ANOVAs between different topics (mean (SD)).

	Automation	Human	Average
Financial news	3.78 (0.72)	4.32 (0.79)	4.05 (0.80)
Earthquake news	4.38 (0.66)	4.77 (0.82)	4.58 (0.76)
Sports news	3.98 (0.86)	4.25 (1.07)	4.11 (0.97)

H3a predicted that readers' actual perceptions of human-written news would be lower than expected. The direction was as we expected, but not statistically significant, $F(1{,}95) = 2.01$, $p = .16$ and thus hypothesis **H3a** was not supported. **H3b** predicted that, for automated news, readers' actual perceptions would be higher than their expectations, which was supported, $F(1{,}94) = 5.61$, $p = .02$ (see Table 5).

H4a predicted readers' that the extent of confirmation would be positively associated with their actual perceptions of the automated news, which was rejected. A Pearson's r was used to test **H4a,** $r = -.355, p = .002$. **H4b** predicted that readers' extent of confirmation would be negatively associated with their prior expectations of automated journalism. A Pearson's r was used to test **H4b**, which was also rejected, $r = -.164, p = .46$.

H5 predicted that readers' extent of confirmation would be positively associated with their satisfaction with automated news. A Pearson's r was used to test **H5**, which was rejected, $r = -.12, p = .308$.

H6 predicted that reader's extent of continuous intention would be positively associated with their satisfaction with automated news. A Pearson's r was used to test **H6**, which was supported, $r = .73, p < .001$.

5 Discussion and Conclusion

Eye-tracking data analysis (RQ) showed a mixed picture. There was no significant difference in terms of normalized article reading duration, fixations durations or counts, and in mean saccade length and duration, which would indicate similar readability levels. However, the maximum saccade length and duration on human-written news were significantly larger than those on automated news. Longer saccades generally indicated easier text, and thus the longer maximum saccadic measures provided a weak evidence for better readability of human-written news. Even so, variables such as eye regressions and directions of other eye movements show the opposite picture. Eye regressions in reading have been correlated with the processing of unfamiliar words or sentences with greater structural or conceptual complexity [37]. There were significantly fewer regressions, and fewer saccades in any direction, on automated news than on human-written news. This result indicated that participants were more likely to encounter unfamiliar words or sentences with more complexed structures when reading human-written news. Overall, the eye movement results provided partial evidence for higher readability of automated news compared to human-written news.

The most interesting finding of this study was that people clearly self-reported that they preferred the readability of human-written news even though eye movements results showed the mixed behavioral differences. For both prior expectations and actual perceptions, readers perceived human-written news significantly more readable than automated journalism. For each type of news, this study showed that readers' actual perceptions were higher than their prior expectations for automated news whereas Haim and Graefe [8] found that automated news did not meet with people's prior expectation. For human-written news, however, the perceived readability from actual reading experience did not meet people's expectations, which was consistent with previous studies [8, 9]. Moreover, there was no significant difference between prior and actual perceptions in terms of human-written news.

Another interesting finding was that the topic of story also produced significant main effects on actual perceptions. For every set of stories, the means of readability of automated news were significantly lower than those of human-written news. These results confirmed the findings in the previous study that people rate human-written news as more readable than automated news [7]. In terms of specific topics, the perception of earthquake news was significantly higher than the other two topics. This finding was not surprising because earthquake automated news was developed at the very early stage of automated journalism, such as the Quakebot developed by *The Los Angeles Times*. After several years of iterating, algorithms of earthquake automated news have been well-developed. Another explanation might be that, compared with sports news and financial news, earthquake news stories are relatively simple and intuitive.

This study offered a new theoretical contribution by testing the full ECT model in the realm of automated news. Earlier study merely tested the first part of ECT model [8, 9]. Although this present study filled in the gap by testing the full ECT model, our finding only partially confirmed ECT hypothesis. By measuring post-consumption behaviour and post-adoption satisfaction after participants read the news, we concluded that readers were still not satisfied with the readability of automated news even though their perceptions exceeded expectations. Their reading experience were neither confirmed with their actual perceptions nor with prior expectations. We partially succeeded in testing ECT model by finding that reader's extent of continuous intention of reading automated news in the future was positively associated with their satisfaction of automated news. Even so, both readers' satisfaction of automated news and their continuous intention of reading automated news were relatively low.

Even though eye movements results provided partial support for higher readability of automated news, people still self-reported they preferred reading human-written news. For instance, readers still preferred the readability of human-written stories even though they typically spent more time reading human-written stories. This could be explained by the possibility that news consumers and researchers might have different definitions of readability. From readers' perspectives, a news story with less complicated vocabularies or less sophisticated grammar doesn't necessarily mean the story is more readable or well-written. The disagreement between people's self-reported reading experience and objective eye-tracking measures should be further studied.

6 Limitations and Future Research

Due to relatively small sample size, we had to generate more data points for data classification. Future studies should to collect more data to better train classification algorithms. Our initial success in finding an array of significant differences in eye-tracking measures between automated news and human-written news and in obtaining a reasonably good classification performance on the small data set indicate a possibility of employing eye-tracking in real-time setting to detect the type of news (automated vs. human-written) articles that are being read. Additionally, participants of this study were relatively young. Previous studies have found different reading behaviours between young and old generations [24]. Future studies should employ more diverse samples.

There are several possible explanations for not completely successful in testing ECT model. First, it is hard to measure readers' satisfaction and post-purchase behaviour because news articles are technically not commodities. Second, participants were not informed which type of news they were consuming until they finish reading all seven articles. This experimental design was different from other previous ECT model testing experiments. Last, the small sample size may also lead to insignificant results in ECT model testing. Future studies should include more participants and test post-purchase behaviours in ECT model.

Acknowledgements. We thank Lan Li, who is a master student in the School of Information at the University of Texas at Austin, for contributing to the data collection.

References

1. Napoli, P.M.: Automated media: an institutional theory perspective on algorithmic media production and consumption. Commun. Theory **24**, 340–360 (2014). https://doi.org/10.1111/comt.12039
2. Graefe, A.: Guide to Automated Journalism. Columbia Journalism School, Tow Center for Digital Journalism, New York (2016)
3. Celeste LeCompte: Automation in the newsroom: How algorithms are helping reporters expand coverage, engage audiences, and respond to breaking news (2015). https://niemanreports.org/articles/automation-in-the-newsroom/
4. Dubay, W.H.: The Principles of Readability. Impact Information, Costa Mesa (2004)
5. Dalecki, L., Lasorsa, D.L., Lewis, S.C.: The news readability problem. Journal. Pract. **3**, 7–8 (2009)
6. Diakopoulos, N.: Algorithmic accountability. Digital Journal. **3**(3), 398–415 (2015). https://doi.org/10.1080/21670811.2014.976411
7. Clerwall, C.: Enter the robot journalist. Journal. Practice **5**, 519–531 (2014). https://doi.org/10.1080/17512786.2014.883116
8. Haim, M., Graefe, A.: Automated news: better than expected? Digital Journal. **5**, 1044–1059 (2017)
9. Jia, C.: Chinese automated journalism: a comparison between expectations and perceived quality. Int. J. Commun. **14**, 2611–2632 (2020)
10. Carlson, M.: The robotic reporter. Digital Journal. **3**(3), 416–431 (2015). https://doi.org/10.1080/21670811.2014.976412
11. van der Kaa, H., Krahmer, E.: Journalist versus news consumer: the perceived credibility of machine written news. BMJ **2**(5147), 305 (2014)
12. Shyam Sundar, S.: Exploring receivers' criteria for perception of print and online news. Journal. Mass Commun. Q. **76**(2), 373–386 (1999). https://doi.org/10.1177/107769909907600213
13. Graefe, A., Haim, M., Haarmann, B., Brosius, H.-B.: Readers' perception of computer-generated news: credibility, expertise, and readability. Journalism **19**(5), 595–610 (2018). https://doi.org/10.1177/1464884916641269
14. Seib, C.: Papers need to work on handling the English language, Austin American-Statesman, p. B25 (1976)
15. Fishkin, S.F.: From Fact to Fiction: Journalism & Imaginative Writing in America. Oxford University Press, Oxford (1985)
16. Flesch, R.F.: A new readability yardstick. J. Appl. Psychol. **32**, 221–233 (1948)
17. Flesch, R.F.: The Art of Readable Writing. Harper, New York (1949)
18. Gedeon, T., Caldwell, S.: Effects of text difficulty and readers on predicting reading comprehension from eye movements. In: 2015 6th IEEE International Conference on Cognitive Infocommunications (CogInfoCom), Gyor, pp. 407–412 (2015). https://doi.org/10.1109/coginfocom.2015.7390628

19. Rayner, K., Chace, K.H., Slattery, T.J., Ashby, J.: Eye movements as reflections of comprehension processes in reading. Sci. Stud. Read. **10**, 241–255 (2006)
20. Martínez-Gómez, P., Aizawa, A.: Recognition of understanding level and language skill using measurements of reading behavior. In: Proceedings of the 19th International Conference on Intelligent User Interfaces 2014, pp. 95–104 (2014). http://dx.doi.org/10.1145/2557500.2557546
21. Gwizdka, J., Hosseini, R., Cole, M., Wang, S.: Temporal dynamics of eye-tracking and EEG during reading and relevance decisions. J. Assoc. Inf. Sci. Technol. **68**(10), 2299–2312 (2017). https://doi.org/10.1002/asi.23904
22. Gwizdka, J.: Differences in reading between word search and information relevance decisions: evidence from eye-tracking. In: Davis, F.D., Riedl, R., vom Brocke, J., Léger, P.-M., Randolph, A.B. (eds.) Information Systems and Neuroscience. LNISO, vol. 16, pp. 141–147. Springer, Cham (2017). https://doi.org/10.1007/978-3-319-41402-7_18
23. Cole, M.J., Gwizdka, J., Liu, C., Bierig, R., Belkin, N.J., Zhang, X.: Task and user effects on reading patterns in information search. Interact. Comput. **23**(4), 346–362 (2011). https://doi.org/10.1016/j.intcom.2011.04.007
24. Shojaeizadeh, M., Djamasbi, S.: Eye movements and reading behavior of younger and older users: an exploratory eye-tacking study. In: Zhou, J., Salvendy, G. (eds.) ITAP 2018. LNCS, vol. 10926, pp. 377–391. Springer, Cham (2018). https://doi.org/10.1007/978-3-319-92034-4_29
25. Sundar, R.P., Becker, M.W., Bello, N.M., Bix, L.: Quantifying age-related differences in information processing behaviors when viewing prescription drug labels. PLoS ONE **7**(6) (2012) http://dx.doi.org.ezproxy.lib.utexas.edu/10.1371/journal.pone.0038819
26. Angela, S., Tanya, B.: The influence of science reading comprehension on South African township learners' learning of science. S. Afr. J. Sci. **115**(1), 72–80 (2019)
27. Bhattacherjee, A.: Understanding information systems continuous: an expectation-confirmation model. MIS Q. **25**(3), 351–370 (2001). https://doi.org/10.2307/3250921
28. Lin, C.S., Sheng, W., Tsai, R.J.: Integrating perceived playfulness into expectation-confirmation model for web portal context. Inf. Manag. **42**(5), 683–693 (2005)
29. Oliver, R.L.: A cognitive model of the antecedents and consequences of satisfaction decisions. J. Mark. Res. **17**, 460–469 (1980). https://doi.org/10.2307/3150499
30. Oliver, R.L.: Cognitive, affective, and attribute bases of the satisfaction response. J. Consum. Res. **20**, 418–430 (1993). http://dx.doi.org/10.1086/209358
31. Oliver, R.L.: Satisfaction. A Behavioral Perspective on the Consumer, 2nd edn. Routledge, London (2015)
32. Salvucci, D.D., Goldberg, J.H.: Identifying fixations and saccades in eye-tracking protocols. In: Proceedings of the 2000 Symposium on Eye Tracking Research & Applications (ETRA 2000), 71–78 (2000). https://doi.org/10.1145/355017.355028
33. Spreng, R.A., Olshavsky, R.W.: A desires congruency model of consumer satisfaction. J. Acad. Mark. Sci. **21**(3), 169–177 (1993)
34. Witten, I.H., Frank, E., Hall, M.A.: Data Mining, Fourth Edition: Practical Machine Learning Tools and Techniques. Morgan Kaufmann Publishers Inc., San Francisco (2016)
35. Chawla, N.V., Bowyer, K.W., Hall, L.O., Philip Kegelmeyer,W.: SMOTE: synthetic minority over-sampling technique. arXiv:1106.1813 Cs (2011)
36. Breiman, L.: Random forests. Mach. Learn. **45**, 5–32 (2011)
37. Rayner, K., Pollatsek, A.: The Psychology of Reading. Lawrence Erlbaum Associates, Mahwah (1989)

NeuroIS to Improve the FITradeoff Decision-Making Process and Decision Support System

Adiel Teixeira de Almeida(✉) and Lucia Reis Peixoto Roselli

Center for Decision Systems and Information Development (CDSID), Universidade Federal de Pernambuco, Recife, PE, Brazil
{almeida,lrpr}@cdsid.org.br

Abstract. NeuroIS approach is used in this research in order to improve the decision-making process and the Decision Support System (DSS), which is a particular kind of information system, for the FITradeoff method. In this research the decision-makers (DMs) behavior is investigated when they are solving Multi-Criteria Decision Making/Aiding problems, considering the holistic evaluation process. In this research, neuroscience experiments were constructed to investigate the holistic evaluation process using graphical and tabular visualizations. These experiments were applied to more than 150 management engineering students. As a result, using an electroencephalogram, the Alpha-Theta Diagram has been proposed, which is a new concept to classify the DMs patterns of behavior, considering Theta (4–8 Hz) and Alpha (8–13 Hz) activities. Based on this diagram, improvements can be suggested to be included in the FITradeoff DSS specially for problems involved in a ranking order context.

Keywords: NeuroIS · Decision Neuroscience · Alpha-Theta Diagram · FITradoff method · Electroencephalogram · Multi-Criteria Decision Making/Aiding (MCDM/A)

1 Introduction

This research is developed using the support of the NeuroIS [1, 2] and Decision Neuroscience approaches to investigate decision-makers (DMs) patterns of behavior, being the decision-making field an important topic to develop studies in NeuroIS [2].

This study is oriented to improve the design of the FITradeoff Decision Support System (DSS), which is a particular kind of information system. This DSS has been constructed for a explicitly Multi-Criteria Decision Making/Aiding (MCDM/A) method [3–6], the FITradeoff method [7]. According to [8] the NeuroIS approach can be used both for support the understanding of DMs, as to orient the design of business and information systems engineering (BISE), which the FITradeoff DSS can be considered an example of BISE.

F. D. Davis et al. (Eds.): NeuroIS 2020, LNISO 43, pp. 111–120, 2020.
https://doi.org/10.1007/978-3-030-60073-0_13

Multi-criteria decision making/aiding (MCDM/A) problems are characterized as problems which presented alternatives evaluated in some criteria or attributes [3–6]. These problems are very common in society routine; some examples are the supplier selection problems [9, 10], the equipment selection problem [11] or maintenance selection polities [12].

In this context, in order to support DMs to solve these problems, many MCDM/A methods, and consequently DSS, have been developed and are presented in the literature. However, it is observed that in the decision-making process performed by these methods fewer considerations about DMs behavioral aspects are presented. Hence, according to [13, 14] the behavior aspects presented in the decision-making process should be considered in order to modulate, i.e. transform the procedure presented in these methods to represent the DMs preferences coherently. Also, [15, 16] considered the inclusion of these aspects as an important topic to advance the researches in MCDM/A area.

Therefore, regarding the consideration of the behavior aspect to modulate MCDM/A methods and its DSS, some NeuroIS studies have been conducted related to the FITradeoff method [17–21]. These studies have been constructed with an eye-tracking and an EEG, which are neurophysiological tools that can advance IS research [22]. Thus, in these studies the neurophysiological tools are used to investigate the DMs behavior when they are performing the holistic evaluation process [17–19], with graphical and tabular visualizations, or the elicitation procedure by decomposition [20, 21], to solve problems in the context of the FITradeoff method [7].

In this paper, an EEG is used to investigate the DMs patterns of behavior when they are performing the holistic evaluation process based on visualization constructed with only two alternatives. Based on the behavior results, considering Theta (4–8 Hz) and Alpha (8–13 Hz) activities, the Alpha-Theta Diagram has been proposed in this research, being possible to identify five patterns of behavior.

2 FITradeoff Method

The Flexible and Interactive Tradeoff (FITradeoff method – [7]) is based on the traditional Tradeoff method [3], presenting the same axiomatic structure of this antecedent method. However, the FITradeoff uses the concepts of partial information [23–26], being not necessary to collect all the DMs preferences for the consequences in a MCDM/A problem. In these problems, the consequences are the evaluation of alternatives in criteria.

Compared to the traditional Tradeoff method [3] the FITradeoff method required less effort for the DMs to perform the decision-making process, since in the FITradeoff only strict preferences are collect. Moreover, in the FITradeoff the DMs do not have to express the exactly point of indifference between the comparison of consequences, which is a requirement in the Tradeoff process. According to [27], this requirement leads to 67% of inconsistencies in the results, which is also confirmed in [20].

The FITradeoff method presents two steps to solve MCDM/A problems. In the first step, the DMs had to order the criteria weights based on their preferences about the alternatives performance scale presented for each criterion. The outcome of this step is an inequality presenting the order of criteria weights.

In the second step the DMs had to express their strict preferences about some consequences in order to reduce the weight space. In other words, some comparisons of consequences are presented to the DM, who had to express his/her strict preference for them. After each preference expressed, inequalities are created being included in a Linear Programming Problem (LPP), which also presents the inequality obtained in the final of the first step.

The FITradeoff method is considered Interactive because, after each preference expressed, the LLP runs updating the set of alternatives in the problem. In other words, the number of alternatives is reduced (for a choice problematic) or the ranking of alternatives has been constructed (for a ranking order problematic) during the decision-making process, based on the dominance relations constructed with the preferences expressed by DMs.

Also, this method is considered Flexible since in it Decision Support System (DSS), for choice problematic, graphical visualizations can be used to support the DMs in the understanding of the problem, being possible to DMs to evaluate the potentially optimal alternatives (POA) in the problem. Moreover, if the DMs desired, based on the holistic evaluation process performed with these POAs, they can select a final alternative into this group, concluding the decision-making process before the mathematical procedure in the LPP has been finished. At the moment, three types of graphics (bar graphic, bubble graphic and spider graphic) are presented in the FITradeoff DSS for choice problematic. The FITradeoff method as available by request at www.fitradeoff.org.

3 NeuroIS Experiments

In order to investigate the holistic evaluation process performed by the DMs, three neuroscience experiments were constructed. The holistic evaluation process is investigated since it provides an important feature in the decision-making process, bringing Flexibility to DMs to proceed in the FITradeoff method.

These experiments were constructed using different configurations of MCDM/A problems. These problems were constructed in Excel and do not present a defined context. The absence of context was considered to allow the generalization of the results. Thus in these experiments, the alternatives were named using letters and the criteria were named using numbers.

In this context, to construct the experiments, these hypothetical problems were illustrated by different forms of visualizations. In other words, problems composed to two, three, four and five alternative evaluated in three, four, five, six and seven criteria were represented by bar graphics, spider graphics, bubble graphic, tables and bar graphic with tables.

Specifically, the first experiment, which was constructed in 2017, was composed for 24 visualizations, presenting 18 bar graphics and other six visualizations (one spider graphic, one bubble graphic, two tables and two bar graphics with tables). The second experiment, constructed in 2018, presented 22 visualizations being 10 bar graphics, 10 tables and two bar graphics with tables. Finally, the third experiment, constructed in 2019, was composed especially for five bar graphics, five spider graphics, and five tables.

Figure 1 illustrates some of the visualizations used in these experiments. It is worth to mention that the tendency line presented in the bar graphic was used to indicate that the criteria weights presented different values, being higher for the first criterion and lower for the last criterion.

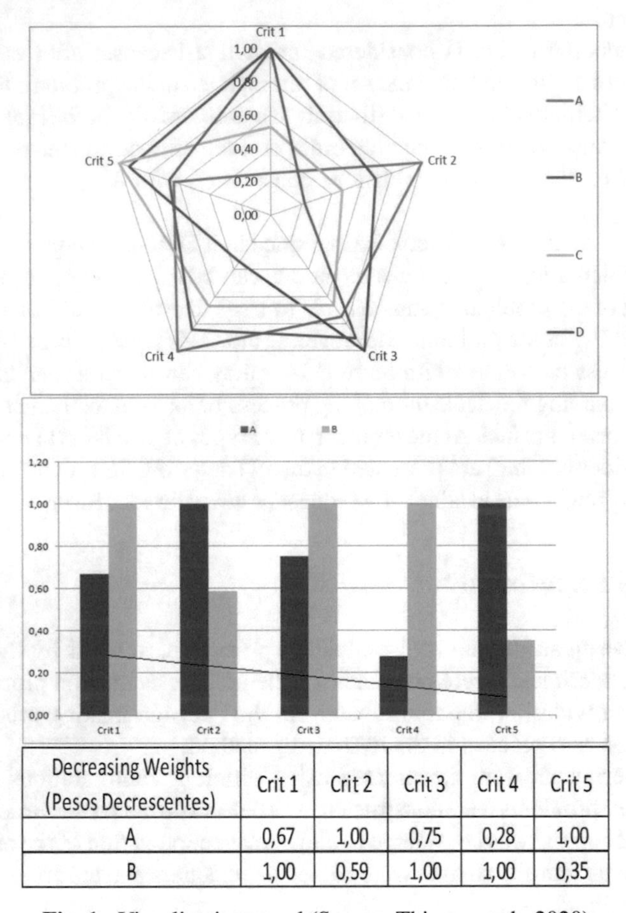

Decreasing Weights (Pesos Decrescentes)	Crit 1	Crit 2	Crit 3	Crit 4	Crit 5
A	0,67	1,00	0,75	0,28	1,00
B	1,00	0,59	1,00	1,00	0,35

Fig. 1. Visualizations used (Source: This research, 2020)

To apply these experiments the first step is concerning to the scheduling of the meetings. The second step is the execution of the experiment, which is divided in 2 phases. In the first phase the explanation of the experiment task is presented, the subjects signed an agreement term and the EEG with 14 channels by Emotiv is coupled in the subjects head. In the second phase, the subjects evaluated the visualizations, do not having limit of time to execute this evaluation.

Also, when the subjects evaluate each one of the visualizations presented in each experiment, the only task required was to select the best alternative in each of them. To select these alternatives the subjects had to apply the Additive Model based in the Multi-Attribute Value Theory (MAVT) concepts [3]. The MAVT [3] theory is used in compensatory methods and corresponds to the aggregation of the criteria in a unique criterion to generate a global value for each alternative. The best alternative is the one which presents the highest global value. In other words, to select correctly the best alternative, the subjects should evaluate the compensation between the performances of each alternative evaluating the tradeoffs between the criteria, based on the criteria weights.

It is worth to mentioning that these experiments were executed in a room inside the NeuroScience for Information and Decision (NSID) laboratory. This room contained a set of table and chair, and a computer to display the visualizations. Also, before the data have been collected, these experiments were approved by the University Ethics Committee.

Finally, the subjects who had participated in these experiments are Management Engineering students from the Federal University of Pernambuco (UFPE). These students attended to the lessons of MCDM/A approach, presenting competence to execute the required task in the experiments. In experiment 1 the sample was composed for 36 students (15 women and 21 men, 16 undergraduate students and 10 graduate students). In experiment 2 for 51 students (25 women and 26 men, of whom 28 were undergraduate students and 23 were graduate students). Finally, in experiment 3 for 78 participants (30 women and 48 men, of whom 34 were undergraduate students and 44 were graduate students).

4 Results

The EEG is used in these experiments in order to investigate the DMs patterns of behavior when they perform the holistic evaluation process. Particularly, considering the subjects who had participated in the third experiment, analyses are constructed to investigate their patterns of behavior. In this context, the research question raised is: Considering the combination of Alpha and Theta activities, the participants may have a particular pattern during the evaluation of the visualizations?

The experiment 3 was considered since it is important to propose suggestions for the FITradeoff DSS, specially the DSS constructed to solve problems involved in a ranking order problematic. It is worth to mentioning, that the experiment 3 used visualizations with only two alternatives since it is important to compare incomparable alternatives, which are common in the ranking order positions during the decision-making process [28].

Also, at the moment, the FITradeoff DSS for ranking order problematic do not pre the holistic evaluation process, being important to investigate this feature in order to include the visualizations in the DSS.

Therefore, based on the EEG results, the Alpha-Theta Diagram is proposed in this research, being the important contribution of this paper. This diagram is a new concept for classify the participant's behavior considering five patterns of DMs behavior.

This diagram is created after the pre-processing data for the EEG (i.e. the data were re-referenced, based on average activity; filtered, to eliminate some higher (59–61 Hz) and lower (0.1 Hz) frequencies; and the Independent Component Analysis (ICA) method was applied). Thus, the power spectrum of the Theta band was obtained by the average of frequencies ranging from 4 to 8 Hz in channel F3, and of the Alpha band by average of frequencies ranging from 8 to 13 Hz in channel P7. The Theta and Alpha power are since in several studies they are associated to the evaluation of the cognitive performance to execute some task [29, 30].

In this context, Theta and Alpha bands were investigated based on the studies [31–35] which commented that cognitive effort can be associated to the increase of Theta activity in frontal channels and the decrease of Alpha activity in parietal channels.

Thus, by the normalization of these bands activities, which was performed to present the values on the same scale, the diagram is constructed. This diagram is generated by the combination of Alpha power and Theta power, which are used to produce four quadrants.

Therefore, based on these quadrants, five patterns of behavior are considered, being four of them corresponding to each one of the quadrants, and the fifth by the combination of these patterns, being considered as a disperse pattern.

Specifically, the upper left quadrant presented the Relaxing behavior, which is characterized by low cognitive effort and engagement (as attention state [31, 34]) by the participant during the evaluation of the visualization (suggested by negative values for Theta activity and positive values for Alpha activity). The upper right quadrant presented the In definition behavior, which is characterized by high cognitive effort and low engagement (suggested by positive values for Theta and Alpha). The lower left quadrant presented the Involvement behavior, which is characterized by low cognitive effort and high engagement (suggested by negative values for Theta and Alpha). Finally, the lower right quadrant presented the Diligence behavior, which is characterized by a high cognitive effort and high engagement (suggested by positive values for Theta and negative values for Alpha).

In this context, in order to investigate the patterns of behavior presented during the execution of the experiment 3, the Alpha-Theta Diagrams were constructed for each subject. Figure 2 illustrates the patterns of behavior indicated in each of the quadrants, these patterns were observed for the subject 31 (who presented an undefined and a relaxed behavior), subject 28 (who presented an undefined behavior), subject 11 (who presented an involved behavior), and subject 1 (who presented a diligent behavior).

Based on the Alpha-Theta Diagrams constructed for each subject it is possible to observe that most of the subjects presented a unique pattern of behavior during the investigation of the visualizations.

Also, it was observed in the experiment executed that the behavior patterns Diligence and Involvement appear 76% of the time. This result is very interesting since it highlights that these desirable behaviors are performed by most of the participants during the evaluation of the visualizations. In other words, this result confirms that most of participants presented cognitive effort and engagement during the required task in the experiment.

Moreover, based on this result, the Alpha-Theta Diagram is replicated for each visualization aggregating the subject's behavior. As result, it is possible to observe that the visualizations were evaluated following the Diligence and Involvement behavior, which is an expected result since most of participants (76%) presented these behaviors during the experiment. Also, it is observed that when subjects evaluate the bar graphics and the tables they presented, in most of the cases, an involved behavior, and when they evaluate the spider graphics, they presented a diligent behavior. Therefore, based

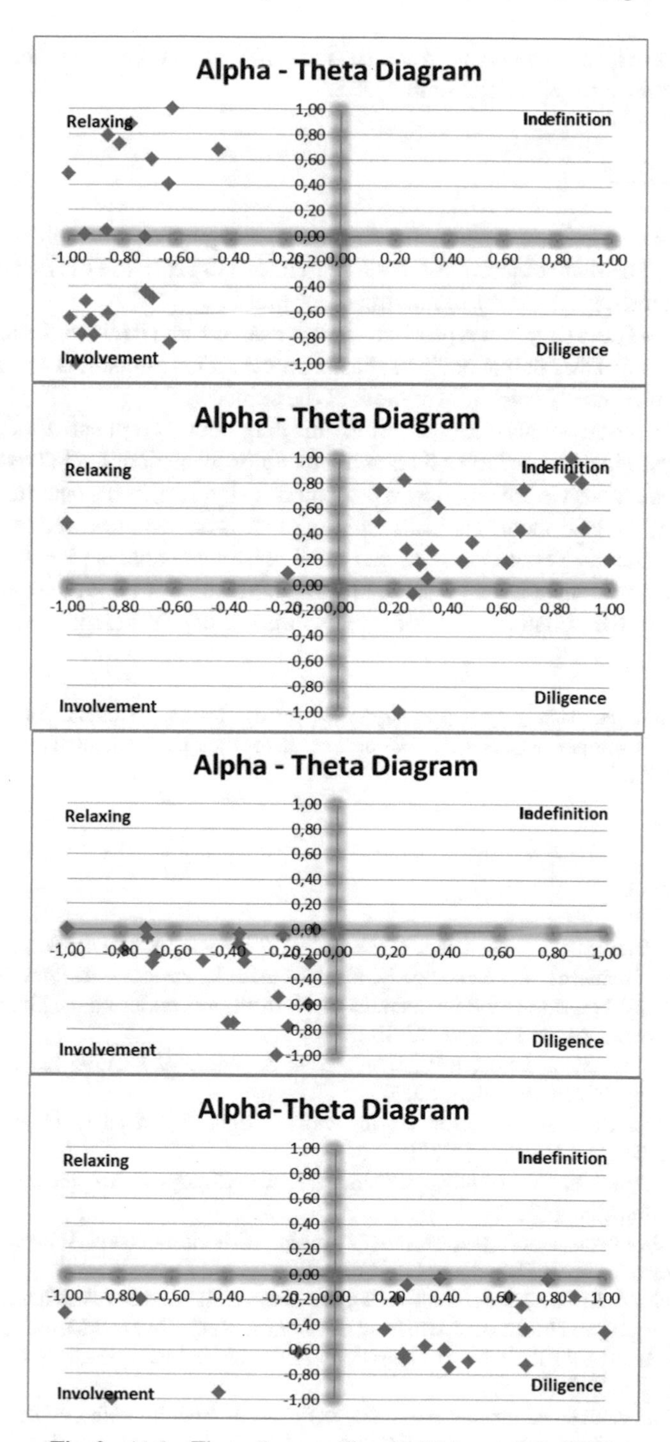

Fig. 2. Alpha-Theta diagram (Source: This research, 2020)

on these patterns, improvements in the FITradeoff DSS can be done, considering the inclusion of these visualizations in this DSS.

5 Conclusion

In this research the holistic evaluation process is investigated using graphical and tabular visualization. The main objective of this research is to improve the FITradeoff Decision Support System, especially for in ranking order problem.

The Alpha-Theta Diagram is proposed in this research and it allows the identification of five patterns of DMs behavior. Thus, based on the EEG results, it is observed a clear definition of patterns of behavior for most of the subjects.

Therefore, based on this clear definition, this diagram was replicated for each visualization designed in the experiment 3 allowing the improvement of the FITradeoff DSS in order to include more visualization with presented desirable patterns (such as Diligence).

Moreover, the FITradeoff decision-making process can be improved by the insights that this tool (Alpha-Theta Diagram) can generate for the analyst, i.e. supporting the analyst in the recommendation presented for the DMs of use or not use some visualization to perform the holistic evaluation process during the FITradeoff decision-making process.

Acknowledgements. This project was supported by the National Council for Scientific and Technological Development (CNPq) and Coordination for the Improvement of Higher Education Personnel (CAPES).

References

1. Riedl, R., Banker, R.D., Benbasat, I., Davis, F.D., Dennis, A.R., Dimoka, A., Gefen, D., Gupta, A., Ischebeck, A., Kenning, P., Müller-Putz, G., Pavlou, P.A., Straub, D.W., vom Brocke, J., Weber, B.: On the foundations of NeuroIS: reflections on the Gmunden Retreat 2009. Commun. Assoc. Inf. Syst. **27**, 15 (2010)
2. Riedl, R., Fischer, T., Léger, P.M., Davis, F.D.: A decade of NeuroIS research: progress, challenges, and future directions (2020)
3. Keeney, R.L., Raiffa, H.: Decision Making with Multiple Objectives, Preferences, and Value Tradeoffs. Wiley, Nova York (1976)
4. Belton, V., Stewart, T.: Multiple Criteria Decision Analysis: An Integrated Approach. Springer, Heidelberg (2002)
5. Figueira, J., Greco, S., Ehrgott, M. (eds.). Multiple Criteria Decision Analysis: State of the Art Surveys. Springer, Heidelberg (2005)
6. de Almeida, A.T., Cavalcante, C., Alencar, M., Ferreira, R., de Almeida-Filho, A.T., Garcez, T.: Multicriteria and Multi-objective Models for Risk, Reliability and Maintenance Decision Analysis. International Series in Operations Research & Management Science, vol. 231. Springer, New York (2015)
7. de Almeida, A.T., de Almeida, J.A., Costa, A.P.C.S., de Almeida-Filho, A.T.: A new method for elicitation of criteria weights in additive models: flexible and interactive tradeoff. Eur. J. Oper. Res. **250**, 179–191 (2016)

8. Loos, P., Riedl, R., Müller-Putz, G.R., Vom Brocke, J., Davis, F.D., Banker, R.D., Léger, P.M.: NeuroIS: neuroscientific approaches in the investigation and development of information systems. Bus. Inf. Syst. Eng. **2**(6), 395–401 (2010)
9. Frej, E.A., Roselli, L.R.P., de Almeida, A.J., de Almeida, A.T.: A multicriteria decision model for supplier selection in a food industry based on FITradeoff method. Math. Probl. **2017**, 1–9 (2017)
10. Barla, S.B.: A case study of supplier selection for lean supply by using a mathematical model. Logist. Inf. Manage. **16**, 451–459 (2003)
11. Lashgari, A., Yazdani-Chamzini, A., Fouladgar, M., Zavadskas, E., Shafiee, S., Abbate, N.: Equipment selection using fuzzy multi criteria decision making model: key study of Gole Gohar iron mine. Eng. Econ. **23**(2), 125–136 (2012)
12. Wang, L., Chu, J., Wu, J.: Selection of optimum maintenance strategies based on a fuzzy analytic hierarchy process. Int. J. Prod. Econ. **107**(1), 151–163 (2007)
13. Korhonen, P., Wallenius, J.: Behavioral issues in MCDM: neglected research questions. In: Multicriteria Analysis, pp. 412–422. Springer, Heidelberg (1997)
14. Hunt, L.T., Dolan, R.J., Behrens, T.E.: Hierarchical competitions subserving multi-attribute choice. Nat. Neurosci. **17**(11), 1613–1622 (2014)
15. Wallenius, J., Dyer, J.S., Fishburn, P.C., Steuer, R.E., Zionts, S., Deb, K.: Multiple criteria decision making, multiattribute utility theory: recent accomplishments and what lies ahead. Manage. Sci. **54**(7), 1336–1349 (2008)
16. Wallenius, H., Wallenius, J.: Implications of world mega trends for MCDM research. In: Ben Amor, S., de Almeida, A., de Miranda, J, Aktas, E. (eds.). Advanced Studies in Multi-Criteria Decision Making. Series in Operations Research, 1st edn., pp. 1–10. Chapman and Hall/CRC, New-York (2020)
17. de Almeida, A.T; Roselli, L.R.P.: Visualization for decision support in FITradeoff method: exploring its evaluation with cognitive neuroscience. In: Lecture Notes in Business Information Processing, vol. 282, pp. 61–73. Springer, Heidelberg (2017)
18. Roselli, L.R.P., Frej, E.A, de Almeida, A.T.: Neuroscience experiment for graphical visualization in the FITradeoff decision support system. In: Chen, Y., Kersten, G., Vetschera, R., Xu, H. (eds.) Group Decision and Negotiation in an Uncertain World. Lecture Notes in Business Information Processing. vol. 315 (2018)
19. Roselli, L.R.P., de Almeida, A.T., Frej, E.A.: Decision neuroscience for improving data visualization of decision support in the FITradeoff method. Oper. Res. Int. J. **19**, 1–21 (2019)
20. Roselli, L.R.P., Pereira, L.S., da Silva, A.L.C.L., de Almeida, A.T., Morais, D.C., Costa, A.P.C.S.: Neuroscience experiment applied to investigate decision-maker behavior in the tradeoff elicitation procedure. Ann. Oper. Res. **289**, 1–18 (2019)
21. Silva, A.L.C.L, Costa, A.P.C.S.: FITradeoff decision support system: an exploratory study with neuroscience tools. In: NeuroIS Retreat 2019, Viena (2019)
22. Dimoka, A., Davis, F.D., Gupta, A., Pavlou, P.A., Banker, R.D., Dennis, A.R., Ischebeck, A., Müller-Putz, G., Benbasat, I., Gefen, D., Kenning, P.H.: On the use of neurophysiological tools in IS research: developing a research agenda for NeuroIS. MIS Q. **36**, 679–702 (2012)
23. Kirkwood, C.W., Corner, J.L.: The effectiveness of partial information about attribute weights for ranking alternatives in multi attribute decision making. Organ. Behav. Hum. Decis. Process. **54**(3), 456–476 (1993)
24. Kirkwood, C.W., Sarin, R.K.: Ranking with partial information: a method and an application. Oper. Res. **33**(1), 38–48 (1985)
25. Punkka, A., Salo, A.: Preference programming with incomplete ordinal information. Eur. J. Oper. Res. **231**(1), 141–150 (2013)
26. Salo, A.A., Hamalainen, R.P.: Preferenceratios in multiattribute evaluation (PRIME)-elicitation and decision procedures under incomplete information. IEEE Trans. Syst. Man Cybernet. Part A Syst. Hum. **31**(6), 533–545 (2001)

27. Weber, M., Borcherding, K.: Behavioral influences on weight judgments in multi-attribute decision making. Eur. J. Oper. Res. **67**, 1–12 (1993)
28. Frej, E.A., de Almeida, A.T., Costa, A.P.C.S.: Using data visualization for ranking alternatives with partial information and interactive tradeoff elicitation. Oper. Res. Int. J. **19**, 909–931 (2019)
29. Klimesch, W.: EEG alpha and theta oscillations reflect cognitive and memory performance: a review and analysis. Brain Res. Rev. **29**, 169–195 (1999)
30. Andreassi, J.L.: Psychophysiology: Human Behavior and Physiological Response, Hillsdale, NJ (1995)
31. de Loof, E., Vassena, E., Janssens, C., de Taeye, L., Meurs, A., Van Roost, D., Verguts, T.: Preparing for hard times: scalp and intracranial physiological signatures of proactive cognitive control. Psychophysiology **56**, 10 (2019)
32. MacDonald, J.S.P., Mathan, S., Yeung, N.: Trial-by-trial variations in subjective attentional state are reflected in ongoing prestimulus EEG alpha oscillations. Front. Psychol. **2**, 82 (2011)
33. Holm, A., Lukander, K., Korpela, J., Sallinen, M., Müller, K.M.I.: Estimating brain load from the EEG. Sci. World J. **9**, 639–651 (2009)
34. Klimesch, W., Schack, B., Sauseng, P.: The functional significance of theta and upper alpha oscillations. Exp. Psychol. **52**(2), 99–108 (2005)
35. Léger, P.M., Davis, F.D., Cronan, T.P., Perret, J.: Neurophysiological correlates of cognitive absorption in an enactive training context. Comput. Hum. Behav. **34**, 273–283 (2014)

Improvements in the FITradeoff Decision Support System for Ranking Order Problematic Based in a Behavioral Study with NeuroIS Tools

Lucia Reis Peixoto Roselli(✉) and Adiel Teixeira de Almeida

Center for Decision Systems and Information Development (CDSID), Universidade Federal de Pernambuco, Recife, PE, Brazil
{lrpr,almeida}@cdsid.org.br

Abstract. This research is performed using NeuroIS tools to improve the information systems (the Decision Support System) constructed for the FITradeoff method. In this study a neuroscience experiment is constructed with the support of an Eye-Tracking in order to investigate the decision-maker (DMs) behavior when they evaluate Multi-Criteria Decision-Making/Aiding (MCDM/A) problems. Therefore, based on the experiment results, a Recommendation Rule is proposed to support the analyst in the advising process performed with the DM and the importance of the holistic evaluation phase for the FITradeoff DSS is highlighted, suggesting that this phase should be included in the DSS constructed for ranking order problematic.

Keywords: NeuroIS · Decision neuroscience · Holistic evaluation · Decision support system · Recommendation rule · FITradeoff method · Multi-Criteria Decision-Making/Aiding (MCDM/A)

1 Introduction

The NeuroIS approach can enable the development of systems based in the investigation of user's behavior when they interact with IS [1–3]. Regarding NeuroIS and Decision Neuroscience, this research is performed to improve the Decision Support System (DSS) developed for the FITradeoff method [4], especially for ranking problematic [5].

In this study the Recommendation Rule is proposed and the holistic evaluation process is highlighted to be applied in the ranking order problematic, being a suggestion for inclusion in this DSS. Concerning to this theme, other papers had already been developed and are presented in the literature in [6–11].

2 DSS for the FITradeoff Method for Ranking Problematic

The Flexible and Interactive Tradeoff (FITradeoff [4]) method is a Multi-Criteria Decision Making/Aiding (MCDM/A) method in the context of Multi-Attribute Value Theory

F. D. Davis et al. (Eds.): NeuroIS 2020, LNISO 43, pp. 121–132, 2020.
https://doi.org/10.1007/978-3-030-60073-0_14

(MAVT) [12, 13] and based in the Tradeoff method [12]. This method was constructed to elicit scaling constants for MCDM/A problem, where MCDM/A problems are problems characterized to present a group of alternatives evaluated in some criteria or attributes of interest [12–14].

The FITradeoff DSSs for choice or ranking problematic [4, 5] present the same steps, the difference between these DSSs is the presence of graphical visualization in the choice problematic version, which allows decision-makers (DMs) to perform the holistic evaluation process.

In the first step of these DSSs, the DMs had to order the scaling constants (weights) for each criterion, following their preferences about the performance scale presented in each criterion. This step is illustrated in Fig. 1 for the supplier selection problem with five alternatives evaluated in seven criteria presented in [15], which is adapted for ranking order problematic [5].

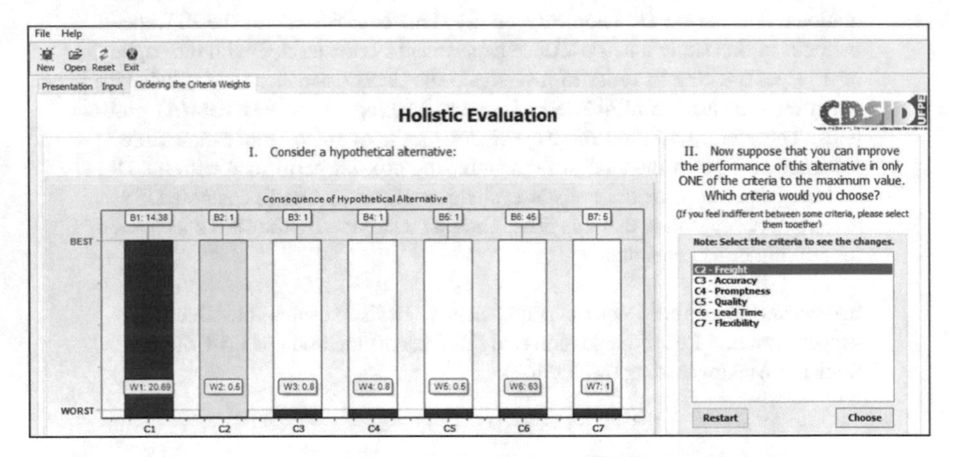

Fig. 1. Step 1 in FITradeoff DSS (Source: FITradeoff DSS, 2016)

In the second step of these DSSs, the performance of some alternatives is compared following the order of criteria weights defined in the last step. Therefore based on the preferences expressed by the DMs, dominance relations for the alternatives are defined.

The second step is illustrated in Fig. 2, in this figure the comparison of 50% of performance in Criteria 1 (Price), i.e. with the price of 17.535, is compared to 100% of performance in Criteria 2 (Freight). For this comparison, the DM was indifferent between the two performances presented.

Therefore, after the expression of this preference, the dominance relations between the alternatives were updated and it was illustrated by the Hasse Diagram in Fig. 3. Based in this diagram it is possible to observe that were defined three positions in the rank. Also, in the Position 3, two alternatives are incomparable; the Supplier 3 incomparable to the Supplier 5 and the Supplier 4 incomparable to the Supplier 5.

Based on this figure, if the DM can compare the incomparable alternatives using the holistic evaluation process, performed with a graphic or a table, he/she can define a dominance relation between these alternatives. Thus, if dominance relation is defined,

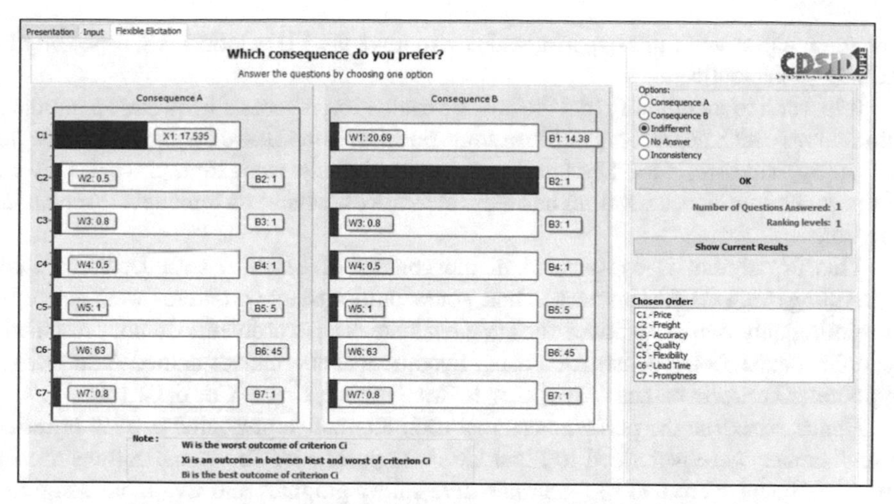

Fig. 2. Step 2 in FITradeoff DSS (Source: FITradeoff DSS, 2016)

a complete order for this problem can be establish, being possible to interrupting the decision process and conclude the FITradeoff method. However, at the moment, for ranking problematic [5], the holistic evaluation is not included yet in the FITradeoff DSS. These DSSs are available by request at www.fitradeoff.org.

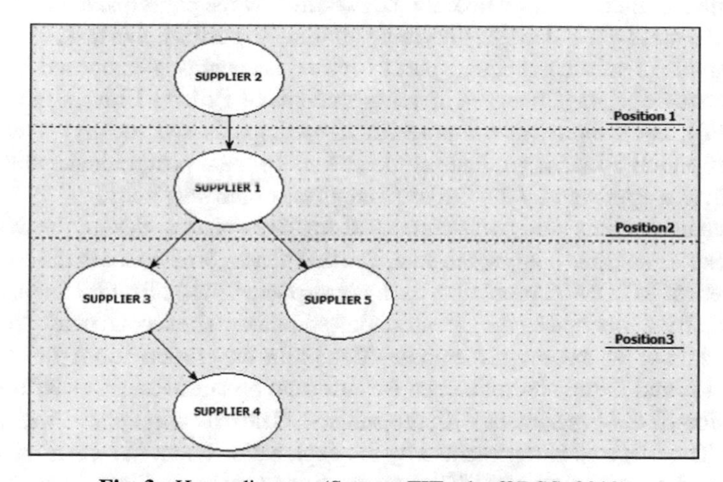

Fig. 3. Hasse diagram (Source: FITradeoff DSS, 2019)

3 Neuroscience Experiment with 2 Alternatives

In this context, in order to investigate how DMs perform the holistic evaluation process, a neuroscience experiment was constructed using the X120 Eye-Tracking by Tobbi Studio.

The main objective of this experiment is to improve the FITradeoff DSS, especially for ranking problematic.

It is worth to mentioning that the holistic evaluation process allows DMs to compare alternatives which are incomparable in some position in the Hasse diagram, as illustrated in Fig. 3. Therefore, if the DMs desire, using this holistic evaluation process, they can define some dominance relations between alternatives, which can remodel the rank order positions.

This experiment is constructed in the context of Multi-Criteria Decision Making/Aiding (MCDM/A) approach. Thus, some multi-attribute problems were generated presenting only two alternatives for DMs evaluate. These problems do not presented a specific context, being constructed using hypothetical alternatives named A and B, and hypothetical criteria named Crit 1, Crit 2, Crit 3, Crit 4, Crit 5, Crit 6, Crit 7.

Hence, based on the performance that each alternative presented in each criterion, visualizations were generated to illustrate these problems. The visualizations used in this experiment were five bar graphics, five spider graphics and five tables. Figure 4 illustrates some of the visualizations used in the experiment; the tendency line, in black color, represented the different weights for the criteria.

Each visualization designed presented a best alternative, which is previously calculated by the experiment developer's team, based on the MAVT concepts [12, 13]. Therefore, the task required for the subjects in this experiment was to select the best alternative in each visualization build, being possible to evaluate which are the visualizations that the subjects correctly perform the holistic evaluation process. It is worth to mentioning that the required task can be executed by the participants since they were attending the class of MCDM/A, providing aptitude to perform this task.

The sample of subjects that take part in this experiment was composed for graduate and postgraduate management engineering students of Federal University of Pernambuco (UFPE). The sample was composed of 78 participants, specifically: 30 women and 48 men, of which 34 were graduate students and 44 were postgraduate students. The experiment was approved by the Ethics Committee of the university.

To synthesizes, this study was composed for three phases: design the experiment, apply the experiment and analyze the data. The first phase corresponds to the construction of hypothetical MCDM/A problems, each correspondent visualization and experiment design (which is supported by the Eye-Tracking software). The second phase corresponds to the schedule of the meetings, the preparation of the experiment room (the same room was used, located in the NeuroScience for Information and Decision - NSID lab) and the execution of the experiment with the subjects. The execution phase was structured in 2 steps. The first step is represented by the explanation about the experiment and the signature of the agreement term. The second step is the execution of the experiment. The last phase is the analyses performed with the eye-tracking data, which investigate the Hit Rate and the pupil diameter variables, being presented in the next section.

4 Experiment Results

In this section the results obtained in the neuroscience experiment are presented, these results are generated using the Hit Rate and the pupil diameter variables. The Hit Rate

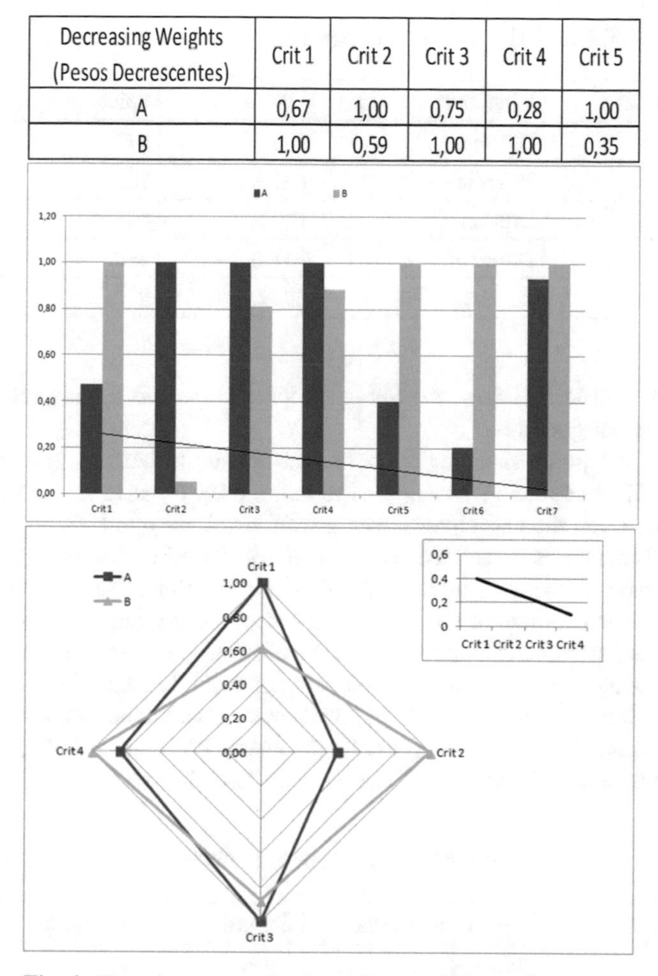

Decreasing Weights (Pesos Decrescentes)	Crit 1	Crit 2	Crit 3	Crit 4	Crit 5
A	0,67	1,00	0,75	0,28	1,00
B	1,00	0,59	1,00	1,00	0,35

Fig. 4. Experiment visualizations (Source: This research, 2020)

(HR) is a variable constructed in [6] and applied in [7–9], which is replicated in this study.

The HR is used to evaluate the visualizations regarding the task required in the experiment. In other words, this variable is used to evaluate which visualizations were used in the correct way, by the subjects, to perform the holistic evaluation and select the best alternative.

To obtain the HR values the Eye-Tracking software was used, this software allows recording the answers provided by each participant for each visualization. Thus, this variable is characterized as the ratio of the correct answers, by the total number of answers (equal to 78 in this experiment). The HR values are presented in Table 1.

Thus, based on Table 1, it is observed that using the bar graphics the subjects obtained a higher perform in selecting the best alternative; i.e. the bar graphics positively perform the required task in the most of the problems built. An exception is observed in the

Table 1. HR value for the experiment with two alternatives.

MCDM/A problem	Weights of criteria	Bar grap	Spider grap	Table
4 Criteria	Different	75%	57%	54%
5 Criteria	Different	91%	78%	93%
6 Criteria	Different	60%	40%	49%
7 Criteria	Different	31%	46%	27%
7 Criteria	Similar	94%	81%	78%

problem with 7 criteria, where the spider graphic presented a substantial HR value higher than the bar graphic.

Regarding the pupil diameter, a coefficient is used to perform the analyses with this variable. This coefficient is selected to being used because it provides a relative comparison between the participants average of the left eye pupil diameter.

This coefficient was calculated for each subject based on the ratio of the average of the left eye pupil diameter by the pupil diameter baseline. This baseline is equal to the pupil diameter captured at the exactly 10 s after the beginning of the visualization evaluation. The participant's pupil diameter coefficients are presented in Appendices A, the sample was reduced after the pre-processing of pupil diameter data.

To synthesize these participant's pupil diameter coefficients the average was calculated once more. Thus, Table 2 presents the average of participant's pupil diameter coefficients for each visualization evaluated in the experiment.

Table 2. Pupil diameter coefficient.

MCDM/A problem	Weights of criteria	Bar grap	Spider grap	Table
4 Criteria	Different	1,006	0,994	0,983
5 Criteria	Different	0,988	0,998	0,976
6 Criteria	Different	0,993	0,995	0,980
7 Criteria	Different	1,002	0,968	0,972
7 Criteria	Similar	0,965	1,007	0,972

Based on the meaning of the pupil diameter coefficient, it is considered that coefficients higher than 1 represents a general increasing in pupil diameter value during the holistic evaluation process. Therefore, based on Table 2, it is observed that the spider graphics presented higher values for the coefficients when compared to the bar graphics and the tables. Also, the bar graphics presented higher values when compared to the tables.

In this context, in order to investigate the significance of these coefficients, a statistical analysis was performed using the non-parametric Wilcoxon signed-rank test with $\alpha = 5\%$ [16].

Hence, based on this analysis, which used the values presented in the Appendices A, significant differences in the coefficient values were observed only for the comparisons where the bar graphics presented higher coefficients. In this context, the increasing in pupil diameter is considered relevant when bar graphics were evaluated by the subjects.

Therefore, connecting this result to the HR result, it is possible to observe that the bar graphics also presented higher HR values. Thus, such results possibly present a causality relation, where the increasing in pupil diameter, which is an indicator of increasing in mental activity [17, 18], possibly produce higher HRs values for the bar graphics.

Finally, an important result is also generated using the HR values. From the discrete probability distributions, it is observed that the experiment required task follows a Bernoulli distribution [16]. Thus, based on HR values, it is possible to obtain the visualizations probability of success in selecting the best alternative (being the HR an estimator of the probability of success).

Also, based on this distribution, the standard deviation can be calculated for a general case, using the probability of success in a range of 0 to 1 (which can be referred to HR values in 0% to 100%). Therefore, the Recommendation Rule is proposed in this study based on the probability of success and the standard deviation, as illustrated in Fig. 5.

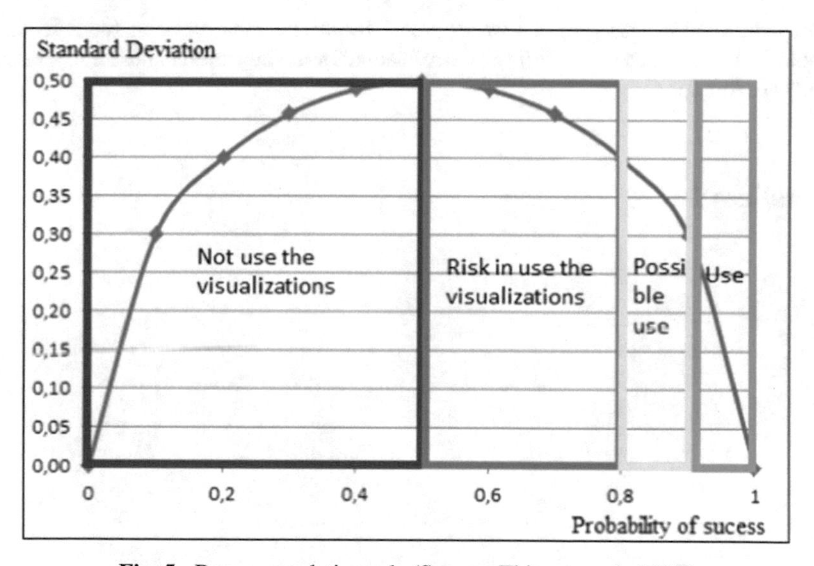

Fig. 5. Recommendation rule (Source: This research, 2020)

5 Conclusion

The HR values are an important insight since they permit the development of the Recommendation Rule to support the analyst in the advertising process with the DMs. Thus, based on this rule the analyst can advise the decision-maker to continue the step 2 of the FITradeoff method or to use a graphical or tabular visualization to define preference relations between the alternatives.

Therefore, based on Fig. 3, if the DM desire to compare the incomparable alternatives, the analyst recommendation should be to not use any of the visualizations to define preference relations between these alternatives (i.e. this problem present seven criteria with different values for the weights, thus the HR values are lower than 50%, and the recommendation in Fig. 5 is "Not use the visualizations").

However, if this problem presented similar values to the criteria weights, the recommendation is the opposite (i.e. "Use the visualizations" to define preference relation), since HR values increased.

Thus, using this former recommendation, the DM can evaluate the visualizations and if he/she defines a preference relation based in the holistic evaluation process this relation should be implemented in the DSS algorithm and in the Hasse Diagram, being improvements for this DSS, which are generated by the holistic evaluation process.

Also, based on the connection of the pupil diameter coefficients and the HR values, it is suggested a relation between these variables during the bar graphics evaluations, when the higher values in HR can be generated by an increasing in mental activity [17, 18] (indicated by the increasing in pupil diameter). Thus, an additional suggestion for improvement in the FITradeoff DSS should be the inclusion of bar graphics in this DSS.

Acknowledgements. This project was supported by the National Council for Scientific and Technological Development (CNPq) and Coordination for the Improvement of Higher Education Personnel (CAPES).

Appendices A

	4 Criteria - different weights			5 Criteria - different weights			6 Criteria - different weights			7 Criteria - different weights			7 Criteria - similar weights		
	Bar grap	Spider grap	Table	Bar grap	Spider grap	Table	Bar grap	Spider grap	Table	Bar grap	Spider grap	Table	Bar grap	Spider grap	Table
76	0,984	1,016	0,993	1,082	0,976	1,016	0,995	1,133	1,001	0,939	0,97	0,976	1,031	1,101	0,982
75	1,072	1,008	0,959	1,076	1,084	0,973	1,047	1,016	1	1,044	0,971	0,964	1,021	1,018	1,006
74	0,996	1,038	0,971	0,992	0,984	1,097	0,974	1,011	0,989	0,958	1,003	1,023	0,997	0,983	0,999
73	0,987	0,972	0,971	0,971	0,979	0,906	0,98	0,976	0,915	0,995	0,975	0,897	0,977	1,004	0,962
72	0,92	0,938	0,917	0,899	0,893	0,9	0,864	0,887	0,911	0,885	0,868	0,867	0,91	0,886	0,835
71	1,223	1,075	1,166	1,078	0,963	1,069	1,034	1,11	1,105	1,153	1,072	1,066	1,034	1,096	1,024
70	0,972	0,918	0,946	0,905	0,953	0,931	0,953	0,934	0,901	1,032	0,919	0,932	0,952	0,884	0,929
69	1,014	0,939	1,032	0,976	0,996	1.015	0,992	0,969	0,959	1,005	0,982	0,996	1,002	0,989	1,078
68	1,034	0,992	1,034	1,043	0,977	0,924	0,985	0,974	1,07	1,017	0,986	0,928	0,993	0,967	0,966
67	1,052	1,021	1,095	1,059	1,021	0,997	1,023	1,073	1,027	1,029	1,095	1,081	1,085	1,09	1,027
66	0,969	1,008	0,968	0,979	1,008	0,967	0,983	1,03	0,974	0,986	0,92	0,986	0,931	0,956	0,972
65	0,985	1,001	1,016	1,006	1,001	1,043	1,003	1,002	1,04	0,997	1,034	0,984	1,066	0,937	1,026
64	1,03	1,027	1,029	1,031	1,027	1,048	0,996	1,018	1,089	1,046	1,017	0,99	0,993	1,047	1,002
63	1,024	0,953	0,988	0,97	0,953	0,967	1,023	0,978	0,926	0,955	1,006	0,998	0,974	0,954	0,979
61	0,987	1,037	1,015	0,965	1,037	0,943	1,054	1,009	0,999	1,041	0,979	0,985	1,097	0,945	0,956
60	1,142	1,066	1,01	1,055	1,066	1,023	1,052	0,991	0,996	1,049	0,997	1,006	1,079	1,01	1,012
59	0,963	1,007	0,961	0,964	1,007	0,982	0,986	0,961	1,006	1,008	0,974	1,006	0,988	0,989	0,915
58	0,894	0,885	0,901	0,998	0,885	0,904	0,892	0,946	0,888	0,946	0,905	0,858	0,938	0,867	0,927
56	0,887	0,951	0,952	0,934	0,951	0,928	0,955	0,975	0,965	1,004	0,927	0,959	0,994	0,953	0,987

(*continued*)

(*continued*)

	4 Criteria - different weights			5 Criteria - different weights			6 Criteria - different weights			7 Criteria - different weights			7 Criteria - similar weights		
	Bar grap	Spider grap	Table	Bar grap	Spider grap	Table	Bar grap	Spider grap	Table	Bar grap	Spider grap	Table	Bar grap	Spider grap	Table
54	1,135	1,03	0,983	1,038	1,03	0,974	1,03	1,015	0,939	1,092	1,016	0,957	1,09	1,041	1,056
53	1,087	1,669	0,95	1,063	1,669	0,958	0,97	1,052	0,958	0,984	0,962	0,936	1,041	0,93	0,967
52	0,996	0,925	1,014	0,974	0,925	0,982	1,014	1,033	0,836	1	0,936	0,947	0,944	0,809	0,852
50	1,01	0,976	1,008	0,98	0,976	0,993	0,974	0,906	0,977	0,977	1,037	1,023	0,974	0,911	0,972
49	0,941	1,02	0,953	0,999	1,02	0,974	0,966	0,981	0,982	1,019	0,96	1,015	0,998	1,007	0,976
47	0,961	0,98	0,925	0,945	0,98	0,99	0,954	0,988	0,911	0,954	0,936	0,953	0,936	0,917	0,941
46	1,092	1,015	1,018	1,037	1,015	0,993	1,185	1,037	1,004	1,107	0,991	1	1,091	1,042	1,025
45	0,962	0,925	0,947	0,969	0,925	1,044	0,989	0,945	0,993	1,02	0,908	1,019	1,042	0,853	0,959
44	1,021	1,029	1,001	0,967	1,029	0,99	1,058	0,967	0,995	1,019	0,961	0,988	0,995	1,026	1,005
43	1,026	0,909	1,001	1,013	0,909	0,991	0,998	0,958	1,054	0,954	0,919	0,942	1,045	0,92	0,984
33	1,005	0,987	0,951	0,971	0,987	0,934	1	0,98	0,926	0,994	0,972	0,928	1,037	1,009	0,932
32	0,997	1,005	0,922	0,979	0,964	0,994	1,005	1,009	1,022	0,984	0,945	0,964	1,027	0,944	0,946
31	0,989	0,937	0,947	1,048	0,965	0,978	0,986	1,047	0,995	0,989	0,948	0,973	0,998	0,934	0,954
28	1,054	0,986	1,034	1,061	0,986	1,017	1,045	0,969	1,01	1,058	1	1,014	1,029	1,014	1,04
27	1,096	0,981	1,088	1,018	1,081	1,041	1,079	1,038	1,031	1,04	1,047	1,072	1,118	1,033	1,044
26	1,061	0,975	1,022	0,961	0,989	0,957	1,027	1,08	1,109	0,962	0,993	0,893	0,996	0,982	1,012
24	0,957	0,921	1,053	0,975	1,002	0,957	0,991	1,013	0,981	1,056	0,902	0,978	1,016	0,997	0,963
23	0,923	0,966	0,974	0,876	1,084	0,901	0,936	0,988	0,986	0,944	0,973	0,968	0,971	0,887	1,023

(*continued*)

(*continued*)

	4 Criteria - different weights			5 Criteria - different weights			6 Criteria - different weights			7 Criteria - different weights			7 Criteria - similar weights		
	Bar grap	Spider grap	Table	Bar grap	Spider grap	Table	Bar grap	Spider grap	Table	Bar grap	Spider grap	Table	Bar grap	Spider grap	Table
22	0,962	0,964	0,928	0,944	0,974	0,962	0,952	0,985	0,903	1,044	0,976	0,931	0,964	0,911	0,915
21	0,884	0,889	0,91	0,856	0,876	0,904	0,94	0,908	0,864	0,89	0,886	0,998	0,898	0,867	0,907
19	0,981	0,998	0,975	0,981	0,973	0,967	0,996	1,032	0,994	1,023	1,047	1,028	1,002	1,036	0,993
17	1,051	0,957	0,951	0,932	0,98	0,972	0,952	1,024	0,963	0,985	0,906	0,869	1,088	0,822	0,912
16	0,967	0,965	0,966	0,962	0,983	0,954	0,982	0,977	0,956	0,977	0,906	0,916	0,976	0,971	0,947
11	0,978	0,935	0,883	0,956	0,955	0,936	0,974	0,951	0,913	0,993	0,944	0,923	1,003	0,949	0,895
2	0,949	0,97	0,939	0,992	0,909	0,959	0,909	0,925	1,035	0,933	0,903	1,004	0,963	0,975	0,986
1	1,041	0,96	0,978	0,97	0,976	0,943	0,992	0,977	1,002	0,985	0,97	0,992	1,02	0,954	0,955

References

1. Riedl, R., Banker, R.D., Benbasat, I., Davis, F.D., Dennis, A.R., Dimoka, A., Gefen, D., Gupta, A., Ischebeck, A., Kenning, P., Müller-Putz, G., Pavlou, P.A., Straub, D.W., vom Brocke, J., Weber, B.: On the foundations of NeuroIS: reflections on the Gmunden Retreat 2009. Commun. Assoc. Inf. Syst. **27**, 15 (2010)
2. Riedl, R., Davis, F.D., Hevner, A.R.: Towards a NeuroIS research methodology: intensifying the discussion on methods, tools, and measurement. J. Assoc. Inf. Syst. **15**, 4 (2014)
3. Loos, P., Riedl, R., Müller-Putz, G.R., Vom Brocke, J., Davis, F.D., Banker, R.D., Léger, P.M.: NeuroIS: neuroscientific approaches in the investigation and development of information systems. Bus. Inf. Syst. Eng. **2**(6), 395–401 (2010)
4. de Almeida, A.T., de Almeida, J.A., Costa, A.P.C.S., de Almeida-Filho, A.T.: A new method for elicitation of criteria weights in additive models: flexible and interactive tradeoff. Eur. J. Oper. Res. **250**, 179–191 (2016)
5. Frej, E.A., de Almeida, A.T., Costa, A.P.C.S.: Using data visualization for ranking alternatives with partial information and interactive tradeoff elicitation. Oper. Res. Int. J. **19**, 909–931 (2019)
6. de Almeida, A.T, Roselli, L.R.P.: Visualization for decision support in FITradeoff method: exploring its evaluation with cognitive neuroscience. In: Lecture Notes in Business Information Processing, vol. 282, pp. 61–73. Springer, Heidelberg (2017)
7. Roselli, L.R.P., Frej, E.A, de Almeida, A.T.: Neuroscience experiment for graphical visualization in the FITradeoff decision support system. In: Chen, Y., Kersten, G., Vetschera, R., Xu, H. (eds.) Group Decision and Negotiation in an Uncertain World. Lecture Notes in Business Information Processing, vol 315 (2018)
8. Roselli, L.R.P., de Almeida, A.T., Frej, E.A.: Decision neuroscience for improving data visualization of decision support in the FITradeoff method. Oper. Res. Int. J. **19**, 1–21 (2019)
9. Roselli, L.R.P., Silva, A.L.C.L, de Almeida, A.T.: Evaluation of graphical visualization behavioral in a multi-criteria decision making context. In: 17th Society of Neuroeconomics, Dublin, Irlanda. (2019)
10. Roselli, L.R.P., Pereira, L.S., da Silva, A.L.C.L., de Almeida, A.T., Morais, D.C., Costa, A.P.C.S.: Neuroscience experiment applied to investigate decision-maker behavior in the tradeoff elicitation procedure. Ann. Oper. Res. **289**, 1–18 (2019)
11. Silva, A.L.C.L, Costa, A.P.C.S.: FITradeoff decision support system: an exploratory study with neuroscience tools. In: NeuroIS Retreat 2019, Viena (2019)
12. Keeney, R.L., Raiffa, H.: Decision Making with Multiple Objectives, Preferences, and Value Tradeoffs. Wiley, New York (1976)
13. Belton, V., Stewart, T.: Multiple Criteria Decision Analysis: An Integrated Approach. Springer, Heidelberg (2002)
14. de Almeida, A.T, Cavalcante, C., Alencar, M., Ferreira, R., de Almeida-Filho, A.T., Garcez, T.: Multicriteria and multi-objective models for risk, reliability and maintenance decision analysis. In: International Series in Operations Research & Management Science, vol. 231. Springer, New York (2015)
15. Frej, E.A., Roselli, L.R.P., de Almeida, A.J., de Almeida, A.T.: A multicriteria decision model for supplier selection in a food industry based on FITradeoff Method. Math. Probl. **2017**, 1–9 (2017)
16. Hines, W.W., Montgomery, D.C.: Probability and Statistics in Engineering and Management Science. Wiley, New York (1990)
17. Laeng, B., Sirois, S., Gredebäck, G.: Pupillometry: a window to the preconscious? Perspect. Psychol. Sci. **7**(1), 18–27 (2012)
18. Porter, G., Troscianko, T., Gilchrist, I.D.: Effort during visual search and counting: Insights from pupillometry. Q. J. Exp. Psychol. **60**(2), 211–229 (2007)

The Impact of Modularization on the Understandability of Declarative Process Models: A Research Model

Amine Abbad Andaloussi[1](✉), Pnina Soffer[2], Tijs Slaats[3], Andrea Burattin[1], and Barbara Weber[4]

[1] Software and Process Engineering, Technical University of Denmark, 2800 Kgs. Lyngby, Denmark
amab@dtu.dk

[2] Department of Information Systems, University of Haifa, 3498838 Haifa, Israel

[3] Department of Computer Science, University of Copenhagen, 2100 Copenhagen, Denmark

[4] Institute of Computer Science, University of St. Gallen, 9000 St. Gallen, Switzerland

Abstract. Process models provide a blueprint for process execution and an indispensable tool for process management. Bearing in mind their trending use for requirement elicitation, communication and improvement of business processes, the need for understandable process models becomes a must. In this paper, we propose a research model to investigate the impact of modularization on the understandability of declarative process models. We design a controlled experiment supported by eye-tracking, electroencephalography (EEG) and galvanic skin response (GSR) to appraise the understandability of hierarchical process models through measures such as comprehension accuracy, response time, attention, cognitive load and cognitive integration.

Keywords: Modularization · Understandability · Declarative process models · DCR graphs · Neurophysiological experiment

1 Introduction

Process digitization begins with a set of process specifications, which are represented as process models and then implemented as part of a process-aware information system (PAIS). Process models serve both enactment and management purposes [1]. They provide a blue-print for process execution – but can also be used to elicit, communicate, and improve the quality of business processes. Designing understandable models is crucial for attaining these purposes.

Processes are represented using languages from the imperative-declarative paradigm (for a literature review, see [2]). Imperative languages clearly depict the different executions supported by the process, and this makes them relatively easy to comprehend.

Work supported by the Innovation Fund Denmark project *EcoKnow* (7050-00034A).

F. D. Davis et al. (Eds.): NeuroIS 2020, LNISO 43, pp. 133–144, 2020.
https://doi.org/10.1007/978-3-030-60073-0_15

However, their support is limited to rigid specifications, which is suitable for repetitive and structured processes, but not for ones where flexibility is an inherent requirement (e.g., knowledge-intensive processes). This need is satisfied by declarative languages, which allow flexible process specifications [1, 3] but are difficult to understand [1].

The use of declarative languages results often in complex models, which are hard to interpret and maintain by humans. Comprehending the human cognitive processes impacting the understandability of process models, paves the way toward the adoption of modeling practices, enhancing the comprehension of declarative models and thus supporting their use for management purposes [4, 5]. In the field of cognitive psychology, existing research has shown the limited capacity of the human working memory [6]. Accordingly, this limited resource must be utilized in a way such that a reader can easily interpret the constraints of the model and extract the required information efficiently. Cognitive load is a common indicator of the use of working memory [7]. Whenever a reader is introduced to a process model, 3 types of load emerge: intrinsic load, extraneous load and germane load [8]. Intrinsic load relates to the inherent complexity of the process, whereas extraneous load raises from the way the process is represented. Germane load, in turn, emerges from the effort invested by the reader to comprehend and reason about the model. While the intrinsic load is changing from one process to another, the extraneous load can be reduced by refining the model representation, hence leaving more capacity for the germane load to emerge and thus an increased ability for the reader to comprehend the process model.

Our research taps into the representation of process models. Considering the intrinsic complexity of processes and the different ways in which entangled constraints could interact in declarative models [9], readers can exceed their working memory capacity and thus limit their understanding of the model. Modularization could reduce the complexity of process models by decomposing them into sub-processes. Modularization has been investigated in computer programming [10], conceptual modeling [11–13] and process modeling [4, 14–16]. With regards to declarative languages, a qualitative study [16] suggests that abstraction and fragmentation are two opposing forces affecting the understandability of modularized process models expressed in the Declare language [17]. Grounded in the theory of cognitive fit [18], we use local and global tasks to perceive the influence of abstraction and fragmentation. As part of our ongoing research, we design a controlled experiment supported by eye-tracking, electroencephalography (EEG) and galvanic skin response (GSR) to investigate end-users' understandability through measures such as comprehension accuracy, response time, attention, cognitive load and cognitive integration. We study modularization in the context of declarative models expressed using the Dynamic Condition Response (DCR graphs) language [19]. We focus on this language in particular because of the availability of industrial-level tools [20] and a wide array of documented real-world applications [21–23]. Section 2 presents the theoretical background, Sect. 3 introduces our research method, and Sect. 4 concludes the paper.

2 Theoretical Background

Modularization and Hierarchy. Modularization denotes the degree to which a system can be devised into independent, composable units [24]. Information hiding is a branch of modularization [24]. In computer programming, it denotes the distinction between the interface and the implementation of a system, which in turn allows refining the implementation without invalidating the interface. In that sense, an interface is seen as an abstraction of the implementation, allowing to reason about the system on a more abstract level [24]. The same principle holds with process models. Hereby, an interface is equivalent to a high-level model providing an overview of the process, while implementations are just like sub-processes describing the low-level details of the process. Information hiding is better supported through hierarchy. Dawkins [25] discusses the notion of "hierarchical reductionism". He used hierarchy to organize complex systems into units, such that each unit of the hierarchy abstracts the details of its subsequent units (placed one level down in the hierarchy) and concretize the details of its former units (placed one level up in the hierarchy). In process modeling, hierarchical reductionism takes information hiding beyond a single abstraction level, making it possible to define sub-processes within a sub-process itself.

Hierarchy was introduced to Declare in [16] and expanded upon in [26]. In DCR, different forms of decomposing models have been introduced [27–29]. For our study we focus on a special type of hierarchy referred to as single-instance sub-processes, where a sub-process is a non-atomic activity which contains embedded activities and constraints that need to be completed before the sub-process can execute [30], similar to what was done for Declare [16, 26].

Impact of Modularization. Modularization has been widely investigated in the literature [10–15]. However, its impact on understandability remains inconclusive and hard to generalize (for a systematic review, see [4]). Hierarchy is claimed to abstract the details in the process model and provide better means to cope with complexity [15, 16]. The notion of complexity has been studied in the computer programming literature [31]. Structural complexity is among the different types of complexity identified in [32]. It denotes the complexity associated with the representation of the artifact (e.g., process model, source code). Structural complexity has been empirically investigated in model comprehension [33]. Using different representations, existing research has shown a significant impact of structural complexity on users' cognitive load (e.g., [34]), comprehension accuracy and response time (e.g., [35]).

Petrusel and Mendling [36] show that users do not typically focus on the entire model but rather limit their attention to only relevant parts of the model. In that vein, abstraction could presumably focus readers' attention on relevant sub-processes. Attention has been investigated in model comprehension. More specifically, recent research has shown how different process representations guide readers' attention towards the relevant parts of the model, and how increased attention accounts for comprehension accuracy [37].

Besides, increased modularization can cause fragmentation and thus requires the reader to continually switch attention between the sub-processes of the model, which leads to the split-attention effect [16, 38]. This effect happens when readers are required to distribute their attention between different sources of information (e.g., different sub-processes) [8]. The split-attention effect can distract readers' ability to focus on relevant aspects, which in turn requires investing additional mental effort when solving a task [8]. In model comprehension, research suggests that splitting the process control-flow and the underlying business rules could presumably influence the reader comprehension accuracy, response time and cognitive load [39].

Additionally, the split attention effect requires the reader to mentally integrate the information extracted from the model sub-processes to understand the process. In model comprehension, recent research linked cognitive integration to comprehension accuracy [37]. In the same research, it has been shown that the visual associations exhibited by readers when making sense of the different components of a model can be used as an indicator for cognitive integration.

Impact of Task Type. The influence of task type on the comprehension of visual representations has been shown in several contexts [40–43]. Following the cognitive fit theory [18], a fit between the task type and the information exposed by the visual representation (e.g., model) is associated with better performance. The impact of the task type has been widely investigated in model comprehension studies [40–43]. Vessey and Galletta [40] identified symbolic tasks (i.e., addressing discrete data values) and spatial tasks (i.e., addressing relationships in data) and proclaimed that tabular representations create better fit for symbolic tasks, while graphical representations make a better fit with spatial tasks. Likewise, Ritchi et al. [41] discerned schema-based tasks (i.e., can be solely completed from the model) and non-schema-based tasks (i.e., addressing aspects beyond the explicit information exposed in the model) and investigated the extent to which graphical and textual representations fit for different tasks. Dun and Grabski [43], in turn, integrated the notion of localization introduced by Larkin and Simon [42] and asserted that the more local the information allowing to solve a particular task, the better is the performance, making the distinction between *local* and *global* tasks a pertinent factor defining the understandability of visual representations. Building upon the tasks' classification of Dun and Grabski [43], we consider the proposed distinction as a relevant dimension for model comprehension.

Interaction Between Modularization and Task Type. In process modeling, the interaction between both factors has been raised in Zugal's literature review [4]. The author suggested that hierarchical models are more efficient for solving local tasks than global tasks. However, he did not provide a clear theoretical understanding of how a fit between the task type and the model representation could affect understandability. To fill this gap, we turn to the theory of cognitive fit to explain how local and global tasks could indeed influence the comprehension of hierarchical process models.

We postulate that modularization supports local tasks through abstraction: the process is divided into sub-processes, and the task addresses specifications within a single sub-process, which in turn creates a good fit between the visual representation of the process and the task at hand. Conversely, modularization complicates global tasks due to fragmentation: the process is divided into sub-processes while the task requires continuous integration of information from different sub-processes, causing a mismatch between the representation of the process and the task at hand.

To further explore the interaction between modularization and task type, we refer to the discussion of structural integration by Gilmore and Green [44], which underlines that understandable representations are those where information can be easily located and transferred to working memory. Gilmore and Green [44] also showed that mental representations preserve some features of the used notation. Hereby, a hierarchical model would produce a more or less "structured" mental model compared to a flat model. Local tasks, in turn, could benefit from the structure of the mental model, which facilitates the retrieval and the transfer of information to working memory. Conversely, additional load could emerge when solving global tasks, as the reader is required to disregard the acquired structure and rather perceive the interplay between activities lying within different sub-processes. Herein, the imposed structure adds additional burden to the reader. It is nonetheless worthwhile to mention that the mental model could also be affected by other factors such as background and experience, which could, in turn, pre-define the way the user acquires and incorporates new information.

3 Research Method

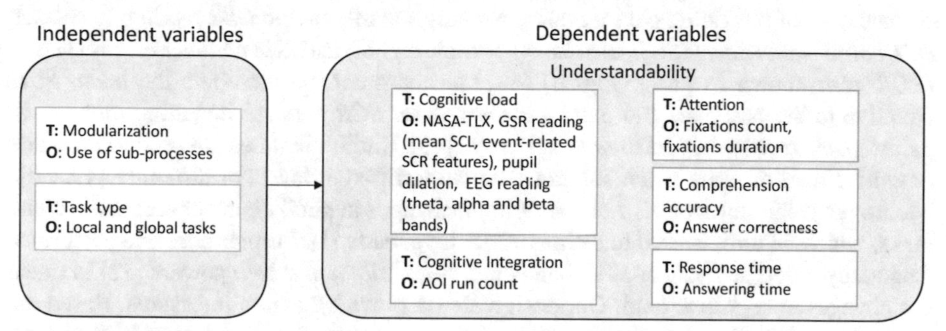

Fig. 1. Envisioned Research Model. T and O refer to the theoretical constructs (T) and their operationalization (O) respectively

Research Model. The theoretical background presented in Sect. 2 suggests that abstraction and fragmentation drive the understandability of modularized process models. Our research aims at providing empirical evidence supporting this proposition. Following a 2×2 factorial design [45], we define *Modularization* (levels: *modularized models versus flat models*) and *Task type* (levels: *local tasks versus global tasks*) as two distinct factors. These factors are expected to impact the user understanding of the model. *Understandability* is a cognitive concept, created in the reader cognition and, thus, not directly tangible [34]. Eventually, it can be estimated only using indirect constructs. Aranda et al. [46] evoke difficulty (i.e., cognitive load), correctness (i.e., comprehension accuracy), and time (i.e., response time) as indicators of understandability. Motivated by the existing literature on model comprehension [36, 37, 39], we additionally consider cognitive integration and attention. Our research model is summarized in Fig. 1.

Following the theoretical foundations set in Sect. 2 and the requirements for a factorial design, we formulate our first *main effect* hypothesis as follows: H_1: **There is a significant difference in the understandability of modularized and flat process models**, while we formulate our second main hypothesis as follows: H_2: **There is a**

significant difference in the understandability of local and global tasks. In addition, we formulate our interaction effect hypothesis as follows H_3**: Modularized models are more understandable for local than for global tasks**.

The theoretical constructs depicted in Fig. 1 are operationalized as follows. With regards to our independent variables, modularization is operationalized using modularized models with sub-processes and flat models without sub-processes, while the task type is operationalized using local and global tasks. Local tasks are meant to make use of abstraction without being affected by fragmentation. They address local aspects requiring to perceive the interplay between activities belonging to the same sub-process. In contrast, global tasks are designed to neglect abstraction and rather cause fragmentation. They address global aspects requiring to perceive the interplay between activities belonging to different sub-processes.

The dependent variables covered by our research model are operationalized using *subjective, neurophysiological, behavioral and performance* measures. To measure cognitive load, we use a subjective rating of cognitive load i.e., *NASA-TLX* [47]. In addition, we use a set of physiological measures. Namely, we rely on the *GSR reading* to extract *skin conductance level* (*SCL*, also know as tonic signal) and *skin conductance response* (*SCR*, also known as phasic signal) [48]. As measures, we compute the *mean SCL* (relative to the baseline) and extract *event-related SCR features* including *number of peaks*, *peak amplitude*, and *area under curve* [49]. Similar features are used to measure cognitive load of users when solving tasks with different levels of difficulty [49, 50]. Moreover, we monitor *pupil dilation* through changes in pupil diameter across different tasks, which, in turn, is used to estimate cognitive load [51]. Furthermore, we perform a frequency-based analysis of EEG bands i.e., *theta, alpha and beta* powers [52] to track the changes in cognitive load. Our design also deploys behavioral measures. Based on the notions of fixation (i.e., the timespan where the eye remains still at a specific position of the stimulus [53]) and areas of interest (AOI, i.e., a grouping of fixations covering a specific area of the stimulus [53]), we use the *AOI run count* (i.e., number of entry and exists to AOIs [37]) to evaluate the participants' cognitive integration, and we use *fixations count and duration* to measure attention. Moreover, we rely on performance measures such as *answer correctness* and *answering time* to analyze the participants' comprehension accuracy and response time.

Material. The material meant for the experiment comprises a set of information-equivalent models represented with and without modularization and a set of local and global tasks allowing to test the impact of abstraction and fragmentation respectively.

Based on the guidelines and recommendations in [54], we define a set of requirements addressing the design of *models* and *tasks*. By following these requirements, we aim at reducing the effects of the confounding factors threatening the validity of our study.

With regards to the models, the insights of Zimoch et al. [54] provide a good starting point to define a uniform visual layout applying to all process models. Herein, we carefully set up a layout where activities are oriented from left to right, depending on their likely order of execution. We also avoid crossing arrows as much as possible, ensure proper spacing between the model's elements and name activities consistently. Besides, we address the complexity of the models to ensure that all of them have the same number of activities and deploy similar constraint patterns. Moreover, similar to [55], we use anonymized models (where activities are labeled with random letters) to avoid the influence of the domain. Last but not least, we append a legend describing the DCR semantics to all models, assuring that participants are able to interpret the DCR notation represented in the models.

As for the design of tasks, we use local tasks addressing the interplay between activities within a single sub-process, and global tasks addressing the interplay between activities located in different sub-processes. Within each task, we ask one question. We design questions to reflect the use of process models in practice. Similar to [5, 16, 44], we formulate questions addressing: the presence and absence of constraints in the model, order of activities and validity of process executions. We make dichotomous questions. Nevertheless, to reduce the chances of guessing answers, we ask participants to justify their answers verbally and allow them to skip answering questions which they are unsure about.

To ensure that our models and tasks are representative, there are a couple of measures which we take into consideration. We design our models in DCR graphs that is a known declarative language with academic and industrial tool-support [20] deployed by several private and public institution in Denmark[1]. Additionally, we rely on the recommendations of experts in DCR graphs to provide models covering a large sub-set of constraint patterns which are frequently used in practice. As for the tasks, we build upon existing literature [5, 16, 44] and use different types of questions reflecting different scenarios where process models are used in practice. Similar to [44], the variety of our questions is not meant to predict differences between different types of questions, but rather to ensure that our findings could be generalized.

We propose different manipulations to cover all the conditions where modularized and flat process models are used to solve local and global tasks. We group the models into sets. Each set S_i ($i \in \mathbb{N}$) is composed of (1) a Process P_i modeled in two variants: one modularized M_{i_m} and one flat M_{i_f} and (2) two tasks: one local T_{i_l} and one global T_{i_g}. Combining models and tasks, each set contains the following: $\{M_{i_m} T_{i_l}, M_{i_m} T_{i_g}, M_{i_f} T_{i_l}, M_{i_f} T_{i_g}\}$. Afterwards, we group the sets into collections. Each Collection is composed of 4 distinct sets, where the respective tasks within each set address presence, absence, order or execution questions. Figure 2 shows an example of a collection.

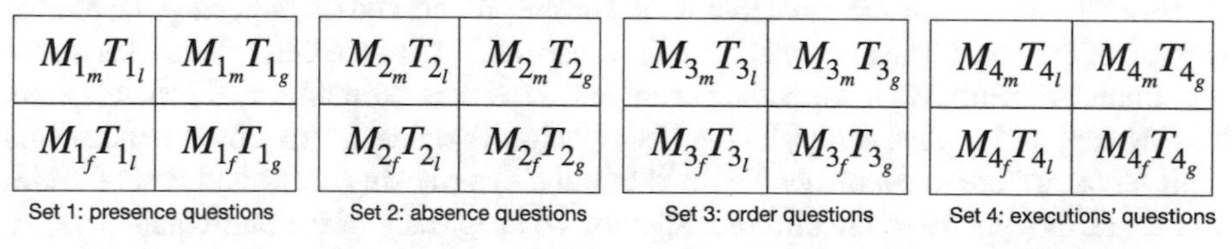

Fig. 2. Example of a collection

Participants. Confounding factors related to the subjects of the study (i.e., participants) represent significant threats to validity if not handled correctly. Following the recommendations in [54], we limit our study to novice participants with no or very limited

[1] See https://dcrsolutions.net

experience with DCR graphs. Doing so, we ensure that the observed effects are due to our manipulations rather than personal factors associated with participants' expertise. Moreover, we perform a screening of all participants checking their physical ability to participate in neuropsychological experiments. Furthermore, to guarantee that all participants are equally trained, we provide a uniform and comprehensive familiarization to all participants, so they have the necessary background about the investigated theory and experiment procedure. By the end of the familiarization, we provide a short quiz to evaluate the technical aptitude of participants.

Experiment Design. We use a within-subject experiment design where each participant is exposed to all conditions. We motivate our design by the idiosyncratic nature of many eye-tracking, EEG and GSR measures [53, 56, 57] (i.e., each participant has her own baseline), which in turn requires a within-subject comparison of the different experiment's conditions. A possible threat to this design might be associated with a learning effect and fatigue during the experiment, which could influence the results of the within-subject comparison. To mitigate these effects, we randomize the experiment's tasks, ensuring that participants receive tasks in different orders.

To ensure a good data quality, we follow existing guidelines on collecting clean eye tracking [53], GSR [56] and EEG [57, 58] data. Before the experiment, a screening form (asking about age range, gender, proficiency in English, vision issues, neurological diseases e.g., epilepsy, attention disorder, handedness and allergies) is sent to check the participant's physical ability to participate in neuropsychological experiments and obtain relative information allowing to determine the most suitable EEG cap size for her. Upon approval, an invitation is sent and information regarding what to avoid (e.g., mascara, eyelash extensions, reflective glasses, artificial hair products, hair pins and clips) are shared. In addition, we ask the participant to wash her hair and dry it completely prior to the experiment day. We prepare and set up electrodes in the EEG recording cap, verify the light conditions and the temperature in the lab before receiving the participant.

We start each experiment session with a familiarization and a quiz. Afterwards, we collect demographic and expertise information from the participant. Next, we seat the participant in front of the eye-tracking station comfortably by (i) adjusting the chair and the table to the participant's preferences, while guaranteeing that the eye tracker can still capture her eyes, (ii) ensuring that feet are flat on the ground, and (iii) adjusting the lumbar support of the chair. Then, we place the EEG cap, and adjust the GSR electrodes on the non-dominant hand. We instruct the participant to breathe normally, avoid chewing and tensing her jaw, keep the hand with the GSR electrodes stable, not to move her head, and limit all other body movements. We calibrate the eye-tracking, GSR and EEG devices and check the quality of the different signals. To keep track of the participant's verbal utterances (i.e., the justifications of her answers), we record the audio for the whole data collection part.

The data collection is composed of a set of trials. During each trial, we show a grey rest screen for 1-2 min and collect new baseline measurements. Afterwards, we select a random task. Here, we display the model and respective question, and then collect the answer from the participant. Next, we display the same model and respective question again, but this time, we ask the participant to justify her answer. By doing so, we can differentiate the initial response time from the time used to justify the answer

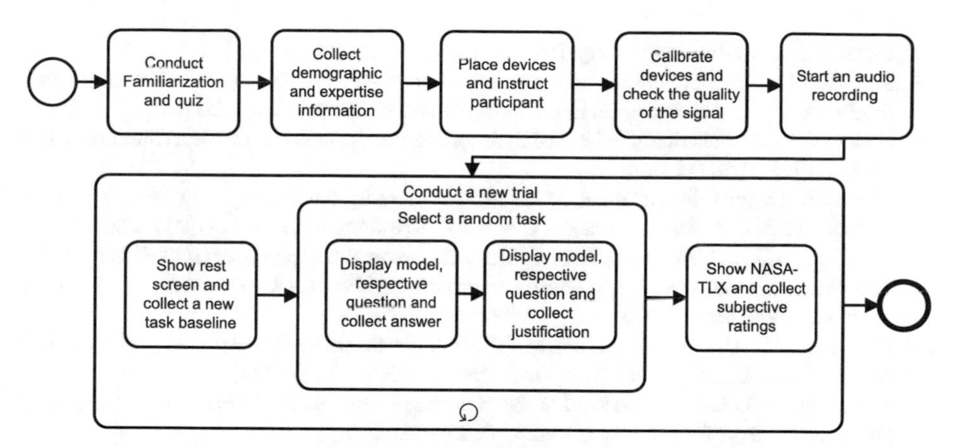

Fig. 3. Data collection procedure

verbally. Finally, we provide the NASA-TLX questionnaire to obtain a subjective rating of cognitive load. Figure 3 depicts a BPMN [59] model summarizing the experiment procedure.

4 Conclusion

This paper describes a research model aimed at investigating the impact of modularization on the understandability of declarative process models. As future work, we are planning to concretize this model and report empirical evidence about the impact of local and global tasks on the understandability of hierarchical process models in DCR graphs.

References

1. Reichert, M., Weber, B.: Enabling Flexibility in Process-Aware Information Systems. Springer, Heidelberg (2012)
2. Andaloussi, A.A., Burattin, A., Slaats, T., Kindler, E., Weber, B.: On the declarative paradigm in hybrid business process representations: a conceptual framework and a systematic literature study. Inf. Syst. **91**, 101505 (2020)
3. Slaats, T.: Declarative and hybrid process discovery: recent advances and open challenges. J. Data Semant. **9**, 1–18 (2020)
4. Zugal, S.: Applying Cognitive Psychology for Improving the Creation, Uderstanding and Maintenance of Business Process Models (2013)
5. Andaloussi, A.A., Burattin, A., Slaats, T., Petersen, A.C.M., Hildebrandt, T.T., Weber, B.: Exploring the understandability of a hybrid process design artifact based on DCR graphs. In: Enterprise, Business Process and Information Systems Modeling, BPMDS 2019, pp. 1–15 (2019)
6. Baddeley, A.: Working memory: theories, models, and controversies. Annu. Rev. Psychol. **63**, 1–29 (2012)
7. Chen, F., Zhou, J., Wang, Y., Yu, K., Arshad, S.Z., Khawaji, A., Conway, D.: Robust Multimodal Cognitive Load Measurement, pp. 13–32. Springer, Cham (2016)

8. Sweller, J.: Cognitive load theory. Psychol. Learn. Motiv. **55**, 37–76 (2011)
9. 9. Pesic, M.: Constraint-based workflow management systems: shifting control to users. PhD thesis, Department of Industrial Engineering & Innovation Sciences (2008)
10. Parnas, D.L.: On the criteria to be used in decomposing systems into moules. Commun. ACM **15**(12), 1053–1058 (1972)
11. Shoval, P., Danoch, R., Balabam, M.: Hierarchical entity-relationship diagrams: the model, method of creation and experimental evaluation. Requirements Eng. **9**(4), 217–228 (2004)
12. Cruz-Lemus, J.A., Genero, M., Piattini. M.: Using controlled experiments for validating UML statechart diagrams measures. In: Software Process and Product Measurement, pp. 129–138. Springer, Heidelberg (2007)
13. Cruz-Lemus, J., Genero, M., Piattini, M.: Toval, A: Investigating the nesting level of composite states in UML statechart diagrams. Proc. QAOOSE **5**, 97–108 (2005)
14. Johannsen, F., Leist, S.: Wand and Weber's decomposition model in the context of business process modeling. Bus. Inf. Syst. Eng. **4**(5), 271–286 (2012)
15. Reijers, H.A., Mendling, J., Dijkman, R.M.: Human and automatic modularizations of process models to enhance their comprehension. Inf. Syst. **36**(5), 881–897 (2011)
16. Zugal, S., Soffer, P., Haisjackl, C., Pinggera, J., Reichert, M., Weber, B.: Investigating expressiveness and understandability of hierarchy in declarative business process models. Softw. Syst. Model. **14**(3), 1081–1103 (2015)
17. Pesic, M., Schonenberg, H., van der Aalst, W.M.P.: DECLARE: full support for loosely-structured processes. In: 11th IEEE International Enterprise Distributed Object Computing Conference (EDOC 2007), pp. 287–287. IEEE (2007)
18. Vessey, I.: Cognitive fit: a theory-based analysis of the graphs versus tables literature. Decis. Sci. **22**(2), 219–240 (1991)
19. Hildebrandt, T.T., Mukkamala, R.R.: Declarative event-based workflow as distributed dynamic condition response graphs. Electron. Proc. Theoret. Comput. Sci. **69**, 59–73 (2011)
20. Marquard, M., Shahzad, M., Slaats, T.: Web-based modelling and collaborative simulation of declarative processes. In: Business Process Management, pp. 209–225. Springer, Cham (2015)
21. Debois, S., Hildebrandt, T., Marquard, M., Slaats, T.: Hybrid Process Technologies in the Financial Sector: The Case of BRFkredit, pp. 397–412. Springer, Cham (2018)
22. Debois, S., Slaats, T.: The analysis of a real life declarative process. In: 2015 IEEE Symposium Series on Computational Intelligence, pp. 1374–1382 (2015)
23. Slaats, T., Mukkamala, R.R., Hildebrandt, T., Marquard, M.: Exformatics declarative case management workflows as DCR graphs. In: Business Process Mangement, pp. 339–354. Springer, Heidelberg (2013)
24. Ostermann, K., Giarrusso, P.G., Kastner, C., Rendel, T.: Revisiting information hiding: reflections on classical and nonclassical modularity. In: ECOOP 2011 – Object-Oriented Programming, pp. 155–178. Springer, Heidelberg (2011)
25. Dawkins, R.: The Blind Watchmaker, p. xiii, p. 332. WW Norton and Company, New York and London (1986). ISBN: 0-393-02216-1
26. Slaats, T., Schunselaar, D.M.M., Maggi, F.M., Reijers, H.A.: The semantics of hybrid process models. In: On the Move to Meaningful Internet Systems: OTM 2016 Conferences, pp. 531–551. Springer, Cham (2016)
27. Hildebrandt, T., Mukkamala, R.R., Slaats, T.: Nested dynamic condition response graphs. In: Proceedings of Fundamentals of Software Engineering (FSEN), April 2011
28. Debois, S., Hildebrandt, T., Slaats, T.: Hierarchical declarative modelling with refinement and sub-processes. In: Business Process Management, pp. 18–33. Springer, Cham (2014)
29. Debois, S., Hildebrandt, T.T., Slaats, T.: Replication, refinement & reachability: complexity in dynamic condition-response graphs. Acta Informatica **55**(6), 489–520 (2018)

30. Seco, J.C., Debois, S., Hildebrandt, T., Slaats, T.: RESEDA: declaring live event-driven computations as reactive semi-structured data. In: 2018 IEEE 22nd International Enterprise Distributed Object Computing Conference (EDOC), pp. 75–84, October 2018
31. Cant, S.N., Jeffery, D.R., Henderson-Sellers, B.: A conceptual model of cognitive complexity of elements of the programming process. Inf. Softw. Technol. **37**(7), 351–362 (1995)
32. Henderson-Sellers, B.: Object-Oriented Metrics: Measures of Complexity. Prentice-Hall, Inc., Upper Saddle River (1995)
33. Houy, C., Fettke, P., Loos, P.: Understanding understandability of conceptual models–what are we actually talking about? In: International Conference on Conceptual Modeling, pp. 64–77. Springer, Heidelberg (2012)
34. Figl, K., Laue, R.: Cognitive complexity in business process modeling. In: International Conference on Advanced Information Systems Engineering, pp. 452–466. Springer, Heidelberg (2011)
35. Sanchez-Gonzalez, L., Ruiz, F., Garcia, F., Cardoso, J.: Towards thresholds of control flow complexity measures for BPMN models. In: Proceedings of the 2011 ACM Symposium on Applied Computing, pp. 1445–1450 (2011)
36. Petrusel, R., Mendling, J.: Eye-tracking the factors of process model com- prehension tasks. In: Advanced Information Systems Engineering, pp. 224–239. Springer, Heidelberg (2013)
37. Bera, P., Soffer, P., Parsons, J.: Using eye tracking to expose cognitive processes in understanding conceptual models. MIS Q. **43**(4), 1105–1126 (2019)
38. Sweller, J., Chandler, P.: Why some material is difficult to learn. Cogn. Instruct. **12**(3), 185–233 (1994)
39. Wang, W., Indulska, M., Sadiq, S., Weber, B.: Effect of linked rules on business process model understanding. Bus. Process Manage. BPM **2017**, 200–215 (2017)
40. Vessey, I., Galletta, D.: Cognitive fit: an empirical study of information acquisition. Inf. Syst. Res. **2**(1), 63–84 (1991)
41. Ritchi, H., Jans, M.J., Mendling, J., Reijers, H.: The influence of business process representation on performance of different task types. J. Inf. Syst. **34**, 167–194 (2019)
42. Larkin, J.H., Simon, H.A.: Why a diagram is (sometimes) worth ten thousand words. Cogn. Sci. **11**(1), 65–100 (1987)
43. Dunn, C., Grabski, S.: An investigation of localization as an element of cognitive fit in accounting model representations. Decis. Sci. **32**(1), 55–94 (2001)
44. Gilmore, D.J., Green, T.R.G.: Comprehension and recall of miniature programs. Int. J. Man Mach. Stud. **21**(1), 31–48 (1984)
45. Wohlin, C., Runeson, P., Host, M., Ohlsson, M.C., Regnell, B., Wesslen, A.: Experimentation in Software Engineering. Springer, Heidelberg (2012)
46. Aranda, J., Ernst, N., Horkoff, J., Easterbrook, S.: A framework for empirical evaluation of model comprehensibility. In: International Workshop on Modeling in Software Engineering (MISE'07: ICSE Workshop 2007). IEEE (2007)
47. Hart, S.G., Staveland, L.E.: Development of NASA-TLX (Task Load Index): results of empirical and theoretical research. In: Advances in Psychology, vol. 52, pp. 139–183. Elsevier (1988)
48. Braithwaite, J.J., Watson, D.G., Jones, R., Rowe, M.: A guide for analysing electrodermal activity (EDA) & skin conductance responses (SCRS) for psychological experiments. Psychophysiology **49**(1), 1017–1034 (2013)
49. Fritz, T., Begel, A., Muller, S.C., Yigit-Elliott, S., Zuger, M.: Using psycho-physiological measures to assess task difficulty in software development. In: Proceedings of the 36th International Conference on Software Engineering, pp. 402–413 (2014)
50. Nourbakhsh, N., Chen, F., Wang, Y., Calvo, R.A.: Detecting users' cognitive load by galvanic skin response with affective interference. ACM Trans. Interact. Intell. Syst. (TiiS) **7**(3), 1–20 (2017)

51. Hess, E.H., Polt, J.M.: Pupil size in relation to mental activity during simple problem-solving. Science **143**(3611), 1190–1192 (1964)
52. Niedermeyer, E., da Silva, F.H.L.: Electroencephalography: basic principles, clinical applications, and related fields. Lippincott Williams & Wilkins, Philadelphia (2005)
53. Holmqvist, K., Nystrom, M., Andersson, R., Dewhurst, R., Jarodzka, H., van de Weijer, J.: Eye Tracking: A Comprehensive Guide to Methods and Measures. OUP Oxford, Oxford (2011)
54. Michael, Z., Rudiger, P., Johannes, S., Manfred, R.: Eye tracking experiments on process model comprehension: lessons learned. In Enterprise, Business-Process and Information Systems Modeling, pages 153–168. Springer, Heidelberg (2017)
55. Pichler, P., Weber, B., Zugal, S., Pinggera, J., Mendling, J., Reijers, H.A.: Imperative versus declarative process modeling languages: an empirical investigation. In: International Conference on Business Process Management, pp. 383–394. Springer, Heidelberg (2011)
56. Imotions. Galvanic skin response: the complete pocket guide (2017)
57. Imotions. Electroencephalography: the complete pocket guide (2019)
58. Muller-Putz, G.R., Riedl, R., Wriessnegger, S.C.: Electroencephalography (EEG) as a research tool in the information systems discipline: Foundations, measurement, and applications. CAIS **37**, 46 (2015)
59. Object Management Group OMG. Business Process Modeling Notation V 2.0 (2006)

The Influence of Negative Emotion as Affective State on Conceptual Models Comprehension

Djordje Djurica(✉) and Jan Mendling

Vienna University of Economics and Business (WU Wien), Vienna, Austria
{djordje.djurica,jan.mendling}@wu.ac.at

Abstract. Comprehension of procedural models presents an essential skill for IS professionals. Although the literature on factors that influence model comprehension is extensive, it is surprising that so far, empirical research about the impact that affective states on model understanding is mainly missing. The purpose of this study is to determine if an affective state of model viewers can have an impact on the understanding of conceptual models. To this end, we develop hypotheses on the effects of emotions, based on Attentional Control Theory and Affective Events Theory. In order to test our hypotheses, we plan to carry out a controlled experiment.

Keywords: Emotions · Conceptual modeling · Model comprehension · Biophysical devices

1 Introduction

Conceptual modeling is an important aid that supports various tasks during system analysis and design. Analysts develop a conceptual model to capture aspects of a real-world phenomenon in such a way that relevant information is later available for the analysis or design of information systems [1]. In order to make the right decisions regarding the design of an information system, it is crucial to fully understand the conceptual model and the business domain in which the system will operate. This makes the research stream on model comprehension highly relevant for designing information systems of good quality [2].

Research on model comprehension is extensive and focuses on many factors that are associated with the model itself and with user characteristics [3]. Papers in this stream focused on the impact of the domain and modeling knowledge [4], learning style [5], cognitive styles [6], and even cultural factors [7]. However, to the best of our knowledge, literature is silent about the impact of affective states on working with conceptual models. It seems that the model reader is implicitly assumed to be a cold-tempered rational analyst who strictly adheres to facts. We know from research on decision-making in business that this assumption does not reflect reality [8, 9]. For these reasons, it is a significant omission of prior research on conceptual modeling that the impact of affective states has

F. D. Davis et al. (Eds.): NeuroIS 2020, LNISO 43, pp. 145–152, 2020.
https://doi.org/10.1007/978-3-030-60073-0_16

not been studied. Its relevance is increasingly acknowledged in other domains and the IS domain [8, 9].

In this research in progress paper, we address this research gap by investigating whether affective states can influence conceptual model comprehension. More specifically, we focus on how negative emotion might influences model understanding. To explain this, we draw on Attentional Control Theory (ACT) [10], Affective Events Theory (AET) [11], and the previous literature about the impact of affective states on cognition. We plan to conduct an experiment in which negative emotion will be induced and where this manipulation will be checked using a biophysical device.

This paper is organized as follows. The next section introduces an overview of previous literature on the influence of user characteristics on conceptual model comprehension, followed by an introduction of basic concepts, a summary of the prior relevant literature, and relevant theories. Further, we derive our hypotheses and present an experimental design. Finally, we close with possible contributions and future steps of our study.

2 Theoretical Background

2.1 Impact of User Characteristics on Conceptual Models Comprehension

Several research streams are relevant to our work. First, there is a research stream investigating conditions that influence how users comprehend a conceptual model. It focuses on the impact of the semantics of the conceptual models on comprehension [12, 13] and, to a lesser extent, on pragmatics [14]. Previous studies that investigate the comprehension of various conceptual models highlight user characteristics as an important factor. The recent paper by Mendling et al. [15] shows many potential aspects of user characteristics that can have an impact on process model comprehension. Some of these aspects are theoretical knowledge [4, 16], duration of practice [17–19], education [17], or familiarity with process modeling [2].

Despite the recognized potential that user characteristics may influence the comprehension of conceptual models, corresponding factors have hardly been taken into consideration as primary factors. Several experiments in this research area have considered them as control variables or covariates, but not as the independent variables [4, 16–19]. It is important to note that non of the mentioned papers took into account the potential effects of the current model viewer's affective state on process model comprehension, which is what our paper aims to rectify.

2.2 Basic Concepts and Prior Research

Affective states is the term often used to describe emotions, moods, affects, and feelings [9]. There is no clear consensus on the definition of either of these constructs, which results in some authors perceiving them as different, but overlapping constructs [20]. In this research, we focus on the emotions as an affective state.

According to Lazarus [21], emotions are short experiences that appear due to cognitive judgment about external stimuli or situations. They are shorter than moods, but more intense, and have a clear trigger [11]. According to the Attention Circumplex Model of Affect by Posner et al. [22], emotions are classified through pleasure and arousal, which represent their bipolar dimensions [23]. Our research aims to discover how negative emotions influence the comprehension of conceptual models. We chose negative emotion since previous research suggests that it reduces the capacity of working memory and negatively influences cognitive performance [10, 24, 25].

Furthermore, research suggests that negative emotions can lead to a decrease in motoric responses resulting in slower reaction times [8, 26, 27]. However, it is important to note that none of the previously mentioned studies, apart from the research by Hibbeln et al. [8] and Bogodistov and Moorman [9], was set in an IS context. Hibbeln et al. [8] attempt to detect negative emotions using mouse cursor movements during a number-ordering task. Their results suggest that negative emotion reduces the speed of the computer mouse during the tasks, resulting in lower efficiency. On the other side, Bogodistov and Moorman [9] conducted an experiment that shed light on how affective states influence the decisions of process users and process designers. Their results suggest that when in the state of fear, process users choose less complex but riskier payment processes, which implies the reduced capabilities of their cognitive system.

2.3 Emotion Recognition Using Physiological Measurements

When we are experiencing the negative emotion of stress, our respiratory system is immediately affected. We breathe quicker and harder in order to distribute oxygen-rich blood throughout our bodies. Our muscles become tense, palms become sweaty, and our heart races, resulting in higher blood pressure [28]. These physiological effects are mediated by the autonomic nervous system. This system consists of sympathetic and parasympathetic divisions that operate concurrently with each other and with the somatic motor system to regulate different types of behavior. The balance between sympathetic and parasympathetic divisions enables our body to maintain stability during the everchanging external conditions [23]. However, emotions can have a significant influence on this balance and result in the number of bodily reactions, as seen in our short example. These reactions can be measured using biophysical devices and used to determine the emotional state of the person [29].

Previous research that focused on emotion recognition using physiological signals presented the number of signals that can be used to deduce a persons' emotional state. The most common ones are: electromyography (EMG), electrodermal activity (EDA), electrocardiogram (ECG), electroencephalogram (EEG), and respiration [30]. In the context of our research, we will focus on detecting negative emotions using EMG, EDA, ECG, and respiration as physiological signals, as these signals are already proven as accurate in the emotion recognition process [31].

2.4 Relevant Theories

The theory that can explain how negative emotions can influence process model comprehension is *Attentional Control Theory (ACT)* by Eysenck et al. [10]. This theory states that negative emotions can result in impaired attentional control, which leads to poorer performance in goal-directed tasks (e.g., searching for specific information on a model) involving a working memory system. The authors assume that this occurs due to an attentional shift from goal-directed to stimulus-directed, increasing awareness of the stimuli that cause negative emotion. ACT was initially developed to describe the influence anxiety has on attentional control but was multiple times extended. It was found to be valid for other types of negative emotions, including sadness, fear, depression, and frustration [8, 32, 33].

Another theory that is important for our research is the *Affective Events Theory (AET)*. This theory suggests that employees can develop an emotional response to many of the events that they encounter at their work, which can impact their subsequent actions [11]. In the context of our research, experiencing negative emotion from one event would transfer this emotion to the subsequent actions, such as process model comprehension tasks.

Drawing on the ACT, AET, and previous research on the influence of negative emotion on cognitive performance and motoric responses, we hypothesize how negative emotion will decrease attentional control and transfer a negative emotional response from one event to a task. This will further influence conceptual model comprehension in terms of both accuracy and efficiency. Therefore, we hypothesize that:

H1: Experiencing negative emotion will decrease conceptual model understanding in terms of accuracy.

H2: Experiencing negative emotion will decrease conceptual model understanding in terms of efficiency.

3 Method

3.1 Experiment Design

In order to answer our hypotheses, we plan to manipulate negative emotion on three levels in a one-factor between-subjects experimental design (high arousal with negative valence, high arousal with positive valence, and a control condition with neutral arousal and valence) to determine the influence of negative emotion on conceptual model comprehension.

In order to capture the valence and arousal dimensions of the participants' emotional state and check if the emotion manipulation was successful, we will use the NeXus-10 MKII device for biophysical measuring and focus on the EMG, EDA, ECG, and respiration rate. We opt for using the biophysical device since it provides us with continuous data about the emotional state of the participant, in comparison to self-reporting surveys that only check at one point of time if the emotion induction was successful.

We also plan to maintain control over potentially confounding factors. To accomplish this, we will use the laboratory environment with individual cubicles and workstations in order to minimize possible distractions. We will also apply constant time-pressure in order to prevent the ceiling effect in regards to accuracy. Further, we will control the layout and size of models, since previous research showed that the graphical layout could influence participants' level of comprehensibility [34, 35].

3.2 Tasks and Materials

In this research, we plan to use process models created using the Business Process Model and Notation (BPMN) from the domains that can be understood by a university student. We will ask our participants comprehension questions that will assess surface-level understanding and show how well users can comprehend the content presented in the process models. These tasks will consist of questions about four different issues that concern the control flow logic of the process models: concurrency, exclusiveness, order, and repetition. We will base these questions on the previous research [12, 13, 18, 35].

The first dependent variable will be *accuracy* and will be evaluated as a number of questions answered correctly. Our questions will relate to formal properties associated with the behaviour represented by process models and, therefore, have precise and objectively correct answers.

The second dependent variable will be *efficiency*, which will represent the amount of time participant needs in order to provide an answer to the task and submit it per correct answer [7].

3.3 Procedure and Manipulation

At the beginning of the experiment, participants will be shown the way to their workstation shielded from distractions and equipped with the biophysical device. At the beginning of the experiment, participants will be randomly assigned to either a negative emotion or a control condition. The experiment will consist of three parts. The first part of the experiment will be an emotion induction procedure, after which all participants will have to provide answers to the same model comprehension tasks. Finally, participants will be debriefed and compensated for their participation.

For inducing negative emotion, we will use an approach from Hibbeln et al. [8]. In this study, the authors induced negative emotions by asking participants to complete an unfairly designed intelligence test taken from Zuckermann [36] with questions that were overly difficult, and impossible to solve during the time provided. Furthermore, Hibbeln et al. [8] implemented a loading delay under time pressure, which already proved to be able to induce negative emotions [37]. Finally, after the test was over, they have informed their participants that they performed poorly on the intelligence test. Authors state that all these mechanisms caused the feelings of unfair treatment and negative emotion. For this reason, we believe that this way of emotion induction is suitable for our research. In the control condition, participants will have to complete the intelligence test, but without emotion induction. Therefore, we do not expect the task to cause any negative emotion.

After the first part of the experiment, all participants will be asked to carefully read the instruction manual and then answer the comprehension questions regarding business process models. The time to provide an answer to each set of questions will be determined in the pretest.

Finally, participants will be debriefed and informed that the intelligence test did measure their intelligence, but that this task was designed to induce negative emotion and to examine the influence this emotion can have on model understanding.

3.4 Participants

In terms of participants for our study, we follow the recommendations of sample selection [38]. We plan to recruit university students from a large business school since we believe that these represent a realistic proxy of the future users of business process models. Previous work on process model comprehension justifies this. Research by [18] showed that there is no correlation between the experience of model use and type of university education with model comprehension. To estimate the required sample size, we will conduct an analysis using the G*Power 3 [39] software and considering the experimental design and nature of the variables that we measure.

4 Ethical Concerns

The proposed experiment will be conducted in accordance with the ethical behavior guidelines recommended by the Association of Information Systems[1]. Participation in the study will be voluntary, and it will be possible to withdraw from participation at any time. The project policy with regard to data protection and privacy issues will take into account national and European data protection directives. Before the start of the studies, we will also contact the Board for Ethical Issues of the Vienna University of Economics and Business (WU Wien). All participants will be fully informed about the planned research use of the data collected. No identifying information will be stored with the research data, so that complete anonymity is guaranteed.

5 Conclusion and Future Steps

To the best of our knowledge, there is no empirical research that focuses on the influence of affective states on the conceptual model comprehension. The results of this research are expected to close this gap and to extend the current body of knowledge about the variables that influence conceptual model comprehension, providing additional insight into how model users' emotional state can influence their task performance. Furthermore, we introduce biophysical devices as a novel method for continuous control of the emotional state of participants. Further research on this topic will focus on neutralizing the effects of negative emotions on model comprehension using primary and secondary notation manipulations.

References

1. Wand, Y., Weber, R.: Research commentary: Information systems and conceptual modeling - a research agenda. Inf. Syst. Res. **13**, 363–376 (2002)
2. Burton-Jones, A., Meso, P.N.: The effects of decomposition quality and multiple forms of information on novices' understanding of a domain from a conceptual model. J. Assoc. Inf. Syst. **9**, 1 (2008)
3. Figl, K.: Comprehension of procedural visual business process models: a literature review. Bus. Inf. Syst. Eng. **59**, 41–67 (2017)

[1] https://aisnet.org/page/AdmBullCResearchCond.

4. Mendling, J., Strembeck, M., Recker, J.: Factors of process model comprehension-Findings from a series of experiments. Decis. Support Syst. **53**, 195–206 (2012)
5. Recker, J., Reijers, H.A., van de Wouw, S.G.: Process model comprehension: the effects of cognitive abilities, learning style, and strategy. Commun. Assoc. Inf. Syst. **34**, 9 (2014)
6. Figl, K., Recker, J.: Exploring cognitive style and task-specific preferences for process representations. Requir. Eng. **21**, 63–85 (2016)
7. Kummer, T.F., Recker, J., Mendling, J.: Enhancing understandability of process models through cultural-dependent color adjustments. Decis. Support Syst. **87**, 1–12 (2016)
8. Hibbeln, M., Jenkins, J.L., Schneider, C., Valacich, J.S., Weinmann, M.: How is your user feeling? MIS Q. **41**, 1–21 (2017)
9. Bogodistov, Y., Moormann, J.: Influence of emotions on IT-driven payment process design: shorter, simpler, and riskier (2019)
10. Eysenck, M.W., Derakshan, N., Santos, R., Calvo, M.G.: Anxiety and cognitive performance: Attentional control theory. Emotion **7**, 336 (2007)
11. Weiss, H.M., Cropanzano, R.: Affective events theory: a theoretical discussion of the structure, causes and consequences of affective experiences at work (1996)
12. Reijers, H.A., Freytag, T., Mendling, J., Eckleder, A.: Syntax highlighting in business process models. Decis. Support Syst. **51**, 339–349 (2011)
13. Mendling, J., Reijers, H.A., van der Aalst, W.M.P.: Seven process modeling guidelines (7PMG). Inf. Softw. Technol. **52**, 127–136 (2010)
14. Khatri, V., Vessey, I.: Understanding the role of IS and application domain knowledge on conceptual schema problem solving: a verbal protocol study. J. Assoc. Inf. Syst. **17**, 2 (2016)
15. Mendling, J., Recker, J., Reijers, H.A., Leopold, H.: An empirical review of the connection between model viewer characteristics and the comprehension of conceptual process models. Inf. Syst. Front. **21**, 1111–1135 (2019)
16. Khatri, V., Vessey, I., Ramesh, V., Clay, P., Park, S.J.: Understanding conceptual schemas: exploring the role of application and IS domain knowledge. Inf. Syst. Res. **17**, 81–99 (2006)
17. Recker, J.: Continued use of process modeling grammars: the impact of individual difference factors. Eur. J. Inf. Syst. **19**, 76–92 (2010)
18. Reijers, H.A., Mendling, J.: A study into the factors that influence the understandability of business process models. IEEE Trans. Syst. Man Cybern. Part A Syst. Hum. **41**, 449–462 (2011)
19. Recker, J., Dreiling, A.: The effects of content presentation format and user characteristics on novice developers' understanding of process models. Commun. Assoc. Inf. Syst. **28**, 6 (2011)
20. Ashkanasy, N.M.: Emotions in organizations: a multi-level perspective. Res. Multi-level Issues **2**, 9–54 (2003)
21. Lazarus, R.S.: Progress on a cognitive-motivational-relational theory of emotion. Am. Psychol. **46**, 819 (1991)
22. Posner, J., Russell, J.A., Peterson, B.S.: The circumplex model of affect: an integrative approach to affective neuroscience, cognitive development, and psychopathology. Dev. Psychopathol. **17**, 715–734 (2005)
23. Lang, P.J., Bradley, M.M., Cuthbert, B.N.: International affective picture system (IAPS): technical manual and affective ratings. NIMH Cent. Study Emot. Atten. **1**, 39–58 (1997)
24. Arnell, K.M., Killman, K.V., Fijavz, D.: Blinded by emotion: target misses follow attention capture by arousing distractors in RSVP. Emotion. **7**, 465 (2007)
25. Blair, K.S., Smith, B.W., Mitchell, D.G.V., Morton, J., Vythilingam, M., Pessoa, L., Fridberg, D., Zametkin, A., Sturman, D., Nelson, E.E., Drevets, W.C., Pine, D.S., Martin, A., Blair, R.J.R.: Modulation of emotion by cognition and cognition by emotion. Neuroimage **35**, 430–440 (2007)
26. Coelho, C.M., Lipp, O.V., Marinovic, W., Wallis, G., Riek, S.: Increased corticospinal excitability induced by unpleasant visual stimuli. Neurosci. Lett. **481**, 135–138 (2010)

27. Murray, N.P., Janelle, C.M.: Anxiety and performance: a visual search examination of the processing efficiency theory. J. Sport Exerc. Psychol. **25**, 171–187 (2003)
28. What happens to your body when you're stressed? https://www.rte.ie/brainstorm/2017/0828/900497-what-happens-to-your-body-when-youre-stressed/
29. Haag, A., Goronzy, S., Schaich, P., Williams, J.: Emotion recognition using bio-sensors: first steps towards an automatic system. In: Lecture Notes in Artificial Intelligence (Subseries of Lecture Notes in Computer Science) (2004)
30. Ali, M., Mosa, A.H., Machot, F. Al, Kyamakya, K.: Emotion recognition involving physiological and speech signals: a comprehensive review. In: Studies in Systems, Decision and Control (2018)
31. Kim, J., André, E.: Emotion recognition based on physiological changes in music listening. IEEE Trans. Pattern Anal. Mach. Intell. **30**, 2067–2083 (2008)
32. Bishop, S.J., Jenkins, R., Lawrence, A.D.: Neural processing of fearful faces: effects of anxiety are gated by perceptual capacity limitations. Cereb. Cortex. **17**(7), 1595–1603 (2007)
33. Sarter, M., Paolone, G.: Deficits in attentional control: cholinergic mechanisms and circuitry-based treatment approaches. Behav. Neurosci. **125**, 825 (2011)
34. Aranda, J., Ernst, N., Horkoff, J., Easterbrook, S.: A framework for empirical evaluation of model comprehensibility. In: International Workshop on Modeling in Software Engineering (MISE 2007: ICSE Workshop 2007), p. 7 (2007)
35. Figl, K., Laue, R.: Cognitive complexity in business process modeling. In: Lecture Notes in Computer Science (Including Subseries Lecture Notes in Artificial Intelligence and Lecture Notes in Bioinformatics) (2011)
36. Zuckerman, M.: The effect of frustration on the perception of neutral and aggressive words. J. Pers. **23**, 407–422 (1955)
37. Ceaparu, I., Lazar, J., Bessiere, K., Robinson, J., Shneiderman, B.: Determining causes and severity of end-user frustration. Int. J. Hum. Comput. Interact. **17**, 333–356 (2004)
38. Compeau, D., Marcolin, B., Kelley, H., Higgins, C.: Generalizability of information systems research using student subjects a reflection on our practices and recommendations for future research. Inf. Syst. Res. **23**, 1093–1109 (2012)
39. Faul, F., Erdfelder, E., Lang, A.G., Buchner, A.: G*Power 3: a flexible statistical power analysis program for the social, behavioral, and biomedical sciences. Behav. Res. Methods **39**, 175–191 (2007)

What Do Users Feel? Towards Affective EEG Correlates of Cybersecurity Notifications

Colin Conrad(✉), Jasmine Aziz, Natalie Smith, and Aaron Newman

Dalhousie University, Halifax, Canada
colin.conrad@dal.ca

Abstract. Security notifications attempt to change risky computer usage behaviour but often fail to achieve their desired effect. Though there are likely many causes for this phenomenon, information systems researchers have posited that emotional reactions to security notifications may play a role in its explanation. This work-in-progress paper descibes a study to create a baseline of electroencephalographic (EEG) and behavioral responses to security notification images by comparing them to known responses to the well-studied International Affective Picture System (IAPS). By creating such a baseline of affective responses to security notification images, future work can explore the effect of passive emotional reactions to security notification designs which would generate insight into effective design practices.

Keywords: Security warnings · Affective processing · Electroencephalography (EEG) · Event-related potential (ERP) · Late positive potential (LPP)

1 Introduction

Security notifications play an important role in the safe operation of computing environments, by informing users of threats and persuading them to change their computer behavior. Security notifications are particularly interesting to information systems researchers because they can fail to evoke desired behavioral change, for reasons that users may not be explicitly aware of. Though there has been considerable recent progress in the design of effective security warnings, future improvements to security information systems may be made by identifying contextual factors that influence secure behavior [1]. Affective considerations, such as degrees of trust, safety, or fear may play a role in unconsciously mediating the relationship between security notifications and behavior [1, 2].

Emotions can be conceptualized as either positive valence (e.g. joy or happiness) or negative valence (e.g. fear, anger or disgust), as well as by their degree of arousal. The late positive potential (LPP), an event-related potential (ERP), has been associated with both high and low valence emotions that also elicit high degrees of arousal [3, 4]. As such, the LPP can potentially be used as a marker of high-arousal emotional responses. In this work-in-progress paper, we describe an experiment to measure an association between

F. D. Davis et al. (Eds.): NeuroIS 2020, LNISO 43, pp. 153–162, 2020.
https://doi.org/10.1007/978-3-030-60073-0_17

the LPP and images of security notifications. Using the International Affective Picture System (IAPS) [5], we will compare LPP responses elicited by standardized negative, positive, and neutral valence stimuli, with those elicited by pictures of computer security notifications and security-unrelated computer images. Motivated by past studies which found that security warnings often fail to produce the desired reactions in users [6], we explore affective correlates of security warning pictures. The outcome of this study will be a baseline of LPP and questionnaire responses which can be used to passively and objectively investigate valence of novel security notification designs.

2 Background and Theoretical Framework

2.1 Security Notifications, Emotions, and Neurophysiology

Computer users are known to resist persuasion by protective messages—such as those given by security notifications—for reasons which they may not be explicitly aware [2]. Until very recently, information systems (IS) research on the subject has focused on either cognitive factors such as habituation or cognitive processing, or on negative emotional factors such as stress or fear [2]. Concerning fear, for instance, protection motivation theory has been identified as useful for explaining desktop security behavior [7–9]. As conceptualized by Rogers [7], this theory holds that there are at least four components that could determine a users' response to a threat: their perception of threat susceptibility, their perception of threat severity, their perception of response efficacy, and their perception of their personal ability to effectively respond [8]. Emotional reactions elicited by security notifications could thus influence perceptions of vulnerability and responses to threats, which ultimately influence behavior.

Though IS researchers have investigated affective factors in the processing of security notifications, much of the past research has been conducted using self-report measures [10, 11]. Such instruments are likely useful for measuring motivation, though they might not effectively measure implicit emotional responses to security notifications. Recognizing limitations to these studies, IS researchers have begun to employ neuroscientific and physiological techniques to investigate security phenomena [2, 9, 12, 13]. Such an approach promises to yield insights into unconscious affective factors which influence motivation for security behavior.

In a 2014 paper, Vance et al. [6] used a combination of EEG and questionnaire measures to predict disregard to security notifications. They found that an attention-related P300 ERP response to a gambling and risk task was a strong predictor of a participant's propensity to disregard security responses, when compared to questionnaire measures. Drawing from this study, we can expand on their findings by exploring a similar ERP measure (i.e., the LPP) to investigate affective factors that may influence this propensity. While Vance et al. [6] investigated attention-related ERP responses to gambling tasks to predict risky behavior, we instead investigate emotion-related ERP responses to security notifications themselves. By doing this, we may identify ERP measures which either better predict threat susceptibility than questionnaires or can later be applied as a passive, real-time measure in an ecologically valid setting.

2.2 EEG and the Late Positive Potential (LPP)

There is an extant literature on measuring emotional responses using ERP. Much of the literature has focused on two ERP components: an early posterior negative component at the 200 ms latency range, and a late positive component (LPP) which often starts at 300 ms and extends to 2000 ms [14, 15]. While the former is thought to affect processes related to cognition of emotions, the LPP may actually consist of an enlarged P300 component which extends well-beyond the normal latency of the P300 response, reflecting an effect of extended task-relevance to an emotion-inducing stimulus [14, 15]. Studies of the LPP have demonstrated an effect that is modulated by the strength of emotions evoked by pictures [16, 17].

Before investigating the impact of affective ERP correlates on security behavior, it is desirable to first have a baseline of responses to stimuli that have been standardized with respect to their emotional valence and arousal levels, for comparison to various security stimuli. The IAPS [5] is a well-studied repository of images which have been indexed based on normative ratings to emotion (valence, arousal, dominance) for study of attention and emotion. Though the IAPS is often used to investigate physiological processes elicited by emotions such as brain oscillations [18], the IAPS has also been used to investigate the emotional effect of art [19], and to validate the measure of emotions in a virtual reality environment [20]. In addition, the IAPS normative ratings are based on responses to the Self-Assessment Manikin (SAM), which is a well-studied affective rating questionnaire system. By combining both physiological and psychological approaches, we may discover gaps between security threat perceptions and unconscious physiological responses. Though we hypothesize that the selected security notifications will exhibit patterns characteristic of neutral photos, the guiding purpose of the study described in this paper is to identify a baseline of affective reactions to security stimuli. Our research question can thus be articulated as follows:

RQ: How do the LPP and self-report measures of valence and arousal differ between security notifications, computer task images, and IAPS stimuli?

3 Methods

3.1 Participants

For the initial study, we will recruit 30 undergraduate students from our university who will be compensated with either $20 cash or course credit. Participants will be excluded if they reported having neurological conditions that could affect EEG (e.g. epilepsy or a recent concussion), uncorrected vision problems, or physical impairments that would prevent them from using a computer keyboard or mouse. In this work-in-progress paper, we report preliminary results from 3 participants.

3.2 Stimuli

Experimental stimuli will include 3 categories of photos from the IAPS database (negative, positive and neutral), as well as two categories of online computer stimuli (security notification images and neutral computer-related images) delivered following a within-subject design, presenting 32 instances of each of the five picture conditions. Security notification images will consist of computer-based images used by antivirus, web browser and firewall systems (e.g. Chrome, McAfee, Norton) while the computer-related images will consist of non-threatening images typical of a computing environment (e.g. a screenshot of a search engine or Wikipedia). All images will be corrected for luminance to control for the effect of luminance on EEG signals [21]. Stimuli will be delivered using PsychoPy [22, 23], which will also be used to mark the onset of each pictures in the EEG recordings via transistor-transistor logic (TTL) codes. A collection of both modern and antiquated security notifications were selected to give a greater range of baseline data (Fig. 1).

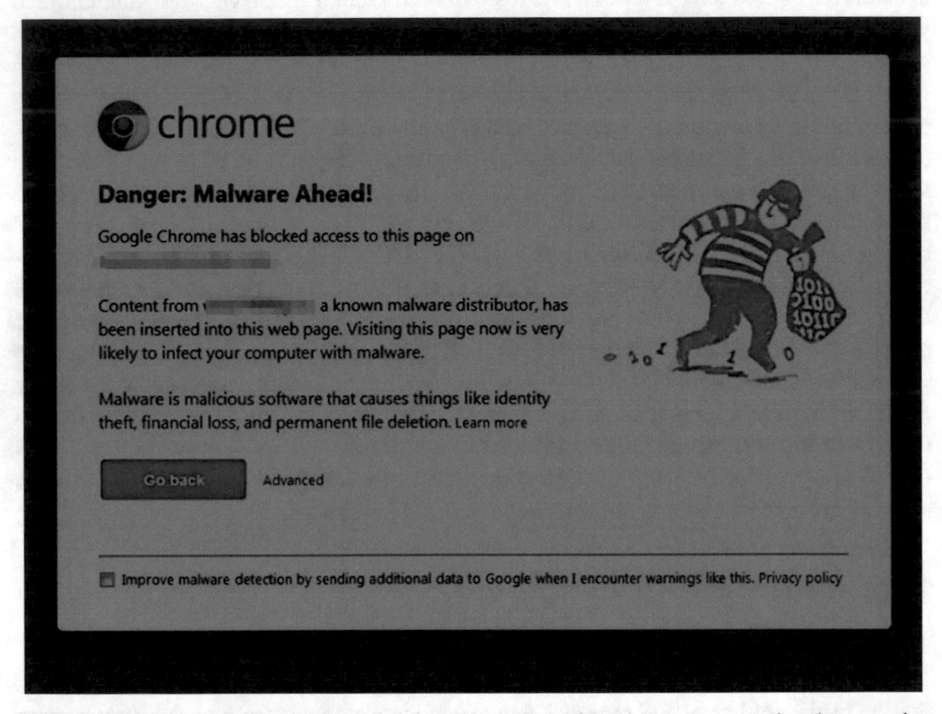

Fig. 1. A sample security warning stimulus. The effect of luminance on emotional processing will be controlled by normalizing all stimuli photos to the luminance baseline provided by the IAPS.

3.3 Questionnaires and Self-assessment

At the outset of the experiment, participants will be asked about their age, gender, perceived skill at using computer systems, years of education and native language. To assess participants' perceived reaction to the photo stimuli, we will use a simplified version of the SAM [5, 24]. The manikin will be presented 2–3 s following the appearance of the stimulus photos and will consist of two 5-point scales which measure degrees of valence and arousal respectively [5]. Following the experiment, participants will be asked questions about their attitudes towards risk and perception of the impact of computer malware [6].

3.4 Procedure

Participants will undergo a consent protocol, will be fitted with an EEG cap, and then brought to a controlled environment. After participating in an initial demographic questionnaire, participants will be presented with a randomized series of images consisting of IAPS photos, security notifications or security-unrelated computer phenomena. Participants will complete the SAM measures for valence and arousal following each picture. Following the study participants will complete the aforementioned post-questionnaire. Each session is expected to take 90 min total. Following the session participants will be debriefed and will receive compensation.

3.5 Data Acquisition

Participants will be fitted with horizontal and vertical electrooculograms (EoG) and 32 scalp electrodes (ActiCap, BrainProducts GmbH, Munich, Germany) positioned at standard locations according to the international 10–10 system and referenced to the midline frontal location (FCz). Electrode impedances will be kept below 20 kΩ at all channel locations throughout the experiment. EEG data will be recorded using a Refa8 amplifier (ANT, Enschende, The Netherlands) at a sampling rate of 512 Hz, bandpass filtered between 0.01 and 170 Hz, and saved using ANT ASAlab.

3.6 Data Processing and Analysis

Data processing and statistical analysis will be conducted in Python using the MNE Python library [23, 25]. Data will be filtered using a 0.1–40 Hz bandpass filter and will be manually inspected for excessively noisy electrodes, which will be removed. The data will be segmented into 2200 ms epochs spanning from 200 ms before the stimulus onset, through the 2000 ms duration of the photo stimuli. Epochs will be manually inspected and those with excessive noise will be removed. Independent component analysis [26] will be used to remove systematic artifacts from the data, including those created by eye blinks, eye movements, and muscle contractions.

Each participant will yield a maximum of 32 epochs for each condition, and the dependent EEG measure will be mean amplitude on each trial between 300 ms and 2000 ms following stimulus outset, which corresponds to the expected window of the LPP component. Statistical analyses will be performed on a region of interest centered around electrode Pz using linear mixed effects modelling [27–30]. Picture condition will be treated as fixed effects while participants, electrode-by-subject, and conditions-by-subject will be treated as random effects. The online security warnings condition will be selected as the fixed effect and the IAPS neutral condition will be specified as the intercept variable. Participants, participants-by-condition, and participants-by-electrode will be specified as random effects. Average SAM responses for each condition will be compared using ANOVA.

4 Preliminary Results

4.1 The LPP Waveform

Preliminary results suggest that there is variance in both EEG and behavioral measures between the conditions. To date, a total of 435 epochs have been analyzed from 3 participants, though collection was suspended due to the COVID-19 pandemic. Results from linear mixed effects analysis found a significant difference between the neutral IAPS and the positive inflection, presumably created by the LPP, elicited by pictures of the security warning condition ($\beta = 2.327$; $t = 2.38$; $p = 0.017$), as well as the positive IAPS condition ($\beta = 2.052$; $t = 2.12$; $p = 0.034$), and the negative IAPS condition ($\beta = 3.199$; $t = 3.28$; $p = 0.001$). Figure 2 visualizes the grand average waveform for three of the five conditions.

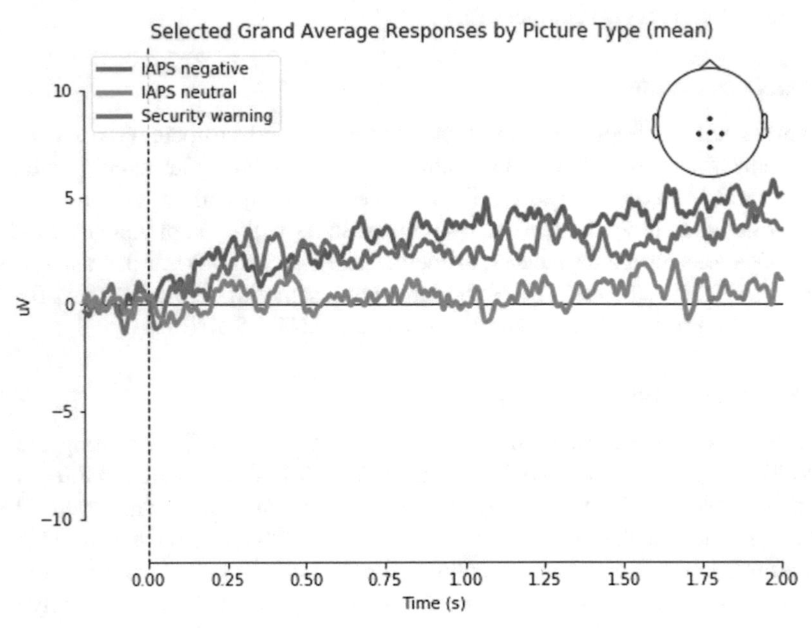

Fig. 2. Selected comparisons of LPP grand average waveforms for the region of interest. Three conditions were selected in order to enhance readability. Variances in response suggest the potential for an effect.

4.2 Valence and Arousal Reports

Preliminary results from ANOVA analysis of the SAM valence responses indicate a statistically significant effect of picture type on valence responses ($F = 36.50$; $p < 0.001$), though not arousal responses ($F = 0.94$; $p = 0.48$). Posthoc analysis using Tukey's test revealed significant differences between security warnings and neutral stimuli ($p = 0.035$), as well as between security warnings and positive stimuli ($p = 0.001$). Responses from computer-related pictures were found to be significantly different positive pictures ($p = 0.011$) but no other conditions. Figure 3 summarizes the mean SAM valence rating responses and Fig. 4 summarizes the SAM arousal ratings.

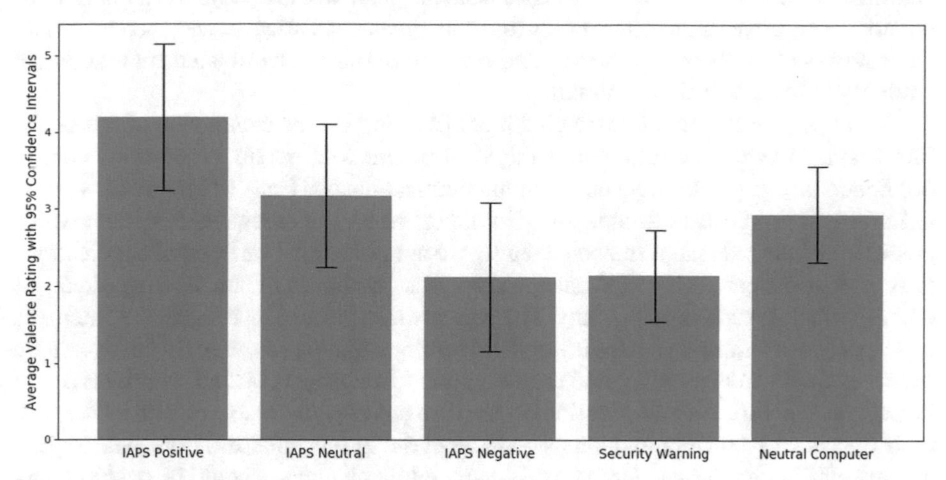

Fig. 3. Comparisons of valence responses from the SAM for each condition. Results from Tukey's test reveal significant differences in valence between IAPS positive stimuli and security warnings, as well as IAPS neutral stimuli and security warnings.

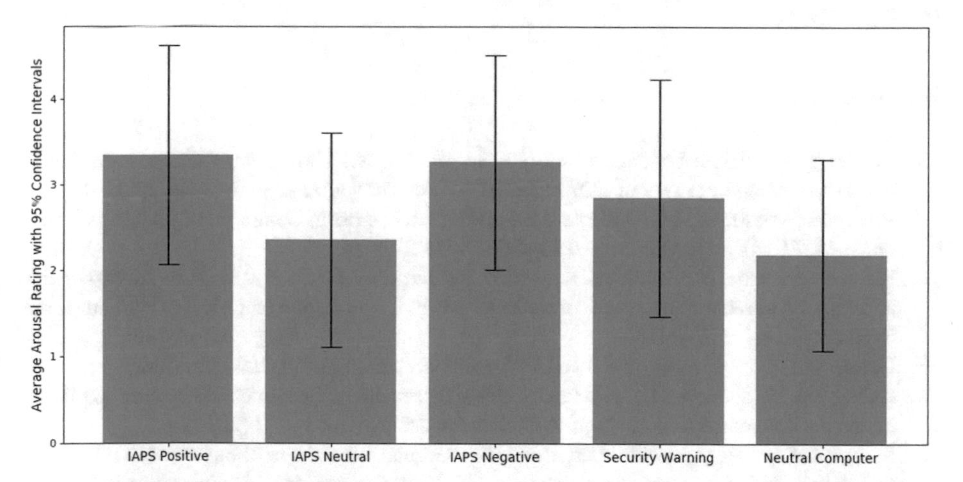

Fig. 4. Comparisons of arousal responses from the SAM for each condition. No significant results were found at this preliminary stage.

5 Discussion and Next Steps

Preliminary results revealed two interesting findings. The first is that there are differences in LPP between security warning stimuli and the IAPS neutral condition. Though the study does not yet have sufficient statistical power to draw conclusions, this suggests that that we may later gather evidence for this trend and may gain insights into differences between the conditions. The second finding is that the security warning condition was significantly different in reported valence from the positive and neutral IAPS conditions, but not the negative. If this effect holds true with greater statistical power, then we would have evidence to believe that the selected security warnings investigated are interpreted similarly to negative valence stimuli.

We anticipate two future research directions following the completion of this study. The first is to conduct a follow-up study of reactions to different varieties of security notifications (e.g. positive and encouraging notifications vs. Fear-evoking notifications) or in different contexts (e.g. contexts of imminent negative consequences vs contexts of possible or future threats). In such a study, we could identify varieties of notifications to further investigate behavioral change outcomes, as well as the moderating effects of cognitive factors such as complexity. The second direction is to investigate the outcomes of responses to stimuli in ecologically valid settings. Similar to the study conducted by Vance et al. [6], future work could use deception to simulate an actual security risk and investigate the differences in reactions to positive and negative valence notifications and their effects on behavioral outcomes—although doing this effectively presents logistical and ethical challenges. Alternatively, such future challenges could be overcome by conducting this study in an office setting in co-operation with industrial partners. The present study nonetheless presents the first steps in extending the work done on security notifications in the information systems field towards a deeper investigation of affective brain processes.

References

1. Reeder, R.W., Porter Felt, A., Consolvo, S., Malkin, N., Thompson, C., Egelman, S.: An experience sampling study of user reactions to browser warnings in the field. In: CHI 2018: Proceedings of the 2018 CHI Conference on Human Factors in Computing Systems, vol. 512, pp. 1–13 (2018). https://doi.org/10.1145/3173574.3174086
2. Brinton Anderson, B., Vance, A., Kirwan, C.B., Eargle, D., Jenkins, J.L.: How users perceive and respond to security messages: a NeuroIS research agenda and empirical study. Eur. J. Inf. Syst. **25**(4), 364–390 (2016)
3. Colombetti, G.: Appraising valence. J. Conscious. Stud. **12**(8–9), 103–126 (2005)
4. Bublatzky, F., Schupp, H.T.: Pictures cueing threat: brain dynamics in viewing explicitly instructed danger cues. Soc. Cogn. Affect. Neurosci. **7**(6), 611–622 (2012)
5. Lang, P.J., Bradley, M.M., Cuthbert, B.N.: International affective picture system (IAPS): affective ratings of pictures and instruction manual. Technical report A-8. University of Florida, Gainesville, FL (2008)
6. Vance, A., Anderson, B., Kirwan, C.B., Eargle, D.: Using measures of risk perception to predict information security behavior: insights from electroencephalography (EEG). J. Assoc. Inf. Syst. **15**(10), 2 (2014)
7. Rogers, R.W.: Attitude change and information integration in fear appeals. Psychol. Rep. **56**(1), 179–182 (1983)
8. Hanus, B., Wu, Y.A.: Impact of users' security awareness on desktop security behavior: a protection motivation theory perspective. Inf. Syst. Manag. **33**(1), 2–16 (2016)

9. Warkentin, M., Walden, E., Johnston, A.C., Straub, D.W.: Neural correlates of protection motivation for secure IT behaviors: an fMRI examination. J. Assoc. Inf. Syst. **17**(3), 1940215 (2016)
10. Johnston, A.C., Warkentin, M.: Fear appeals and information security behaviours: an empirical study. Manag. Inf. Syst. Q. **34**(3), 549–556 (2010)
11. Guo, K.H., Yuan, Y., Archer, N.P., Connelly, C.E.: Understanding nonmalicious security violations in the workplace: a composite behavior model. J. Manag. Inf. Syst. **28**(2), 203–236 (2011)
12. Vance, A., Jenkins, J.L., Anderson, B.B., Bjornn, D.K., Kirwan, C.B.: Tuning out security warnings: a longitudinal examination of habituation through fMRI, eye tracking, and field experiments. Manag. Inf. Syst. Q. **42**(2), 355–380 (2018)
13. Kirwan, B., Anderson, B., Eargle, D., Jenkins, J., Vance, A.: Using fMRI to measure stimulus generalization of software notification to security warnings. In: Davis, F.D., Riedl, R., vom Brocke, J., Léger, P.M., Randolph A.B., Fischer, T. (eds.) Information Systems and Neuroscience. Lecture notes in Information Systems and Organisation, pp. 93–99. Springer, Cham (2020)
14. Luck, S.: An Introduction to the Event-Related Potential Technique, 2nd edn. (2014)
15. Hajcak, G., MacNamara, A., Olvet, D.M.: Event-related potentials, emotion, and emotion regulation: an integrative review. Dev. Neuropsychol. **35**(2), 129–155 (2010)
16. Hajcak, G., Olvet, D.M.: The persistence of attention to emotion: brain potentials during and after picture presentation. Emotion **8**(2), 250–255 (2008)
17. Brown, S.B., van Steenbergen, H., Band, G.P., de Rover, M., Nieuwenhuis, S.: Functional significance of the emotion-related late positive potential. Front. Hum. Neurosci. **6**, 33 (2012)
18. Güntekin, B., Başar, E.: A review of brain oscillations in perception of faces and emotional pictures. Neuropsychologia **58**, 33–51 (2014)
19. Gerger, G., Leder, H., Kremer, A.: Context effects on emotional and aesthetic evaluations of artworks and IAPS pictures. Acta Physiol. (Oxf) **151**, 174–183 (2014)
20. Marín-Morales, J., Higuera-Trujillo, J.L., Greco, A., Guixeres, J., Llinares, C., Scilingo, E.P., Alcañiz, M., Valenza, G.: Affective computing in virtual reality: emotion recognition from brain and heartbeat dynamics using wearable sensors. Sci. Rep. **8**(1), 1–15 (2018)
21. Bradley, M.M., Hamby, S., Löw, A., Lang, P.J.: Brain potentials in perception: picture complexity and emotional arousal. Psychophysiology **44**(3), 364–373 (2007)
22. Peirce, J.W.: Generating stimuli for neuroscience using PsychoPy. Front. Neuroinform. **2**, 10 (2009)
23. Conrad, C., Agarwal, O., Woc, C.C., Chiles, T., Godfrey, D., Krueger, K., Marini, V., Sproul, A., Newman, A.: In: Davis, F.D., Riedl, R., Vom Brocke, J., Léger, P.-M., Randolph, A., Fischer, T. (eds.) Information Systems and Neuroscience, pp. 287–293 (2020)
24. Leventon, J.S., Bauer, P.J.: Emotion regulation during the encoding of emotional stimuli: effects on subsequent memory. J. Exp. Child Psychol. **142**, 312–333 (2016)
25. Gramfort, A., Luessi, M., Larson, E., Engemann, D.A., Strohmeier, D., Brodbeck, C., Parkkonen, L., Hämäläinen, M.S.: MNE software for processing MEG and EEG data. Neuroimage **86**, 446–460 (2014)
26. Delorme, A., Makeig, S.: EEGLAB: an open source toolbox for analysis of single-trial EEG dynamics including independent component analysis. J. Neurosci. Methods **134**(1), 9–21 (2004)
27. Davidson, D.J.: Functional mixed-effect models for electrophysiological responses. Neurophysiology **41**(1), 71–79 (2009)
28. Newman, A.J., Tremblay, A., Nichols, E.S., Neville, H.J., Ullman, M.T.: The influence of language proficiency on lexical semantic processing in native and late learners of English. J. Cogn. Neurosci. **24**(5), 1205–1223 (2012). https://doi.org/10.1162/jocn_a_00143

29. Tremblay, A., Newman, A.J.: Modelling non-linear relationships in ERP data using mixed-effects regression with R examples. Psychophysiology **52**(1), 124–139 (2014). https://doi.org/10.1111/psyp.12299
30. Conrad, C., Newman, A.J.: Measuring the impact of mind wandering in real time using the P1-N1-P2 auditory evoked potential. In: Davis, F., Riedl, R., vom Brocke, J., Léger, P-M, Randolph, A. (eds.) Information Systems and Neuroscience. Lecture notes in Information Systems and Organisation, pp. 37–45. Springer (2018). https://doi.org/10.1007/978-3-030-01087-4_5

Detecting Mind Wandering Episodes in Virtual Realities Using Eye Tracking

Michael Klesel[1,2], Michael Schlechtinger[1(✉)], Frederike Marie Oschinsky[1], Colin Conrad[3], and Bjoern Niehaves[1]

[1] University of Siegen, Siegen, Germany
{michael.klesel,michael.schlechtinger,frederike.oschinsky, bjoern.niehaves}@uni-siegen.de
[2] University of Twente, Enschede, The Netherlands
[3] Dalhousie University, Halifax, Canada
colin.conrad@dal.ca

Abstract. Virtual Reality (VR) allows users to experience their environment differently and more immersively than traditional information systems (IS). Therefore, it is important to also study cognitive processes in VR settings. In this proposal, we focus on the concept of mind wandering, which is an emerging concept in IS research that can be studied using neurological measures such as eye tracking. Current literature suggests that mind wandering is a complex concept with different dimensions, namely deliberate and spontaneous mind wandering. While previous literature has provided initial evidence on the feasibility of eye tracking to approximate mind wandering, this study seeks to investigate how well eye tracking performs when it comes to a more nuanced perspective on mind wandering applied in an VR setting.

Keywords: Mind wandering · Deliberate · Spontaneous · Virtual reality · Eye tracking

1 Introduction

For decades, information systems (IS) researchers have acknowledged the importance of cognitive processes during technology use. Constructs such as cognitive absorption [1] or IT-mindfulness [2] have widely been applied and have uncovered significant effects in IS-related contexts. With the rise of NeuroIS, the importance of cognitive aspects in technology-related settings has again been emphasized.

This study focuses on mind wandering, which is a cognitive concept that has only recently gained significant attention in psychology and neuroscience [3]. Mind wandering refers to episodes where our mind shifts to internal thoughts. While mind wandering can have severe negative effects [4], there are also an increasing number of studies that have demonstrated positive aspects of mind wandering, including a higher degree of creativity [5, 6].

F. D. Davis et al. (Eds.): NeuroIS 2020, LNISO 43, pp. 163–171, 2020.
https://doi.org/10.1007/978-3-030-60073-0_18

Several studies have investigated the concept of mind wandering in different scenarios, with various measurement techniques. However, little is known about mind wandering episodes in virtual reality (VR). Since a major driver of VR technologies relates to the fact that they affect and potentially even enhance our cognition, investigating mind wandering episodes in VR promises to generate further insights. To stress this argument, Thornhill-Miller and Dupont [7] "highlight[s] virtual reality (VR) as perhaps the safest, most fully developed of the emerging technologies of cognitive enhancement and as an underused tool for the enhancement of creativity in particular" (p. 102).

To better understand the relationship between VR and the concept of mind wandering, this study proposes an experiment to further investigate mind wandering in VR. The remainder is structured as follows: First, we briefly review the concept of mind wandering and how it is measured (Sect. 2). In Sect. 3, we propose the experimental setting that allows us to investigate mind wandering in VR. We conclude by reflecting on potential insights and future directions of this research.

2 Related Work

2.1 Mind Wandering

IS research often assumes that technology users are continuously focused [1, 8, 9]. However, empirical evidence shows that peoples' thoughts frequently proceed in a seemingly haphazard manner and effortlessly jump from one topic to another [10–12]. For up to half of their waking time, minds are not tethered to the actual moment or task, but easefully disconnected from the external environment [13].

Mind wandering is commonly described as a shift of attention away from a primary task toward dynamic, unconstrained spontaneous thoughts [4, 14] and as the mind's capacity to move away aimlessly from external happenings [15]. According to Christoff et al. [10], mind wandering can be defined as: "a mental state, or a sequence of mental states, that arise relatively freely due to an absence of strong constraints on the contents of each state". While mind wandering has widely been considered a failure of attention and control [16–20], recent studies highlight its advantages, including more effective brain processing, pattern recognition, and creativity [5, 12, 21, 22]. Specifically, mind wandering can help consider future events, solve problems, and create new ideas, e.g., at the digital workplace. It predominantly occurs during a resting state, task-free activity, and non-demanding circumstance [10, 12, 23, 24].

Since mind wandering can be a decisive factor for how users process information when using technology, IS researchers have started to acknowledge its relevance [25–28]. Sullivan et al. [26] were first to show that mind wandering influences functional outcomes of interacting with technology (i.e., creativity). They developed a domain-specific definition for technology-related mind wandering, being "task-unrelated thought which occurs spontaneously and the content is related to the aspects of computer systems" [26]. Moreover, Oschinsky et al. [25] revealed a significant difference between hedonic system use and utilitarian system use when it comes to mind wandering. Their study showed that the design of a system influences mind wandering, which in turn is known to affect antecedents of IT behavior and thus actual IT use.

There is a potential relationship between mind wandering and cognitive load, which has been investigated in the IS discipline. Representations of goal-states can be cued by goal-related stimuli under high cognitive load [3]. On the contrary, episodes of spontaneous thought are connected to low-level attention and uncontrolled, automatic thinking. As long as mind wandering is taking place, we seem to lack the ability to terminate or suspend it – we are fully immersed and yet relaxed and calm. The important difference of focused thinking under high cognitive load and the potential trigger of mind wandering episodes under low cognitive load is not yet sufficiently explored in the domain of NeuroIS research, and it is possible that there is an inverse relationship between the two constructs.

Because the interest in mind wandering has significantly increased in psychological and neuroscientific as well as IS research [22], different measurement scales have been proposed. However, the operationalization of mind wandering in IS-related conditions is still immature and incomplete [25–27, 29]. For instance, only little research exists that investigates the neurophysiological measures (e.g., EEG) in the domain of IS research (i.e., NeuroIS). Since self-report measurement does not seem to be the most efficient and appropriate way to assess the appearance of mind wandering experiences, refining the corresponding measurement instruments continues to be an important goal for research in this area [12]. We seek to contribute to closing this gap and propose the inclusion of and triangulation with objective data through eye tracking.

2.2 Eye Tracking and Mind Wandering

We conducted a literature review to identify how previous studies have measured mind wandering. For this study, we focus on the underlying type of technology (computer vs. VR) as well as the measurement of mind wandering (self-reported and using eye tracking). An overview of previous studies is given in Table 1.

Table 1. Studies on mind wandering and eye tracking.

Technology		Measurement		Example references
Computer	Virtual reality	Self-report	Eye tracking	
✓		✓		[25, 26]
	✓	✓		[30]
✓		✓	✓	[31–48]
	✓	✓	✓	(this study)

Table 1 highlights a variety of mind wandering findings which were collected by using self-reports and eye tracking. A large proportion of this literature deals with the risks of automobile crashes due to driver mind wandering. For example, He et al. (2011) highlighted deficits in vehicle control while mind wandering [39]. Others emphasize the increased chance of mind wandering due to the emergence of autonomous driving systems and offered suitable predictors [38, 40].

Mind wandering was also assessed in the context of attention while performing reading and learning tasks. Bixler et al. (2014–2016) aim for a fully automated mind wandering detection system using a machine learning model. To approach this goal, the researchers pseudo-randomly probed participants to report mind wandering episodes while performing computerized reading tasks. Meanwhile, the machine learning model tried to predict mind wandering due to gaze data followed up by a learning process based on the self-reported data [31–34]. Our findings indicated, that large chunks of eye tracking literature centers around utilizing objective data to create neural networks or machine learning models [41, 42]. Other researchers also probing for mind wandering in attention-tasks, familiarized test subjects with massive open online courses. Establishing on prior knowledge on objective mind wandering detection equipment, Zhao et al. (2017), successfully detected mind wandering with a common webcam [48].

Most of the discussed research used eye tracking devices in the form of cameras below or above the computer monitor (e.g., Tobii eye tracking devices) to record mind wandering. It is clear that eye tracking has a number of advantages over other methods for mind wandering research. However, there is a gap when it comes to the investigation of mind wandering in VR. In the remainder of this paper, we will describe an experiment which seeks to bridge this gap.

3 Methods

3.1 Participants and Materials

30 participants will be recruited at two different universities located in Canada and Germany to participate in a mailroom sorting task. Stimuli delivery and eye tracking will be conducted using HTC Vive PRO Eye SRanipal SDK, will be developed using the Unity engine and delivered using SteamVR. Participants will be screened for normal or corrected-to-normal eyesight, use of upper limbs and proficiency in English or German. Participants will be informed that we are investigating mind wandering in a simulated work environment. We will seek approval from our university's research ethics board and each session will last for 30 min in a controlled setting. At the completion of each session participants will receive CAD $15 or 15€ depending on where they conducted the experiment.

3.2 Procedure

Participants will undergo a consent protocol, complete an initial demographic questionnaire and will then be fitted with the HTC Vive PRO Eye VR-system. Participants will then take part in a virtual corporate mail room sorting task where they are given a series of addressed virtual envelopes and asked to place them in the appropriate bin. Participants will be asked to repeatedly retrieve a letter using the VR wand, read the address, and determine which of 16 bins to place it. The virtual letters will contain a selection of information consisting of addressee, title, department and address. Bins will be arranged according to department and will be clearly labelled at the base of each bin. Participants will not be required to walk during the routine.

3.3 Questionnaires and Physiological Measures

At three points throughout the experiment participants will be prompted with an experience sample where they will be asked about their degree of experienced mind wandering immediately preceding the sample [49]. Following the experiment, participants will complete a questionnaire about perceived degree of mind wandering throughout and its degree of spontaneity [50]. Task engagement times will be recorded by the software using events that record the time of letter retrieval and letter delivery, as well as task success (operationalized as the proportion of successful tasks/total number of tasks) and eye tracking engagements with task objects. During the time between each retrieval and delivery, eye fixation counts and fixation durations on 17 areas of interest will be recorded by the VR software.

3.4 Data Analysis

One of the challenges of eye tracking in a VR environment is that the environment is fluid and involves user-directed motion. This task was selected because though it creates a realistic simulation, it also constrains motion considerably and the equipment is optimized for such tasks. Eye fixation targets will consist of Unity objects which are pre-designed and modified for this VR environment. When eye fixations lock on to one of the programmed objects, a method will be called which records eye fixations and durations during which they are fixated on the object. Each participant is expected to yield between 5000 and 7000 trials which each correspond to a retrieval/delivery window. Analysis will be conducted on trials with time windows that completely precede the 30 s before a mind wandering probe samples. Trials will be labeled afterwards based on whether participants reported being in a state of mind wandering. The result is a largely automated process and manual intervention is only required to add data about the mind wandering state.

Two linear mixed effects investigations will be conducted on the resulting data. In the first investigation, fixation counts and fixation durations (for both target and non-target areas) as well as task duration will be investigated as fixed effects. Reported mind wandering will be investigated as the intercept variable. The reported mind wandering and on-task states will be treated as random effects to account for differences in number of trials and variances in reported mind wandering. This will identify variables which influence mind wandering. In the second investigation, the same variables will be investigated, though the mind wandering condition will be included as a fixed effect and task success as the intercept variable. Finally, multivariate linear regression will be used to assess the effects of the ex post measures on task success rates.

3.5 Outlook

As noted by Thornhill-Miller and Dupont [7], VR can be a promising technology to enhance cognitive processes. Consequently, this study seeks to extend current insights in terms of how to stimulate (or reduce) mind wandering episodes in technology-related settings. With a better understanding of the cognitive processes at play in everyday business tasks, we can uncover new insights into how to design our environments. Virtual reality promises to help create realistic, yet controlled environments which make new research directions possible. The results from this project can also inform organizations how to use VR to design processes that could be affected by mind wandering.

Perhaps the most promising way that this work can be further developed is to design and implement adaptive systems. Adaptive systems change based on a users' mental or physical state with the goal of improving an information system. When complete, we would have demonstrated eye-tracking correlates of mind wandering, which might be implemented to create such environments. In the future, we may extend this work to investigate how mind wandering interventions can change behavior, and whether these changes have implications to the productivity of organizations.

Acknowledgements. We would like to acknowledge that this research is part of the aSTAR research project, which is funded by the Federal Ministry of Education and Research of the Federal Republic of Germany (BMBF, promotional reference 02L18B010), the European Social Fund and the European Union.

References

1. Agarwal, R., Karahanna, E.: Time flies when you're having fun. Cognitive absorption and beliefs about information technology usage. MIS Q. **24**, 665–694 (2000)
2. Thatcher, J.B., Wright, R.T., Sun, H., Zagenczyk, T.J., Klein, R.: Mindfulness in information technology use. Definitions, distinctions, and a new measure. MIS Q. **42**, 831–847 (2018)
3. Fox, K.C.R., Christoff, K. (eds.): The Oxford Handbook of Spontaneous Thought. Mind-Wandering, Creativity, and Dreaming. Oxford University Press, New York (2018)
4. Smallwood, J., Schooler, J.W.: The restless mind. Psychol. Bull. **132**, 946–958 (2006)
5. Baird, B., Smallwood, J., Mrazek, M.D., Kam, J.W.Y., Franklin, M.S., Schooler, J.W.: Inspired by distraction. Mind wandering facilitates creative incubation. Psychol. Sci. **23**, 1117–1122 (2012)
6. Agnoli, S., Vanucci, M., Pelagatti, C., Corazza, G.E.: Exploring the link between mind wandering, mindfulness, and creativity. A multidimensional approach. Creat. Res. J. **30**, 41–53 (2018)
7. Thornhill-Miller, B., Dupont, J.-M.: Virtual reality and the enhancement of creativity and innovation: under recognized potential among converging technologies? J. Cogn. Educ. Psychol. **15**, 102–121 (2016)
8. Addas, S., Pinsonneault, A.: E-mail interruptions and individual performance. Is there a silver lining? MIS Q. **42**, 381–405 (2018)
9. Devaraj, S., Kohli, R.: Performance impacts of information technology. Is actual usage the missing link? Manag. Sci. **49**, 273–289 (2003)
10. Christoff, K., Irving, Z.C., Fox, K.C.R., Spreng, R.N., Andrews-Hanna, J.R.: Mind-wandering as spontaneous thought. A dynamic framework. Nat. Rev. Neurosci. **17**, 718–731 (2016)
11. Killingsworth, M.A., Gilbert, D.T.: A wandering mind is an unhappy mind. Science **330**, 932 (2010)
12. Smallwood, J., Schooler, J.W.: The science of mind wandering. Empirically navigating the stream of consciousness. Annu. Rev. Psychol. **66**, 487–518 (2015)
13. Golchert, J., Smallwood, J., Jefferies, E., Seli, P., Huntenburg, J.M., Liem, F., Lauckner, M.E., Oligschläger, S., Bernhardt, B.C., Villringer, A., et al.: Individual variation in intentionality in the mind-wandering state is reflected in the integration of the default-mode, fronto-parietal, and limbic networks. NeuroImage **146**, 226–235 (2017)

14. Andrews-Hanna, J.R., Irving, Z.C., Fox, K.C.R., Spreng, R.N., Christoff, K.: The neuroscience of spontaneous thought. An evolving, interdisciplinary field. In: Fox, K.C.R., Christoff, K. (eds.) The Oxford Handbook of Spontaneous Thought. Mind-Wandering, Creativity, and Dreaming, pp. 143–163. Oxford University Press, New York (2018)
15. Giambra, L.M.: Task-unrelated thought frequency as a function of age. A laboratory study. Psychol. Aging **4**, 136–143 (1989)
16. Drescher, L.H., van den Bussche, E., Desender, K.: Absence without leave or leave without absence. Examining the interrelations among mind wandering, metacognition and cognitive control. PLoS ONE **13**, 1–18 (2018)
17. Baldwin, C.L., Roberts, D.M., Barragan, D., Lee, J.D., Lerner, N., Higgins, J.S.: Detecting and quantifying mind wandering during simulated driving. Front. Hum. Neurosci. **11**, 1–15 (2017)
18. Mooneyham, B.W., Schooler, J.W.: The costs and benefits of mind-wandering. A review. Can. J. Exp. Psychol. **67**, 11–18 (2013)
19. Smallwood, J., Fishman, D.J., Schooler, J.W.: Counting the cost of an absent mind. Mind wandering as an underrecognized influence on educational performance. Psychon. Bull. Rev. **14**, 230–236 (2007)
20. Zhang, Y., Kumada, T., Xu, J.: Relationship between workload and mind-wandering in simulated driving. PLoS ONE **12**, 1–12 (2017)
21. Smeekens, B.A., Kane, M.J.: Working memory capacity, mind wandering, and creative cognition. An individual-differences investigation into the benefits of controlled versus spontaneous thought. Psychol. Aesthet. Creat. Arts **10**, 389–415 (2016)
22. Fox, K.C.R., Beaty, R.E.: Mind-wandering as creative thinking. Neural, psychological, and theoretical considerations. Curr. Opin. Behav. Sci. **27**, 123–130 (2019)
23. Buckner, R.L., Vincent, J.L.: Unrest at rest. Default activity and spontaneous network correlations. NeuroImage **37**, 1091–1096 (2007)
24. Northoff, G.: How does the brain's spontaneous activity generate our thoughts? In: Fox, K.C.R., Christoff, K. (eds.) The Oxford Handbook of Spontaneous Thought. Mind-Wandering, Creativity, and Dreaming, pp. 55–70. Oxford University Press, New York (2018)
25. Oschinsky, F.M., Klesel, M., Ressel, N., Niehaves, B.: Where are your thoughts? On the relationship between technology use and mind wandering. In: Proceedings of the 52nd Hawaii International Conference on System Sciences, Honolulu, Hi, USA, pp. 6709–6718 (2019)
26. Sullivan, Y., Davis, F., Koh, C.: Exploring mind wandering in a technological setting. In: Proceedings of the 36th International Conference on Information Systems, pp. 1–22. Fort Worth, United States of America (2015)
27. Wati, Y., Koh, C., Davis, F.: Can you increase your performance in a technology-driven society full of distractions? In: Proceedings of the 35th International Conference on Information Systems, Auckland, New Zealand, pp. 1–11 (2014)
28. Conrad, C., Newman, A.: Measuring the impact of mind wandering in real time using an auditory evoked potential. In: Davis, F.D., Riedl, R., Vom Brocke, J., Léger, P.-M., Randolph, A.B. (eds.) Information Systems and Neuroscience, pp. 37–45 (2019)
29. Sullivan, Y., Davis, F.: Self-regulation, mind wandering, and cognitive absorption during technology use. In: Proceedings of the 53rd Hawaii International Conference on System Sciences, Honolulu, Hi, USA, pp. 4483–4492 (2020)
30. Lin, C.-T., Chuang, C.-H., Kerick, S., Mullen, T., Jung, T.-P., Ko, L.-W., Chen, S.-A., King, J.-T., McDowell, K.: Mind-wandering tends to occur under low perceptual demands during driving. Sci. Rep. **6**, 1–11

31. Bixler, R., D'Mello, S.: Toward fully automated person-independent detection of mind wandering. In: Hutchison, D., Kanade, T., Kittler, J., Houben, G.-J. (eds.) User Modeling, Adaptation, and Personalization. Proceedings of the 22nd International Conference, UMAP 2014, Aalborg, Denmark, 7–11 July 2014, pp. 37–48. Springer, Cham (2014)
32. Bixler, R., D'Mello, S.: Automatic gaze-based detection of mind wandering with metacognitive awareness. In: Ricci, F. (ed.) User Modeling, Adaptation, and Personalization. Proceedings of the 23th International Conference, UMAP 2015, Dublin, Ireland, 29 June–3 July 2015, pp. 31–43. Springer, Cham (2015)
33. Bixler, R., D'Mello, S.: Automatic gaze-based user-independent detection of mind wandering during computerized reading. User Model. User Adapt. Interact. **26**(1), 33–68 (2015)
34. Faber, M., Bixler, R., D'Mello, S.K.: An automated behavioral measure of mind wandering during computerized reading. Behav. Res. Methods **50**(1), 134–150 (2017)
35. Foulsham, T., Farley, J., Kingstone, A.: Mind wandering in sentence reading: decoupling the link between mind and eye. Can. J. Exp. Psychol. Revue canadienne de psychologie experimentale **67**, 51–59 (2013)
36. Franklin, M.S., Broadway, J.M., Mrazek, M.D., Smallwood, J., Schooler, J.W.: Window to the wandering mind. Pupillometry of spontaneous thought while reading. Q. J. Exp. Psychol. **66**, 2289–2294 (2013)
37. Gwizdka, J.: Exploring eye-tracking data for detection of mind-wandering on web tasks. In: Davis, F.D., Riedl, R., Vom Brocke, J., Léger, P.-M., Randolph, A.B. (eds.) Information Systems and Neuroscience, vol. 29, pp. 47–55. Springer, Cham (2019)
38. Hatfield, N., Yamani, Y., Palmer, D.B., Yahoodik, S., Vasquez, V., Horrey, W.J., Samuel, S.: Analysis of visual scanning patterns comparing drivers of simulated L2 and L0 systems. Transp. Res. Rec. **2673**, 755–761 (2019)
39. He, J., Becic, E., Lee, Y.-C., McCarley, J.S.: Mind wandering behind the wheel: performance and oculomotor correlates. Hum. Factors **53**, 13–21 (2011)
40. Huang, G., Liang, N., Wu, C., Pitts, B.J.: The impact of mind wandering on signal detection, semi-autonomous driving performance, and physiological responses. Proc. Hum. Factors Ergon. Soc. Annu. Meet. **63**, 2051–2055 (2019)
41. Hutt, S., Krasich, K., Mills, C., Bosch, N., White, S., Brockmole, J.R., D'Mello, S.K.: Automated gaze-based mind wandering detection during computerized learning in classrooms. User Model. User Adapt. Interact. **29**(4), 821–867 (2019)
42. Hutt, S., Mills, C., White, S., Donnelly, P.J., D'Mello, S.K.: The eyes have it: gaze-based detection of mind wandering during learning with an intelligent tutoring system. In: Proceedings of the 9th International Conference on Educational Data Mining (2016)
43. Reichle, E.D., Reineberg, A.E., Schooler, J.W.: Eye movements during mindless reading. Psychol. Sci. **21**, 1300–1310 (2010)
44. Robison, M.K., Gath, K.I., Unsworth, N.: The neurotic wandering mind: an individual differences investigation of neuroticism, mind-wandering, and executive control. Q. J. Exp. Psychol. **2006**(70), 649–663 (2017)
45. Smilek, D., Carriere, J.S.A., Cheyne, J.A.: Out of mind, out of sight: eye blinking as indicator and embodiment of mind wandering. Psychol. Sci. **21**, 786–789 (2010)
46. Steindorf, L., Rummel, J.: Do your eyes give you away? A validation study of eye-movement measures used as indicators for mindless reading. Behav. Res. Methods **52**(1), 162–176 (2019)
47. Unsworth, N., Robison, M.K.: Tracking arousal state and mind wandering with pupillometry. Cogn. Affect. Behav. Neurosci. **18**(4), 638–664 (2018). https://doi.org/10.3758/s13415-018-0594-4
48. Zhao, Y., Lofi, C., Hauff, C.: Scalable mind-wandering detection for MOOCs: a webcam-based approach. In: Data Driven Approaches in Digital Education, pp. 330–344. Springer, Cham (2017)

49. Wammes, J.D., Smilek, D.: Examining the influence of lecture format on degree of mind wandering. J. Appl. Res. Mem. Cogn. **6**, 174–184 (2017)
50. Carriere, J.S.A., Seli, P., Smilek, D.: Wandering in both mind and body. Individual differences in mind wandering and inattention predict fidgeting. Can. J. Exp. Psychol./Revue Canadienne De Psychologie Experimentale **67**, 19–31 (2013)

Feeling the Pain of Others in Need: Studying the Effect of VR on Donation Behavior Using EEG

Anke Greif-Winzrieth(✉), Michael Knierim, Christian Peukert, and Christof Weinhardt

Institute of Information Systems and Marketing, Karlsruhe Institute of Technology (KIT), Karlsruhe, Germany
{anke.greif-winzrieth,micheal.knierim,christian.peukert, christof.weinhardt}@kit.edu

Abstract. Virtual reality (VR) enables people to engage in experiences that reach far beyond physical reality. This has inspired humanitarian organizations (among others the United Nations) to use VR technology to raise the awareness of humanitarian crises by virtually transporting people to the regions affected. As a consequence, these immersive experiences may lead to a change in the readiness to donate. As scientific evidence for this effect is still rare we propose an experimental design which aims at investigating how immersion affects donation behavior. In particular, neurophysiological measurement (EEG) shall shed light on the influence of immersion on emotional and motivational processes. First results from a convenient sample of young men indicate that donation behavior is linked to the dynamics of frontal alpha asymmetry changes.

Keywords: VR · NeuroIS · EEG · Frontal alpha asymmetry · Donations

1 Introduction

Since the advent of high-quality, low-cost VR systems, humanitarian organizations like the United Nations have started to use immersive 360° videos "to inspire viewers towards increased empathy, action and positive social change" [1]. Convergingly, some research suggests that higher levels of immersion in VR may facilitate empathy [2, 3], and lead to higher intentions to volunteer for charitable purposes. However, the immersion effects do not necessarily seem to translate into higher donations to charities [4]. A main caveat of most of this previous work is a focus on hypothetical donation decisions and a sole assessment of intentions to donate. To examine how immersion affects emotional and motivational processes and resulting donation behavior, we herein propose an experimental design using neurophysiological (EEG), and behavioral measures (donation to a local humanitarian organization). Electroencephalographic (EEG) investigations constitute a particularly prominent method of interest in NeuroIS research, likely due to relatively low cost, portability and high temporal resolution [5]. We focus on frontal

F. D. Davis et al. (Eds.): NeuroIS 2020, LNISO 43, pp. 172–180, 2020.
https://doi.org/10.1007/978-3-030-60073-0_19

alpha asymmetry (FAA) as an established measure for approach/avoidance motivation [6] which has also been proposed as a predictor of monetary donations to charity after a charities promotional video [7]. Yet, the study of the effectiveness of 360° VR videos in comparison to 360° desktop videos has not been examined. To support the experiment proposal, we report pilot study results that indicate substantial changes in FAA in a sample of young men, a population group that has previously been found to show lower emotional responding to charity-supporting desktop videos [8].

2 Theoretical Background: VR and Frontal Alpha Asymmetry

VR systems are capable to deliver a feeling of being present in a virtual world [9]. Thereby, a VR system's degree of immersion (i.e., the technical capabilities) largely influences the perceived presence [10]. To stress that one feels more present in a virtual space rather than in one's own physical location, the term *telepresence* is established [11]. Researchers make for instance use of the increased feeling of telepresence to treat real-world phobias in a virtual environment [12]. Furthermore, purely virtual experiences can lead to similar physiological responses as if one would encounter the same situation in reality [13]. It is therefore interesting to follow a NeuroIS research approach to gain further insights on the neural correlates of immersion.

The EEG provides a feasible and cost-effective approach to study cognitive-affective processes by measuring electrical discharges (post-synaptic potentials) from large clusters of neurons at the scalp level [5, 14]. While this means that subcortical structures (e.g. the limbic system) cannot be observed through the EEG, oscillating electrical discharges have been linked to emotional and motivational processes when organized in functionally differing frequency bands [5]. Of particular interest has been the observation of alpha band activity (typically 8–13 Hz) over the prefrontal cortex [14]. With alpha being considered a marker of cortical idling [15], it is understood that asymmetric frontal alpha (i.e. different alpha levels on left and right hemispheres) shows connections to state and trait level motivational and emotional experiences [6, 14]. This metric is termed *Frontal Alpha Asymmetry* (FAA) [16]. A substantial body of research demonstrates that relatively greater left than right frontal power characterizes approach-oriented situations (e.g. jealousy, anger, or self-control) and/or individuals (e.g. high dispositional anger or high trait optimism) [6, 14]. On the other hand, greater right than left frontal power is considered to reflect withdrawal-related motivational traits and states (e.g. sadness, fear, but also empathy) or internalizing personality traits (e.g. depression or anxiety) [14]. FAA scores are thus employed by scholars worldwide to study constructs like temperament and personality, and various types of motivation and emotion processes [6, 14]. FAA has also been used in the context of charity campaign engagement, but mainly with a focus on desktop stimuli and female subjects [7, 8].

3 Proposed Experimental Design

To study the effect of VR on donation behavior, we propose the following experiment:

Treatments. Subjects are randomly assigned to either the DESKTOP or the VR treatment (between subjects). As stimulus, we use the 360° video *Clouds over Sidra* [1], introducing impressions of the daily life of a 12-year-old Syrian refugee girl. Subjects in the DESKTOP treatment watch the video on a full HD 24" desktop screen and use the mouse to change the viewing angle. Subjects in the VR treatment watch the video in an Oculus Go head mounted display (HMD) where the viewing direction is adjusted to the head orientation.

Procedure. Figure 1 depicts the experimental procedure. Subjects are seated on a swivel chair allowing for 360° body rotation in a soundproofed and air-conditioned cubicle equipped with a computer, mouse, keyboard and speakers. The experimenter then sets up the EEG system and subjects follow the instructions on the screen, guiding them through the first resting phase. Next, the experimenter sets up the HMD (VR) or the headphones (DESKTOP) and instructs the subject how to start the video. After the video the experimenter removes the HMD or the headphones and the subject is guided through the second resting phase. The screen following the resting phase explains the donation decision. The donation is implemented as described there and the decision is made in private as well as anonymous (the experimenter who interacts with the subject delivers the money in a closed envelope and does not know the payment and the amount donated). This is particularly stressed in the instructions to avoid social desirability and experimenter demand effects. The session ends with a short survey.

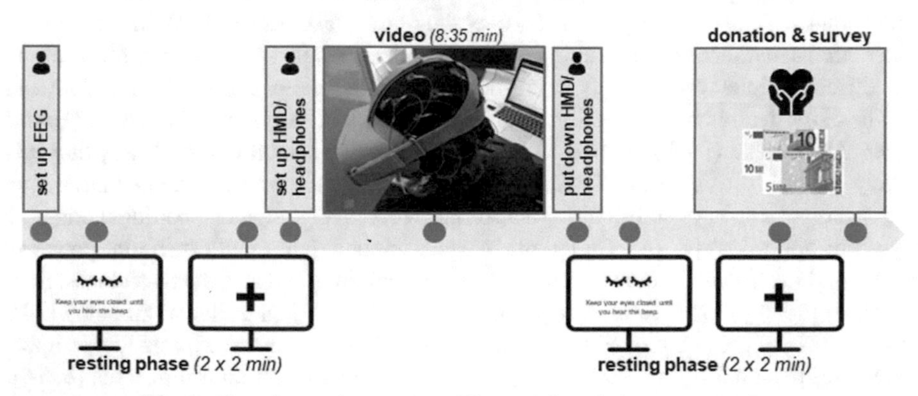

Fig. 1. Experimental procedure. Picture taken during a test session.

Behavioral Measure. To capture the effects of watching the video on monetary decisions, subjects are asked to split €15 between themselves and an organization supporting local refuges, particularly children. Subjects can donate any amount between €0 and €15 in steps of €1 and receive their share in cash after the experiment.

EEG Measure. EEG data is collected following recent recommendations for FAA research [14]. An Emotiv Flex 32-channel gel-based EEG headset (Ag/AgCl electrodes, 128 Hz sampling rate, 14 bit resolution) is used with the reference (CMS) electrode on the left earlobe and the ground (DRL) electrode on the right earlobe. Following the 10–10 system, electrodes are evenly distributed and data is analyzed for the prefrontal locations (Fp1, F3, F7, F8, F4, Fp2). To improve the FAA recording quality, resting state data from a 2 min period before and after the video stimulus is collected [14].

Further Variables. We further collected demographic variables (gender, age, education, job, financial situation, income), and prior experience with VR devices.

4 Pilot Study

We conducted a pilot study implemented in oTree [17] with 5 subjects in the VR Treatment. Homogeneity in the small sample was increased by screening subjects for being male, right-handed, in their early twenties, having full (corrected) vision and full color vision, and to be free of general health impediments for the last week. All participants were students with a monthly disposable income of less than €500 and two of them had used VR headsets more than five times prior to their participation in the experiment. The others had used such devices only once (2) or never before (1).

Data Analysis The EEG processing was fully automated to reduce the influence of unreliable researcher decisions [14, 18]. The pipeline includes the steps (in order): line noise removal (50 & 100 Hz), robust common average re-referencing [18], detrending and denoising (1 Hz and 40 Hz FIR filter), outlier trimming (>500 mV/250 ms), paroxysmal artefact removal (artefact subspace reconstruction [19]), and independent component removal using the ADJUST toolbox in EEGLab [20]. Afterwards, Morlet wavelet decomposition was employed to extract frequency powers. For each subject individualized alpha power was extracted (−2 to +2 Hz around the peak alpha frequency observed at O1 & O2 electrodes) [21]. Alpha band scores were within-subject z-scored to normalize the power changes across subjects. The FAA score was then calculated from decibel normalized frontal electrode pairs by subtracting the left hemisphere from the right hemisphere for homologous pairs (i.e. Fp2-Fp1, F4-F3, F8-F7) and averaging the differences scores. In this way, higher asymmetry scores indicate relatively greater left frontal activity (e.g. higher approach motivation) [14]. For all aggregations median averaging was used to reduce the potential impact of outliers in the data [22].

Preliminary Results. First we assessed median and individual changes in FAA from before to after the presentation of the VR video as summarized in Fig. 2 (left). Overall, the presentation of the video appears to have a reducing, albeit not unanimous effect on FAA scores. This indicates higher avoidance motivation, perceived lack of self-control, sadness or anxiety – or stated otherwise: at least no tendency to want or feel like approaching or changing the current situation. This observation indicates initial support to the idea, that the VR video could alter emotional and motivational experiences. Importantly, the effect seems present even in this sample of young men, a target population that has previously been found to be less susceptible to emotional response from desktop video stimuli in the context of charity campaign donations [7, 8]. Including donations (coloring of the lines in Fig. 2 (left) reveals that subjects with strong FAA reductions tended to donate very little or nothing while one of the subjects who showed a slight increase in FAA, donated almost all of his payment. This finding is in line with previous work that finds positive correlations between FAA scores and donations to charity campaigns in young women watching a desktop video of a child in need [7].

Second, we assessed the progression of FAA scores over the course of the VR video. A gradual reduction in FAA medians appears in the first four fifths of the video, followed by an increase in the last fifth when the video presents a donation appeal. Therefore, the overall reduction in FAA could be considered as a state of greater avoidance motivation caused by a sense of powerlessness ("it is sad, but what can I do?") that possibly changes when the opportunity for action is made salient ("it is sad, but I *can* do something!"). Drawing from this consideration, we derive the hypothesis that a salient presentation of opportunities for action could increase individuals' approach motivation towards challenging situations. To test this hypothesis, a larger experiment could manipulate (show/hide) the presentation of such action opportunities.

The late increase in FAA that seems to be present in most subjects does however not appear to predict donation heights. We consider two possible explanations: On the one hand, the timing of the opportunity for action could have an influence as to whether more action is actually taken (here: if more money is donated). For example, accumulation of avoidance motivation in earlier stages could be overpowering the late shift towards approach. We plan to vary the timing of the call for action (donation appeal) to see, if an early hint to the opportunity to act might increase approach motivation before being confronted with the problem. On the other hand, the progressions reveal that while an overall shift from before to after the video might be subtle, during the video strong changes in FAA can occur. We propose that these might be linked to donations: for subject C, who donated the most, we observe the overall highest increase (cf. segment 1) and for subject B, who donated the second most we observe a remarkable increase (cf. segment 5). Thus, we derive the hypothesis that the maximum increase in the FAA score during the video could predict donation heights. Again, the varied timing of action opportunity provision would be a possible option to manipulate and further investigate these experience patterns.

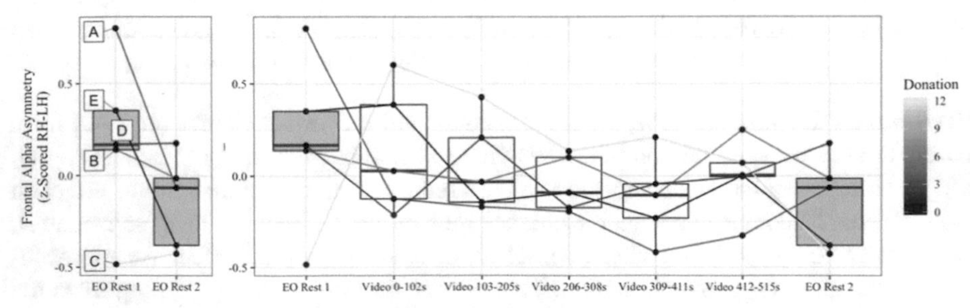

Fig. 2. Changes in FAA scores from before to after the VR video stimulation (left) and across five equally spaced units during the VR video (right).

5 Discussion and Future Research

Altogether, it needs to be critically appraised that the small sample for this first explorative study is not sufficient to conclude the presence of substantial effects. Whether or not FAA is influenced strongly by VR video stimulation (and more so than from standard desktop screen videos) and how FAA patterns are related to experience and behavior requires larger samples. Given the present observations, it would appear though as highly valuable to include within-subject observations into statistical inferences. For example, linear mixed models with individual random intercepts and slopes might be an important approach to study the relationships between FAA and experience (e.g. motivation and emotion) and behavior (e.g. donation) outcomes.

An important challenge for a more elaborate experiment design is that visual differences between VR and desktop modalities may arise and could cause confounding effects in the FAA feature. To account for this challenge, we have reviewed more related work that conducts virtual/mixed reality research together with EEG, to inform ourselves about how visual confounds might have been accounted for previously. So far, it appears that no general strategy exists, as other work does not seem to take special measures to overcome it (e.g. [23] or [24] – albeit both with a large screen display VR solution). Furthermore, only recent work appears to have investigated potential artefacts from head-mounted VR systems [25]. This work has not found VR-related artefacts, or at least not in lower frequencies (only >90 Hz). However, in order to account for the possibility that visual artefacts might confound FAA observations, we propose an extended experimental design that includes additional within-subject stimuli (neutral and affectively challenging – as recommended by [14]). Neutral stimuli relate to static eyes-open resting conditions (fixation cross) and dynamic resting state conditions (e.g. a video of a fishtank as used in [26]). Affective stimuli relate to dynamic videos (e.g. videos of facial expressions). To account for potential affective spill-over effects, these additional stimuli will be separated by washout phases (using simple eyes open resting phases or time production tasks – e.g. [27]). The repeated measurements can then be used to compare FAA across different stimuli in order to assess whether or not general differences exist between VR and desktop treatments. Furthermore, the repeated measurements allow for better within-subject standardization (e.g. z-scoring) that could further improve the inter-individual comparability of findings overall and in the donation elicitation stimulus in particular.

In order to further assess the effect of VR on approach/avoidance motivation, the experiment should additionally collect self-reported measures to check whether results are consistent with the EEG measurement. As proposed by [28], these measures should include both, state (e.g. positive and negative affect (PANAS) [29]) and trait variables (e.g. behavioral inhibition/activation (BIS-BAS) [30]). As an increased feeling of presence in VR as compared to less immersive devices (e.g. desktop screen) is widely used to explain the technology's effects on humans' psyche [31], the study should also assess telepresence (e.g. as proposed by [32]).

6 Conclusion

Within this work-in-progress paper, we propose and discuss an experimental design to study the effect of VR on donation behavior. The results of our pilot study indicate that in the VR treatment donation behavior may be linked to emotional and motivational processes as shown in the dynamics of frontal alpha asymmetry changes. However, so far we have only studied the VR treatment with a small sample and the full data collection for both treatments remains to be done. A more elaborate design for a larger scale study should further control for potentially confounding effects in the FFA caused by the visual differences between the desktop and VR devices and include further self-reported measures to carry out triangulation. Our results may guide the design of donation campaigns and especially shed light on the question whether the application of VR technology leads to a better outcome – a common assumption that has hardly been proven so far.

References

1. United Nations Virtual Reality (UNVR), http://unvr.sdgactioncampaign.org/vr-films. Accessed 28 Feb 2020
2. Schutte, N.S., Stilinović, E.J.: Facilitating empathy through virtual reality. Motiv. Emot. **41**(6), 708–712 (2017). https://doi.org/10.1007/s11031-017-9641-7
3. Shin, D.: Empathy and embodied experience in virtual environment: to what extent can virtual reality stimulate empathy and embodied experience? Comput. Human Behav. **78**, 64–73 (2018). https://doi.org/10.1016/j.chb.2017.09.012
4. Kandaurova, M., Lee (Mark), S.H.: The effects of Virtual Reality (VR) on charitable giving: the role of empathy, guilt, responsibility, and social exclusion. J. Bus. Res. **100**, 571–580 (2019). https://doi.org/10.1016/j.jbusres.2018.10.027
5. Müller-Putz, G.R., Riedl, R., Wriessnegger, S.C.: Electroencephalography (EEG) as a research tool in the information systems discipline: foundations, measurement, and applications. Commun. Assoc. Inf. Syst. **37**, 911–948 (2015). https://doi.org/10.17705/1cais.03746
6. Briesemeister, B.B., Tamm, S., Heine, A., Jacobs, A.M.: Approach the good, withdraw from the bad—a review on frontal alpha asymmetry measures in applied psychological research. Psychology **4**, 261–267 (2013). https://doi.org/10.4236/psych.2013.43a039
7. Huffmeijer, R., Alink, L.R.A., Tops, M., Bakermans-Kranenburg, M.J., Van IJzendoorn, M.H.: Asymmetric frontal brain activity and parental rejection predict altruistic behavior: moderation of oxytocin effects. Cogn. Affect. Behav. Neurosci. **12**, 382–392 (2012). https://doi.org/10.3758/s13415-011-0082-6
8. Martinez-Levy, A., Cherubino, P., Cartocci, G., Modica, E., Rossi, D., Mancini, M., Trettel, A., Babiloni, F.: Gender differences evaluation in charity campaigns perception by measuring neurophysiological signals and behavioural data. Int. J. Bioelectromagn. **19**, 25–35 (2017). www.ijbem.org
9. Witmer, B.G., Singer, M.J.: Measuring presence in virtual environments: a presence questionnaire. Presence Teleoperators Virt. Environ. **7**, 225–240 (1998). https://doi.org/10.1162/105474698565686
10. Cummings, J.J., Bailenson, J.N.: How immersive is enough? A meta-analysis of the effect of immersive technology on user presence. Media Psychol. **19**, 272–309 (2016). https://doi.org/10.1080/15213269.2015.1015740

11. Steuer, J.: Defining virtual reality: dimensions determining telepresence. J. Commun. **42**, 73–93 (1992)
12. Balan, O., Moise, G., Moldoveanu, A., Moldoveanu, F., Leordeanu, M.: Automatic adaptation of exposure intensity in VR acrophobia therapy, based on deep neural networks. In: Proceedings of the 27th European Conference on Information Systems (ECIS), pp. 1–14. Stockholm & Uppsala, Sweden (2019)
13. González-Franco, M., Peck, T.C., Rodríguez-Fornells, A., Slater, M.: A threat to a virtual hand elicits motor cortex activation. Exp. Brain Res. **232**(3), 875–887 (2013). https://doi.org/10.1007/s00221-013-3800-1
14. Smith, E.E., Reznik, S.J., Stewart, J.L., Allen, J.J.B.: Assessing and conceptualizing frontal eeg asymmetry: an updated primer on recording, processing, analyzing, and interpreting frontal alpha asymmetry. Int. J. Psychophysiol. **111**, 98–114 (2017). https://doi.org/10.1016/j.physbeh.2017.03.040
15. Allen, J.J.B., Coan, J.A., Nazarian, M.: Issues and assumptions on the road from raw signals to metrics of frontal EEG asymmetry in emotion. Biol. Psychol. **67**, 183–218 (2004). https://doi.org/10.1016/j.biopsycho.2004.03.007
16. Davidson, R.J., Schwartz, G.E., Saron, C., Bennett, J., Goleman, D.J.: Frontal versus parietal EEG asymmetry during positive and negative affect. Psychophysiology **16**, 202–203 (1979)
17. Chen, D.L., Schonger, M., Wickens, C.: oTree—an open-source platform for laboratory, online, and field experiments. J. Behav. Exp. Financ. **9**, 88–97 (2016). https://doi.org/10.1016/j.jbef.2015.12.001
18. Bigdely-Shamlo, N., Mullen, T., Kothe, C., Su, K.-M., Robbins, K.A.: The PREP pipeline: standardized preprocessing for large-scale EEG analysis. Front. Neuroinform. **9**, 1–20 (2015). https://doi.org/10.3389/fninf.2015.00016
19. Mullen, T.R., Kothe, C.A.E., Chi, M., Ojeda, A., Kerth, T., Makeig, S., Jung, T.-P., Cauwenberghs, G.: Real-time neuroimaging and cognitive monitoring using wearable dry EEG. IEEE Trans. Biomed. Eng. **62**, 2553–2567 (2015). https://doi.org/10.1109/TBME.2015.2481482. Real-time
20. Mognon, A., Jovicich, J., Bruzzone, L., Buiatti, M.: ADJUST: An automatic EEG artifact detector based on the joint use of spatial and temporal features. Psychophysiology **48**, 229–240 (2011). https://doi.org/10.1111/j.1469-8986.2010.01061.x
21. Klimesch, W.: EEG alpha and theta oscillations reflect cognitive and memory performance: a review and analysis. Brain Res. Rev. **1**, 169–195 (1999)
22. Cohen, M.X.: Analyzing Neural Time Series Data: Theory and Practice. MIT press, Cambridge (2014)
23. Kober, S.E., Kurzmann, J., Neuper, C.: Cortical correlate of spatial presence in 2D and 3D interactive virtual reality: an EEG study. Int. J. Psychophysiol. **83**, 365–374 (2012). https://doi.org/10.1016/j.ijpsycho.2011.12.003
24. Slobounov, S.M., Ray, W., Johnson, B., Slobounov, E., Newell, K.M.: Modulation of cortical activity in 2D versus 3D virtual reality environments: an EEG study. Int. J. Psychophysiol. **95**, 254–260 (2015). https://doi.org/10.1016/j.ijpsycho.2014.11.003
25. Hertweck, S., Weber, D., Alwanni, H., Unruh, F., Fischbach, M., Latoschick, M.E., Ball, T.: Brain activity in virtual reality: assessing signal quality of high-resolution EEG while using head-mounted displays. In: IEEE Conference on Virtual Reality and 3D User Interfaces (VR), pp. 970–971 (2019)
26. Piferi, R.L., Kline, K.A., Younger, J., Lawler, K.A.: An alternative approach for achieving cardiovascular baseline: viewing an aquatic video. Int. J. Psychophysiol. **37**, 207–217 (2000)
27. Honma, M., Kuroda, T., Futamura, A., Shiromaru, A., Kawamura, M.: Dysfunctional counting of mental time in Parkinson's disease. Sci. Rep. **6** (2016). https://doi.org/10.1038/srep25421

28. Rodrigues, J., Müller, M., Mühlberger, A., Hewig, J.: Mind the movement: frontal asymmetry stands for behavioral motivation, bilateral frontal activation for behavior. Psychophysiology **55**, e12908 (2018). https://doi.org/10.1111/psyp.12908
29. Watson, D., Clark, L.A., Tellegen, A.: Development and validation of brief measures of positive and negative affect: the PANAS scales. J. Pers. Soc. Psychol. **54**, 1063–1070 (1988). https://doi.org/10.6102/zis242
30. Carver, C.S., White, T.L.: Behavioral inhibition, behavioral activation, and affective responses to impending reward and punishment: the BIS/BAS scales. J. Pers. Soc. Psychol. **67**, 319 (1994)
31. Schuemie, M.J., van der Straaten, P., Krijn, M., van der Mast, C.A.P.G.: Research on presence in virtual reality: a survey. Cyber Psychol. Behav. **4**, 183–201 (2001). https://doi.org/10.1089/109493101300117884
32. Kim, T., Biocca, F.: Telepresence via television: two dimensions of telepresence may have different connections to memory and persuasion. J. Comput. Commun. **3**, JCMC325 (2006). https://doi.org/10.1111/j.1083-6101.1997.tb00073.x

Using NeuroIS Tools to Understand How Individual Characteristics Relate to Cognitive Behaviors of Students

Tanesha Jones[1], Adriane B. Randolph[1(✉)], Kimberly Cortes[2], and Cassidy Terrell[3]

[1] Department of Information Systems, Kennesaw State University, Kennesaw, GA, USA
tjone148@students.kennesaw.edu, arandol3@kennesaw.edu

[2] Department of Chemistry and Biochemistry, Kennesaw State University, Kennesaw, GA, USA
klinenbe@kennesaw.edu

[3] University of Minnesota Rochester, Rochester, MN, USA
terre031@r.umn.edu

Abstract. NeuroIS tools have increasingly been used to examine cognitive behaviors in educational settings. Here we present results of ongoing work applying neurophysiological tools to examine the cognitive load of student learners in the context of chemistry education. In particular, we investigate how individual characteristics relate to the Pope Engagement Index for students interacting with an information system for visualizing molecules. Characteristics such as meditation, levels of athleticism, and medication affecting alertness were found to significantly and positively correlate with cognitive load.

Keywords: Cognitive load · Individual characteristics · Pope engagement index · EEG · Chemistry student learners

1 Introduction

Increasingly, neuroIS tools are being used in "neuro-education" to better understand student learners and their cognitive processes [1–3]. Researchers are able to use tools such as electroencephalography (EEG) in conjunction with traditional psychometric tools to describe a learner's full-body experience, including their feelings and levels of engagement that may otherwise be difficult to articulate. In fact, neuroIS tools may be able to help better pinpoint when such mental changes take place and thus offer more clarity on what to change in a learning environment [3].

Within neuroIS, cognitive load has received particular focus as a construct of interest and has been measured using EEG and eye-tracking tools [4, 5]. Here, we also focus on cognitive load, as measured using the popular Pope Engagement Index (PEI) calculated from surface EEG recordings [6]. In particular, we investigate the relationship of cognitive load with the individual characteristics of chemistry students from a university in a metropolitan midwestern city as part of a federally-funded grant project[1] in the United States.

[1] This work was funded by the National Science Foundation under Grant Number 1711425.

F. D. Davis et al. (Eds.): NeuroIS 2020, LNISO 43, pp. 181–184, 2020.
https://doi.org/10.1007/978-3-030-60073-0_20

2 Methodology

At the start of the session, students completed a survey about their individual characteristics that was adopted from brain-computer interfacing, a sub-area of neuroIS [7]. These individual characteristics were not limited to age, gender, and race, but also included differences in self-perceived levels of athleticism, hand dexterity, medication intake, smoking status, prior biometric tool use, and video game experience. We also obtained cognitive measures of spatial ability using the Purdue Visual Rotation Test (PVRoT) [8] and the Hidden Figures Test (HFT) [9].

Then, students engaged in a simulated learning environment where an instructor was present to explain the lesson and then remained while the student worked through exercises on their personal laptop. The students engaged in six different activities using PyRx, an opensource tool for visualizing chemical molecules [10]. This tool was purported to have an "easy-to-use user interface" (https://pyrx.sourceforge.io/home) yet some students still experienced difficulty with its setup. Students used PyRx to visualize proteins and match different ligands to the protein sites, much like finding the right key for a lock.

A 16-channel, research-grade BioSemi ActiveTwo[2] bioamplifier system recorded students' electrical brain activity during the activities. This system was run on a Windows laptop. The electrode cap was configured according to the widely used 10–20 system of electrode placement [11]. Active electrodes were placed on the cap to allow for the recording of brain activations down-sampled to 256 Hz using a Common Average Reference (CAR). The sixteen recorded channels were: frontal-polar (Fp1, Fp2), frontal-central (FC3, FCz, FC4), central (C3, Cz, C4), temporal-parietal (TP7, TP8), parietal (P3, Pz, P4), and occipital (O1, Oz, O2).

The recorded data was later analyzed using using the EEGLab plugin (https://sccn.ucsd.edu/eeglab/index.php) to Matlab to ascertain band powers and calculate cognitive load according to the PEI best represented by the calculation of (combined beta power) / (combined alpha power + combined theta power) [6]. A separate PEI was calculated for each of the six activities performed. RStudio Cloud (https://rstudio.cloud.com) was then used to find correlations between the student learners' individual characteristics and their PEI per activity.

3 Preliminary Results

Results for two of the six activities are presented here where we could compare the same students across both activities. Out of the original ten participants, the same six students completed both of these activities. The average age was 21 (ranged 20–22 year) with 2 males and 4 females. Correlations were found significant at the level where alpha equaled 0.05.

Activity 1: Docking the Ligand. This activity entailed students selecting the ligands and isolating the possible binding sites of the protein in the software. For this activity, we

[2] https://cortechsolutions.com/product-category/eeg-ecg-emg-systems/eeg-ecg-emg-systems-activetwo/.

found strong positive correlations to the PEI with Meditation and Athleticism. Table 1 summarizes the correlations found for this activity. This indicates that a student who regularly engaged in meditation and had higher levels of self-rated athleticism experienced higher cognitive load.

Table 1. Significant correlations of individual characteristics to the Pope Engagement Index for the Docking the Ligand activity.

	Correlation	P-value
Meditation	1.0000	0.0000
Athleticism	0.9147	0.0106

Activity 2: Visualizing the Docked Ligand. This activity entailed selecting and viewing a particular ligand bound to the protein in PyRx and zooming in to see more detail. For this activity, we found a strong positive correlation only with AffectiveMeds to the cognitive load experienced by the student as measured by their PEI. This indicates that taking medication that made the student more alert correlated with higher levels of cognitive load experienced for this activity. Particularly, the correlation coefficient was 0.8832 with a p-value of 0.0197.

Summary. Meditation, Athleticism, and AffectiveMeds were the three characteristics found with strong positive correlations to students' PEI for the two different activities. Common convention implies that individuals would translate the calm and focus often gained from meditating, engaging in athletic activities, and from stimulating medication to achieve lower cognitive load; it seems something more is yet to be revealed. Through scatterplot analysis, one participant may be dominating the results, however we felt it more important to retain all data points due to the distinctly small sample size of six students. Although a definite conclusion is impossible at this stage, we are encouraged to seek understanding of the full picture of our students when pairing them with such learning activities. These results indicate that some individual characteristics may have more influence than others on a student's cognitive load experienced in this setting.

4 Conclusion

This study provides an example of how neuroIS tools may be used in an educational setting to better understand the cognitive processes of students. In particular, we investigated the relationship between individual characteristics and the cognitive load of chemistry students as measured by the Pope Engagement Index calculated from EEG recordings. This work-in-progress paper presents a snapshot from a larger study that spans three years and is still under analysis.

Although results cannot be generalized to a wider population due to low sample size, these efforts indicate the importance of gaining a comprehensive view of students to better understand the impacts on their learning environment. Certainly, more data should better reveal which characteristics have particular saliency in this setting. Further, we may find distinction in our future results by dissecting our reliance on the original calculation of the PEI as a measure for cognitive load. Lastly, we will seek to delineate the nature of the tasks as having external and internal attentional components in line with newer research examining the relationship of alpha waves to cognitive load, where alpha is a key component of PEI calculations.

References

1. Charland, P., et al.: Measuring implicit cognitive and emotional engagement to better understand learners' performance in problem solving. Zeitschrift für Psychologie (2017)
2. Charland, P., et al.: Assessing the multiple dimensions of engagement to characterize learning: a neurophysiological perspective. J. Visual. Exp. JoVE **101**, e52627 (2015)
3. Randolph, A., et al.: Application of NeuroIS tools to understand cognitive behaviors of student learners in biochemistry. In: Davis, F., Riedl, R., vom Brocke, J., Léger, P.M., Randolph, A., Fischer, T. (eds.) Information Systems and Neuroscience, pp. 239–243. Springer, Cham (2020)
4. Fischer, T., Davis, F.D., Riedl, R.: NeuroIS: a survey on the status of the field. In: Davis, F., Riedl, R., vom Brocke, J., Léger, P.M., Randolph, A. (eds.) Information Systems and Neuroscience, pp. 1–10. Springer, Cham (2019)
5. Riedl, R., Fischer, T., Léger, P.-M.: A decade of NeuroIS research: status quo, challenges, and future directions. In: Thirty Eighth International Conference on Information Systems, South Korea (2017)
6. Pope, A.T., Bogart, E.H., Bartolome, D.S.: Biocybernetic system evaluates indices of operator engagement in automated task. Biol. Psychol. **40**(1), 187–195 (1995)
7. Randolph, A.B., Moore Jackson, M.M.: Assessing fit of nontraditional assistive technologies. ACM Trans. Access. Comput. **2**(4), 1–31 (2010)
8. Bodner, G.M., Guay, R.B.: The Purdue visualization of rotations test. Chem. Educ. **2**(4), 1–17 (1997)
9. French, J.W., Ekstrom, R.B., Price, L.A.: Manual for kit of reference tests for cognitive factors (revised 1963). Educational Testing Service, Princeton, NJ (1963)
10. Dallakyan, S., Olson, A.J.: Small-molecule library screening by docking with PyRx. In: Hempel, J., Williams, C., Hong, C. (eds.) Chemical Biology, pp. 243–250. Springer, Cham (2015)
11. Homan, R.W., Herman, J., Purdy, P.: Cerebral location of international 10–20 system electrode placement. Electroencephalogr. Clin. Neurophysiol. **66**(4), 376–382 (1987)

Beyond System Design: The Impact of Message Design on Recommendation Acceptance

Antoine Falconnet[1], Wietske Van Osch[1], Joerg Beringer[2], Marc Fredette[1], Sylvain Sénécal[1], Pierre-Majorique Léger[1], and Constantinos K. Coursaris[1](✉)

[1] HEC Montréal, Montréal, Canada
{antoine.falconnet,vanosch,marc.fredette,ss,pml,coursaris}@hec.ca
[2] Blue Yonder, Coppell, USA
joerg.beringer@jda.com

Abstract. The current paper reports on the results of a pilot study to explore the impact of message design on users' likelihood to accept system-generated recommendations as well as their intention to use the recommendation system (RS). We aim to extend the RS literature, which has hitherto focused on system design elements, but has generally overlooked the importance of message design, a key element in facilitating effective attention and information processing, particularly in the context of managerial decision-making.

Keywords: Managerial decision-making · Recommendation systems · Message design · Acceptance and use intention · Eye tracking

1 Introduction

Recommendation systems (RS) are increasingly popular tools for augmenting the human process of decision-making. Studies of RS have focused largely on design implications at the system level [1, 2], with limited research on RS interface design [3]. During these studies, interaction data from a user's interaction with the recommendations are collected and subsequently juxtaposed against attributes of artifacts stored in large repositories to subsequently produce and present recommendations. To date, however, there has been no research exploring the impact of the recommendation message's design on the user's interaction with and behavior toward the recommendation, such as accepting it (e.g., for web-based resource recommendations, clicking on the linked recommendation; for behavioral recommendations, clicking on an 'accept' button), rejecting/ignoring it, or requesting additional information in support of the recommendation.

Particularly in the context of managerial decision-making, it is not the system alone, but also the nature of the message content that will drive the manager's perceptions of trust and likelihood to accept a decision recommendation [4]. In this context, as information processing of recommendations would be done on the basis of both content elements (Areas of Interest or AOIs) and interaction elements (e.g., buttons), a granular

F. D. Davis et al. (Eds.): NeuroIS 2020, LNISO 43, pp. 185–190, 2020.
https://doi.org/10.1007/978-3-030-60073-0_21

analysis of both design elements simultaneously is warranted. Hence, this study sets out to explore what would make users of a management dashboard trust and accept decision recommendations generated by the built-in decision support system with minimal cognitive effort and time required. Specifically, our research aims to answer the following research questions:

RQ1. What is the effect of message design (choices) on a user's cognitive and emotional responses to system-generated recommendations?

RQ2. What is the effect of message design (choices) on a user's behaviors vis-à-vis the likelihood to accept system-generated recommendations?

RQ3. What is the effect of message design (choices) on a user's attitude toward the recommendation and the recommender system?

2 Related Work

2.1 Previous Work on Recommender Systems

Extant research on RS has provided extensive support regarding the effects of the use of Recommendation Agents (RAs) on user perceptions of ease of use, control, trust and system effectiveness [1] and has shown that these four user evaluation outcomes have an impact on intentions for future use. Furthermore, prior work demonstrated that RA type and RA use influence users' decision-making effort and quality; it also showed that recommendation content influences users' evaluations of the RAs and subsequent decision-making; however, the effects of the nature of the content on the user's perceptions of both the recommendation and the system as well as the subsequent intention to accept future recommendations and adopt (use) the system is unknown.

Finally, only limited empirical research has begun to explore the effects of RS use on adoption intentions [8] and having used psychometric methods for additional validation. In this research, we aim to build on their [2] validated model and extend it by exploring the effects of message design using mixed-methods.

2.2 Message Design Components

The importance of message design is well-understood in domains ranging from advertising [5, 6] and marketing [7] to health information [8, 9]. Indeed, the domain of message design is highly interdisciplinary and existing research on message design shows the complexity of such design [7]. Underpinning message design are four groups of design principles, namely functional (e.g., problem definition), administrative (e.g., information access or cost), aesthetic (e.g., harmony), and cognitive principles (e.g., facilitating attention) [4].

In this study, given our focus on managerial decision-making, we focus on four message components that emerge from two groups of design principles, namely functional and cognitive. Functional principles deal with providing requisite information (e.g., about the problem or solution) and do so in a clear manner, as complicated language in message design is known to impair understanding [4]. Cognitive principles concern message design so as to facilitate attention and effective information processing [4]. The structure of messaging—i.e., the sequence in which information is presented is also part of these cognitive principles [4]. We manipulate three aspects of message design associated with the above functional and cognitive principles of providing information clarity, and facilitating attention and information processing, namely: the (lack of) specificity of the problem and/or description, and the sequencing of information (presenting problem or solution first). We also manipulate decision-making complexity given its importance in creating understandable language [10, p. 169].

2.3 Research Model and Hypotheses

Based on an integration of the RS and message design literatures discussed above, we propose 18 hypotheses reflected in Fig. 1 below.

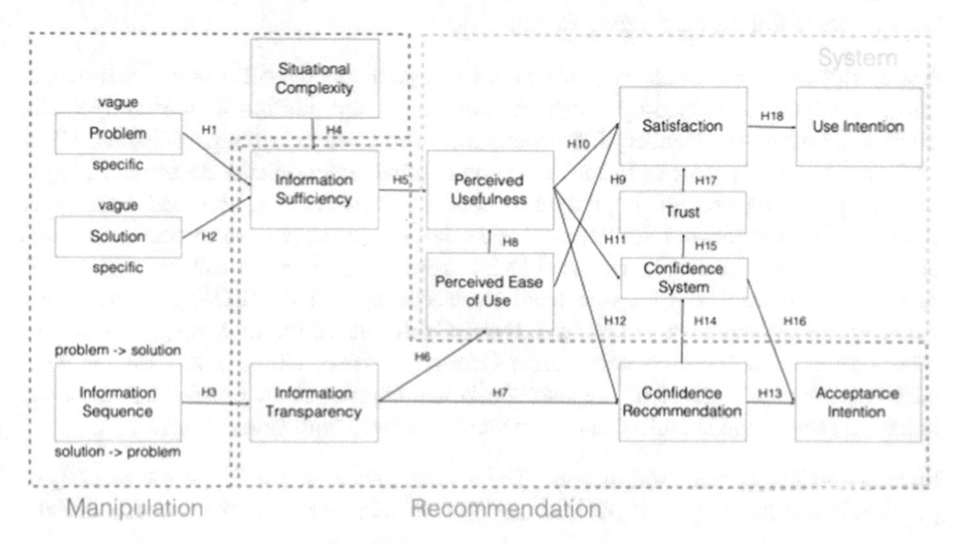

Fig. 1. Proposed research model

3 Methodology

Experimental Design: A four-factor, each with two levels (i.e., $2 \times 2 \times 2 \times 2$), between-within research design was employed in this study. Factors involved (i) Information Sequence (Problem-to-Solution or reverse); Information Specificity comprised of (ii) Problem Specificity (Vague vs. Specific) and (iii) Solution Specificity (Vague vs. Specific); and (iv) Decision Complexity (Simple Decision vs. Complex Decision). Participants were presented with 3 stimuli per treatment condition, for a total of 48 stimuli.

Participants: The pilot study gender-balanced sample involved six (6) participants, with ages between 23 and 26 years old (M = 24.33 years), all attending a large Business School in North America. They had normal or corrected-to-normal vision without glasses, and had never been diagnosed with epilepsy, or other health, neurological and psychiatric conditions. Participants were offered a $20 gift card as compensation.

Experimental Procedure and Stimuli: Using a scenario where participants acted as a restaurant manager, decisions related to inventory or order delivery situations had to be made on the basis of system recommendations. Decision-making varied in terms of the construction of the recommendation message in relation to its information sequence and specificity, but also in terms of the situation's complexity. Successive screenshots ($n = 48$) showing situations (i.e., a problem and a solution recommended by the recommender system) in text were used as stimuli. Two buttons («CONFIRM» and «DETAILS») corresponding to the two decision options available were shown below each recommendation. Participants had to either Confirm the recommendation as-is or request additional Details if unsure. The Details themselves would not be shown to the participant (something they were aware of during the briefing stage).

Apparatus and Measures: A comprehensive approach in the collection of neurophysiological data was enabled by a sophisticated, integrated multi-system setup [11, 12]. Tobii x60 (Tobii AB, Danderyd, Sweden) was used to capture the participants' USB-keyboard-entered responses to each decision-making situation and the associated eye-tracking providing gaze and pupil dilation data as a proxy for cognitive load. Facereader (Noldus, Wageningen, the Netherlands) and a desktop-based built-in webcam were used to record the emotional valence based facial expressions of participants throughout the experiment. Arousal was inferred from electrodermal activity (EDA) collected via a MP-150 Biopac Bionomadic (Santa Barbara, California). Participant responses to self-reported measures were collected via a Qualtrics (Provo, Utah, United States) web-based survey administered on the same desktop computer as the presented stimuli and in sequence with the stimuli. Constructs were measured with single items.

Survey and Instrument Validation: The questionnaire used in this study consists of previously validated scales [2, 13] measuring constructs shown in the research model.

4 Preliminary Results and Ongoing Work

Responding RQ1, preliminary analysis of three sets of neurophysiological data is presented below. First, valence was positively associated with Information Sufficiency ($b = .007$, $p < .0001$), solution specificity ($b = .008$, $p < .0001$), and usefulness ($b = .007$, $p < .0001$); similarly, arousal was negatively related to transparency ($b = -.052$, $p < .05$). Cognitive load was greater for vague rather than specific problems ($b = -.035$, $p < .0001$) and for vague rather than specific solutions ($b = -.029$, $p < .0001$). Also, the arousal slope associated with simple decisions was roughly one-half of the arousal slope for complex decisions; hence, an additive effect on arousal may be experienced in complex decision situations. Lastly, gaze tracking visualizations show users re-reading the solution when information is sequenced in the solution-to-problem format.

Regarding RQ2, Information Specificity was found to drive users to accept the recommendation significantly more so for messages providing problem specificity (H4: $b = 1.3157$, $p < 0.0001$) and solution specificity (H4: $b = 1.8626$, $p < 0.0001$).

Answering RQ3, all hypothesized relationships received strong statistical support (Hypotheses 1 through 18) and are presented in three sets of results below. First, information specificity (i.e., problem and solution specificity) impacted information sufficiency (respectively, H1: $b = .895$, $p < .001$; H2: $b = 1.423$, $p < .001$). Situational complexity and information sequence, respectively, negatively affected information sufficiency (H4: $b = -.548$, $p < 0.001$) and information transparency (H3: $b = -.687$, $p < .001$). Second, effects were shown for: information sufficiency on usefulness (H5: $b = .808$, $p < .001$); information transparency on ease of use (H6: $b = .458$, $p < .001$) and recommendation confidence (H7: $b = .934$, $p < .001$); and recommendation confidence on intention to accept the recommendation (H13: $b = .891$, $p < .001$), which was also affected by confidence in the system (H16: $b = .935$, $p < .001$). Third, ease of use impacted usefulness (h8: $b = .634$, $p < .001$) and satisfaction (H9: $b = .792$, $p < .001$). Usefulness influenced recommendation confidence (H12: $b = .780$, $p < .001$), system confidence (H11: $b = .843$, $p < .001$), and system satisfaction (H10: $b = .810$, $p < .001$). Recommendation confidence affected system confidence (H14: $b = .833$, $p < .001$), in turn, system trust (H15: $b = .77$, $p < .001$), and ultimately satisfaction (H17: $b = .873$, $p < .001$). Lastly, satisfaction affected the intention to use the RS (H18: $b = .949$, $p < .001$).

5 Discussion and Concluding Remarks

This paper reports the results of a pilot exploring the effects of three message design components—specificity of problem and solution descriptions and information sequence—on users' perceptions and attitudes towards the recommendation and the RS as a whole. Despite limitations in terms of sample size and research design (lack of counterbalancing), findings offer strong support of the importance of message design as a critical element in facilitating information processing, particularly in the context of managerial decision-making. Findings help extend RS literature by shifting the focus from system to message design. It underscores the importance of this overlooked design aspect, which is not just a critical antecedent of users' recommendation acceptance rather also of their attitude toward and intention to use the system. A complete neurophysiological data analysis from the study will be presented at the conference.

References

1. Xiao, B., Benbasat, I.: E-commerce product recommendation agents: use, characteristics, and impact. MIS Q. **31**(1), 137–209 (2007)
2. Pu, P., Chen, L., Hu, R.: A user-centric evaluation framework for recommender systems. In: Proceedings of the fifth ACM Conference on Recommender Systems, pp. 157–164 (2011)
3. Bigras, É., Léger, P.M., Sénécal, S.: Recommendation agent adoption: how recommendation presentation influences employees' perceptions, behaviors, and decision quality. Appl. Sci. **9**(20), 4244 (2019)
4. Petterson, R.: Introduction to message design. J. Vis. Literacy **31**(2), 93–104 (2012)
5. Coursaris, C.K., Sung, J., Swierenga, S.J.: Effects of message characteristics, age, and gender on perceptions of mobile advertising–an empirical investigation among college students. In: 2010 Ninth International Conference on Mobile Business and 2010 Ninth Global Mobility Roundtable (ICMB-GMR), pp. 198–205. IEEE, June 2010

6. Coursaris, C.K., Sung, J., Swierenga, S.J.: Exploring antecedents of SMS–based mobile advertising perceptions. Int. J. Electron. Financ. **6**(2), 143–156 (2012)
7. Coursaris, C.K., Van Osch, W., Balogh, B.A.: A social media marketing typology: classifying brand Facebook page messages for strategic consumer engagement. In: ECIS 2013, p. 46, June 2013
8. Spoelstra, S.L., Given, C.W., Sikorskii, A., Coursaris, C.K., Majumder, A., DeKoekkoek, T., Schueller, M., Given, B.A.: Feasibility of a text messaging intervention to promote self-management for patients prescribed oral anticancer agents. In Oncol Nurs Forum, vol. 42, no. 6, pp. 647–657, November 2015
9. Spoelstra, S.L., Given, C.W., Sikorskii, A., Coursaris, C.K., Majumder, A., DeKoekkoek, T., Schueller, M., Given, B.A.: Proof of concept of a mobile health short message service text message intervention that promotes adherence to oral anticancer agent medications: a randomized controlled trial. Telemed. e-Health **22**(6), 497–506 (2016)
10. Malamed, C.: Visual Language for Designers: Principles for Creating Graphics That People Understand. Rockport Publishers, Beverly (2009)
11. Léger, P.M., Sénecal, S., Courtemanche, F., Ortiz de Guinea, A., Titah, R., Fredette, M., Labonte-LeMoyne, E.: Precision is in the eye of the beholder: application of eye fixation-related potentials to information systems research. Assoc. Inf. Syst. (2014)
12. Léger, P.M., Courtemanche, F., Fredette, M., Sénécal, S.: A cloud-based lab management and analytics software for triangulated human-centered research. In: Information Systems and Neuroscience, pp. 93–99. Springer, Cham (2019)
13. Davis, F.D.: Perceived usefulness, perceived ease of use, and user acceptance of information technology. MIS Q. **13**(3), 319–340 (1989)

Nudging to Improve Financial Auditors' Behavior: Preliminary Results of an Experimental Study

Jean-François Gajewski[1], Marco Heimann[1], Pierre-Majorique Léger[2], and Prince Teye[1(✉)]

[1] Université de Lyon, Jean Moulin, iealyon, Magellan, Lyon, France
{jean-francois.gajewski,marco.heimann}@univ-lyon3.fr, prince.teye1@univlyon3.fr
[2] HEC Montréal, Montreal, Canada
pierre-majorique.leger@hec.ca

Abstract. This study investigates the impact of adapting Audit Management Information System (AMIS) user interface using nudges on the attentional behavior of auditors during the identification and diagnosis of audit evidence indicative of aggressive financial reporting. Specifically, in this preliminary phase of our multi-step research project, we investigate the visual behaviors of nudged vs. non-nudged auditors during evidence review. We test our predictions using eye-tracking, in a controlled experiment where participants are tasked with performing an audit of financial reporting in a AMIS. Results prove that nudged conditions are associated with longer average fixation duration, fixation counts and revisits of accounts with aggressive reports. By identifying the visual attention differences of nudged and non-nudged conditions, we highlight how contextually adapting user interface can draw on nudges to effectively enhance audit performance.

Keywords: Eye-tracking · Behavioral auditing · Nudges

1 Introduction

Audit management information systems (AMIS) are central to performing quality audits [1]. Despite legislation and regulations put in place to ensure the quality of audits, auditors remain human beings. As such they are subject to behavioural biases which could adversely affect the quality of audits. Since AMIS provides the context for auditor decisions, their design is paramount.

A recent suggestion to address these shortcomings of our brains is to use nudges [2]. Nudges consist in better designing the choice environment by attracting the person's attention in order to incite the person to take decisions in a predictable way. In the case of audit, nudges will consist in contextually altering the design of AMIS in a predictable and systematic way in order to improve the audit quality. According to Sunstein [3], successful nudges do not merely modify the environment but draw on a

F. D. Davis et al. (Eds.): NeuroIS 2020, LNISO 43, pp. 191–197, 2020.
https://doi.org/10.1007/978-3-030-60073-0_22

sound understanding of the cognitive processes underlying the behavioral change which is intended.

The main objective of this study is to explore the differences in the attentional characteristics induced by nudges in an aggressive reporting detection task. In order to highlight these differences, we use eye-tracking measures to explore visual attention differences induced by modifications of the AMIS. In a further step of our research program we intend to link those differences to successful identification of aggressive audit items. Aggressive financial reporting is the use of optimistic projections in the accounting standards to create financial statements that present a rosier picture of a company than is actually the case. These actions are taken to give the investment community a falsely enhanced view of a business, or for the personal gain of management.

This study enables us to understand how individual auditors analyse financial information during audits under various conditions and which conditions are best for aggressive reporting detection. The study aims at bridging the gap between the application of nudges to AMIS, and behavioral auditing. Our research so far offers both theoretical and practical contributions. With the rise of digitalization heralding the shift in auditing from mostly hardcopy materials to digital trails, the importance of understanding the traits of a skeptical auditor as per his attention measured with his eye movements on digital platforms cannot be overemphasized. The interaction of the auditor with AMIS interface plays a role in the quality of the work performed. Given that visual attention is to be kept at an acceptable level all throughout the audit engagement, the use of nudges as portrayed by this paper serves as an efficient tool to achieve desired levels. Then again, this paper tests a practical application of nudge theory as championed by Nobel winner Richard Thaler and Cass Sunstein to the field of auditing for which very little work has been done. The paper also shows that despite behavioral biases to which auditors may be subject, nudges could be used to serve as effective tools to remedy negative impacts which may arise there from. The next phase of our research will involve the use of professional auditors with work experience to test these hypotheses.

2 Prior Research and Hypotheses Development

Various studies have shown that auditors are subject to various cognitive biases which could interfere in audit [3–4]. Some of these cognitive biases are optimism bias [5], confirmation bias [6] and anchoring bias [4]. Given the potential negative impacts of cognitive biases, some studies have been done on mitigating the impacts of cognitive biases [5].

Previous studies of visual attention show that, on average, ineffective searches of target information are correlated with a higher number of fixations of the stimulus [7, 8], and that erroneous detections are associated with longer and more frequent fixations [3, 7, 9, 10]. Furthermore, chances for successful detection decrease with time [9], which could be due to cognitive resource depletion through stimulus encoding processes. In auditing, research suggests that financial auditors with high levels of professional skepticism exhibit a higher degree for information search when confronted with more aggressive reporting [11]. This translates to more time spent on the examination of audit evidence [12, 13].

Following Sunstein [2] we test two nudges, i.e. adaptation of the AMIS interface to reduce the negative effects of cognitive biases on the detection of aggressive reporting by auditors. A review of the nudge literature and an examination of applications across different domains conclude that nudges are effective in pushing people to choose desired results [2, 14]. The first nudge levels auditor's tendency to act similarly to other auditors and draws on social norms [14]. In the field of social psychology, social norms can be viewed as a tool to guide behavior in a certain situation or environment as "mental representations of appropriate behaviour". They can be cultural products (including values, customs, and traditions). The second nudge uses the justification technique to reduce the use of cognitive shortcuts (heuristics) by auditors. The justification technique consists in requiring participants to give a reasoned explanation of their judgements [6].

As a matter of fact, nudges may increase visual attention, because justification and social norms may induce more rational behaviour. This type of behaviour is likely to push the auditors to put more attention on the items of financial reporting. Based on eye-tracking and nudge literature, we propose three hypotheses:

H1 *Nudges Lead to More Visual Attention in Examining All Stimuli.*

H2 *Nudges Lead to More Visual Attention Paid to Aggressive Financial Reporting Items.*

H3 *Using More Nudges Will Lead to More Visual Attention in the Examination of Stimuli.*

3 Research Method

In this preliminary phase of our research project, we conducted a within-subject experiment with 12 participants with different ages and experience. All participants were undergraduate students in an important business school in North America and had attended several courses in auditing. All our participants were offered a $30 gift card as a compensation for their participation. The compensation was fixed for all participants. The experimental design was approved by the Institutional Review Board.

3.1 Research Design and Protocol

In order to confirm our hypotheses, we conducted a computerized experiment in which we tasked our participants with examining pieces of audit evidence adapted from Phillips [15]. The experiment lasted 27 min and followed a 2 (social norm) times 2 (justification) within subject semi-randomized design. By semi-randomized we mean that, to avoid cross contamination of the nudges, the experimental conditions were always presented in the following order: the first audit represented the control condition with no nudge followed by the other three conditions, which were the manipulations of the social norms nudge, justification nudge and a combination of both. These three manipulations were randomized for each participant. Thus, all participants were exposed to the control condition (without nudge) before being exposed to the nudge conditions to avoid endogeneity. Also, all the subjects read the nudges before the auditing task.

Consequently, each participant had to inspect, in the fictitious interface of an AMIS, four distinct sets of 14 items of auditing evidence emanating from 14 distinct accounts in the financial statements (for a total of 56 items of audit evidence classified under 14 distinct accounts of the financial accounts). Each set of audit evidence contained 3 items which represented aggressive financial reporting.

To mitigate the effect of prior knowledge of the business audited and the anchoring effect from the first attempt, different fictitious businesses were used in the four attempts. The stimuli were created using simple scenarii well-known to all individuals majoring in accounting. The ability to identify the aggressive financial reporting items was pretested and the results were satisfactory. To prepare the participants and mitigate the effect of learning through the trial, the experiment started with a short presentation of the company in question as well as basic information needed for the audit such as materiality and audit year. Again, the items of audit evidence were randomized in each of the four audits for all participants to eliminate the learning effect. After having examined all the 14 items of the audit evidence at their own pace, participants manually moved to the next page where they evaluated the level of aggressiveness of the entire financial reporting of the company in question and then performed a free recall task which involved the identification of the financial reporting items adjudged aggressive. After that, the Hurtt professional skepticism scale was administered. The Hurtt scale measures *ex ante* an individual's level of trait professional skepticism [16]. Demographic data was further collected for control purposes.

3.2 Apparatus and Measures

Eye-tracking (Red 250, SensoMotoric Instruments GmbH, Teltow, Germany) was used to gather the behavioral measures throughout the experiment, at a sampling frequency of 60 Hz. The number of fixations, which is the stabilization of the eye on an object [16], and their duration were gathered for each area of interest (AOI), as the literature tends to agree that fixation is related to the cognitive processing of visual information [16, 17]. The time before the first fixation in an AOI and the total view time of a stimulus were also collected. 14 AOIs were placed on the page for the fourteen financial account items. Due to the randomization on the page of account examination, AOIs could have different representations per participant and per attempt. The AOIs were also set for the nudges. For each participant, the eye-tracker was calibrated to a maximum average deviation of 0.5°, using a 9-points predefined calibration grid [18].

4 Preliminary Results

We briefly present several preliminary results from our study. Hypothesis 1 states that nudges lead to more visual attention in examining all stimuli. Fixation duration, similar to number of fixations and dwell time, represents the relative engagement with the object. The greater the value, the greater the level of visual attention [19]. A linear regression with random intercept model and a two-tailed level of significance is performed to compare the fixation count for the control condition and the manipulated conditions. Results suggest that audit tasks performed with nudges are associated with higher fixation counts than audit tasks performed without nudges: fixation counts seem lower for control condition than the justification nudge (t-value 13.69, $p < .0001$), fixation counts seem lower for the no-nudge condition compared with the nudge of social norms (t-value -1.88, p 0.0609), fixation counts seem lower for the no nudge condition compared with the condition where the two nudges of social norms and justification were combined (t-value -5.91, $p < .0001$).

Hypothesis 2, stipulates that nudges lead to more attention paid to aggressive financial reporting items. A major aim of the experiment was to nudge participants to increase their level of skepticism in such a way that more attention is paid aggressive financial reporting elements. Aside from the validation of the first hypothesis which showed a significant engagement in the examination of all stimuli, the target stimuli (aggressive financial reporting elements) received more visual attention from participants across nudged conditions compared with the non-nudged condition. This test was conducted using the Poisson regression with random intercept variable. The results showed that fixation counts were lower for non-target AOIs than target AOIs (t-value -5.13, $p < .0001$).

In order to test Hypothesis 3, we apply a linear regression with random intercept model and a two-tailed level of significance of Fixation Counts. The conditions of the social norms nudge and justification nudge are compared individually with the combination of both nudges. Results showed that fixation counts seem higher for the justification condition than for the combined condition (t-value 11.79 $p < .0001$). Fixation counts seem lower for the social norms condition than for the combined condition (t-value -4.06, $p < .0001$). This shows that hypothesis 3 is only partially fulfilled. Results seem to suggest that the justification nudge seems to be more effective. This is also corroborated by the number of revisits which is higher for the justification condition than all other conditions as tested with the Poisson regression with random intercept model.

5 Discussion and Conclusion

Our preliminary results suggest that H1, H2 are supported with H3 partially supported. Significant links between nudges and more visual attention in the examination of audit evidence (H1), and between nudges and more visual attention to aggressive financial reporting items (H2) are supported. Thus, this study presents evidence that the level of professional skepticism can increase significantly through the use of nudges.

Our research so far offers both theoretical and practical contributions. On the theoretical side, our results support the idea that increased visual attention induced by nudges is associated with higher detection of aggressive financial reporting. One explanation for that could be found in the reduction of the cognitive load behind the justification of the decisions [20, 21]. However, we find no preliminary pupillometric evidence. An alternative is that justification increases perceived accountability which requires auditors to go deeper in the analysis of items. In practice this finding could help adapt the design of AMIS through behaviorally informed nudges.

Limitations of this exploratory phase of our study center on the models used as stimuli for our experiment. While we tried to minimize the impact of individual knowledge variation in accounting by using basic audit evidence from a manufacturing firm which could be well-known to all potential participants, complexities which are more pronounced in today's financial reporting and audit fields are not taken into consideration. Another limitation can be found in our sample. A bigger and more equally distributed sample will allow more complex statistical analyses and provide more significant findings. Moreover, a dual sample with experts and non-experts could be more relevant to better analyze auditors' behavior.

Another limitation lies in the analysis of professional skepticism. This study could be extended by relating professional skepticism to the use of nudges.

Finally, no measure was taken to evaluate the 'stopping rule', which is when a subject decides to terminate his information search because he judges that he has enough information to complete his task [22–24]. The next step of our research will evaluate this concept in order to better understand eye-tracking data, especially the measures linked to the view time.

References

1. Pratap, K., Gregory, D.: Market Guide for Audit Management Solutions (2019). https://www.gartner.com/en/documents/3903083/market-guide-for-audit-management-solutions
2. Thaler, R., Sunstein, C.: Nudge: Improving Decisions About Health, Wealth, and Happiness. Rev. and expanded ed. Penguin Books, New York (2009)
3. Sunstein, C.: Human Agency and Behavioral Economics: Nudging Fast and Slow. Palgrave Macmillan, London (2017)
4. Knapp, M.C., Knapp, C.A.: Cognitive Biases in Audit Engagements: Certified Public Accountant. CPA J. **82**, 40 (2012)
5. Shanteau, J.: Cognitive heuristics and biases in behavioral auditing: review, comments and observations. Account. Organ. Soc. **14**(1), 165 (1989)
6. Hilton, D.: The psychology of financial decision making: applications to trading, dealing and investment analysis. J. Psychol. Financ. Markets **2**, 37–53 (2001)
7. McMillan, J.J., White, R.A.: Auditors' belief revisions and evidence search: the effect of hypothesis frame, confirmation bias, and professional skepticism. Account. Rev. **68**(3), 443–465 (1993)
8. Holmqvist, K., Nyström, M., Andersson, R., Dewhurst, R., Jarodzka, H., Van de Weijer, J.: Eye Tracking: A Comprehensive Guide to Methods and Measures. OUP, Oxford (2011)
9. Goldberg, J.H., Kotval, X.P.: Computer interface evaluation using eye movements: methods and constructs. Int. J. Ind. Ergon. **24**(6), 631–645 (1999)
10. Van Waes, L., Leijten, M., Quinlan, T.: Reading during sentence composing and error correction: a multilevel analysis of the influences of task complexity. Read. Writ. **23**(7), 803–834 (2009)
11. Henderson, J.M., Hollingworth, A.: The role of fixation position in detecting scene changes across saccades. Psychol. Sci. **10**(5), 438–443 (1999)
12. Fullerton, R., Durtschi, C.: The Effect of Professional Skepticism on the Fraud Detection Skills of Internal Auditors, 11 November 2004. SSRN. https://ssrn.com/abstract=617062. https://dx.doi.org/10.2139/ssrn.617062
13. Nkansa, P.C.: Professional Skepticism and Fraud Risk Assessment: An Internal Auditing Perspective. The University of Memphis, Ann Arbor (2016)
14. Dolan, P., Hallsworth, M., Halpern, D., King, D., Metcalfe, R., Vlaev, I.: Influencing behaviour: the MINDSPACE way. J. Econ. Psychol. **33**(1), 264–277 (2012)

15. Phillips, F.: Auditor attention to and judgments of aggressive financial reporting. J. Account. Res. **37**(1), 167–189 (1999)
16. Hurtt, R.K.: Development of a scale to measure professional skepticism. Auditing **29**(1), 149–171 (2010)
17. Yusuf, S., Kagdi, H., Maletic, J.I.: Assessing the comprehension of UML class diagrams via eye tracking. In: 15th IEEE International Conference, Program Comprehension, ICPC 2007, pp. 113–122 (2007)
18. Fehrenbacher, D.D., Djamasbi, S.: Information systems and task demand: an exploratory pupillometry study of computerized decision making. Decis. Support Syst. **97**, 1–11 (2017)
19. Just, M.A., Carpenter, P.A.: Eye fixations and cognitive processes. Cogn. Psychol. **8**(4), 441–480 (1976)
20. Rayner, K.: Eye movements in reading and information processing: 20 years of research. Psychol. Bull. **124**(3), 372 (1998)
21. Boutin, K. D., Léger, P. M., Davis, C. J., Hevner, A. R., Labonté-LeMoyne, É.: Attentional characteristics of anomaly detection in conceptual modeling. In: Information Systems and Neuroscience, pp. 57–63 (2019)
22. Riedl, R., Léger, P.M.: Fundamentals of NeuroIS. Studies in Neuroscience, Psychology and Behavioral Economics. Springer, Heidelberg (2016)
23. Nickles, K.R., Curley, S.P., Benson, P.G.: Judgment-Based and Reasoning-Based Stopping Rules in Decision Making Under Uncertainty, vol. 7285. University of Minnesota (1995)
24. Browne, G.J., Pitts, M.G.: Stopping rule use during information search in design problems. Organ. Behav. Hum. Decis. Process. **95**(2), 208–224 (2004)

Improving Driving Behavior with an Insurance Telematics Mobile Application

Perrine Ruer[1](✉), Sylvain Senecal[1], Mathieu Brodeur[1], Frédérique Bouvier[1], Alexander Karran[1], Marc Fredette[1], Thibaud Chatel[2], and Pierre-Majorique Leger[1]

[1] Tech3Lab, HEC, Montreal, Canada
{perrine.ruer,sylvain.senecal,mathieu.2.brodeur, frederique.bouvier,alexander-john.karran,marc.fredette, pierre-majorique.leger}@hec.com

[2] Insurance and Research Team, Desjardins Group, Montreal, Canada
thibaud.chatel@dgag.ca

Abstract. The research presented in this manuscript assesses the impact of the introduction of a mobile phone distraction feature included in a car insurer telematics mobile application. We analyze car drivers' behaviours and cognitive states using EEG and vehicular telemetry during a simulated driving task. Two groups of qualified drivers were compared: one group were provided awareness of feature introduction and one group with no awareness of the feature introduction. As a whole, our results suggest that the awareness of app feature updates may have led to the creation and internalization of better driving habits during a 3-month period and thereafter, an improvement in behavioral outcomes in the simulated environment. This research contributes to road safety research by assessing the cognitive and behavioural impact of telematics application features on driving behaviour and has practical implications for the automotive insurance sector.

Keywords: Driving behavior · EEG · Insurance telematics · Mobile application

1 Introduction

Advances in technology are changing the automobile insurance sector. In the past, auto insurance used personal (e.g., age, gender, prior driving experience) and vehicle information to identify probable risky clients. Nowadays, more information is available with GPS localization such as location, time of the day, traffic congestion, or mileage. New technologies have been developed based on GPS data combined with driving behavior habits [1–3] and folded in a new field referred to as insurance telematics. In the automotive insurance industry, telematics technologies are usually in the format of a fixed device inside the car (i.e., hardware) or a mobile application on the driver's smartphone (i.e., software) [1]. These technologies collect and analyze driving habit data (speed compliance, sudden stop, etc.) that traffic authorities report as "at-risk drivers" [2]. With this information, auto insurers calculate a score based on several trips defining the policyholder's safety profile and rating [1]. Car insurance companies now couple profiles

F. D. Davis et al. (Eds.): NeuroIS 2020, LNISO 43, pp. 198–203, 2020.
https://doi.org/10.1007/978-3-030-60073-0_23

and ratings with telematics technologies to offer financial incentives to policyholders to encourage safer driving behaviors [2, 3].

Our research question examined how a telematic mobile application can change the driver's behavior. Specifically, we focus on how the introduction of a new feature (driver distraction measurement) into the mobile application influences driver behaviors. We conducted a longitudinal laboratory NeuroIS experiment to test drivers' behaviors before and after the introduction of the new driving distraction feature, which notifies the driver if they touch the phone screen (for instance, if the driver tries to send a text message). Before the introduction of the new feature, the mobile app provided personalized feedback on the user's driving and provided a score after five journeys. The mobile app measures speed, fast acceleration and hard braking. The new feature targets distraction caused by mobile phone usage while driving, a significant cause of road accidents. Its goal is to promote change driving habits in order for people to restrict cell phone use while driving.

In this study, we assess drivers' driving behaviors and cognitive activity in a simulated driving task using electroencephalography (EEG) before and after the introduction into the mobile application of the new distraction feature. The results suggest that when drivers are consciously aware of the introduction of the distraction feature, their driving behavior and cognitive state while driving appear to be less risky in the simulated environment when tested three months later. These findings have implications for the auto insurance industry and contribute to the literature on policyholders' behaviors and the influence of the telematics app.

2 Methodology

We conducted a laboratory-based two-phase test-retest study. Both data collections, i.e., Phases 1 and 2, followed an identical protocol. All participants were current users of the Desjardins Insurance mobile app. Participants were required to drive in a fixed driving simulator utilising control devices (force-feed-back steering wheel, gear lever, gas and brake pedals) as well as visual and auditory rendition devices. After phase 1 participants were split into two groups, group 1 were provided knowledge of an upcoming app update and group 2 were not. The study was approved by our institution's research ethics committee, and written informed consent was obtained from all subjects prior to participation. All participants were compensated upon completion of both phases of the experiment.

Participants. Nineteen participants with normal to corrected-to-normal vision, who had used the existing telematics mobile (Desjardins Insurance) app for >3 months were recruited through a research organization. Five participants were excluded due to simulator sickness in phase 1 or other technical problems with tools during data collection. Thus, a total of fourteen drivers participated in both phases (7 males, 6 females[1]; aged 20–43, $\mu = 32.4$ years, SD = 8.2 years). All participants possessed a valid driving license ($\mu = 10.9$ years, SD = 7.7 years), with an average of 50 to 100 km driven per week.

[1] One participant did not state his/her gender.

Apparatus and Measures. The driving simulator scenarios were generated using City Car Driving software. The road environment was displayed on three screens, providing a 120° field of vision. Telemetry data were recorded within the simulation software. Three types of data were recorded: simulator data (pedals, gearbox and steering wheel positions, speed and position of the simulator), traffic data (position, speed, braking) and scenario data.

EEG data were recorded and sampled at 1,000 Hz from 32 Ag-AgCl preamplified electrodes mounted on the actiCap and with a brainAmp amplifier (Brainvision, Morrisville). Syncronization of EEG data and behavioral data was performed using guidelines from [4, 5] and analysed using EEGLAB (San Diego, USA) and Brainvision (Morrisville, USA).

Procedure. Upon arrival at the laboratory, participants signed an informed consent form after which participants were asked to complete a psychometric battery consisting of (age, gender, years of education, driving habits, behaviors frequency, distractions while driving, assessment of other drivers and oneself). After completion, EEG sensors were placed on the participant. Instructions were then given to all participants to drive like they would in real life respecting speed limits, and the road safety code. Participants are then asked to perform four tasks, which ascend in order of difficulty (i.e. additional stimuli). The simulated driving scenario lasted approximately 25 minutes per participant with each subtask lasting approx. 5 min. After the driving task, participants completed another psychometric battery consisting of a questionnaire and a semi-structured interview concerning their driving experience. After completion of the first phase of the study, at the end of the interview, participants were split into two groups one group $n =$ 7 were informed that a driver distraction measure involving mobile phone usage was to be implemented within the Desjardins Insurance application and the other group were not so informed. For the purposes of analysis, the two groups were named "aware" and "unaware").

All participants were invited to perform the same experiment three months after the first phase. During the three-month interval, the mobile app was updated with the introduction of the new feature. The experimental procedure was identical during the second phase. During the post experiment interview, in order to verify that whether participants were conscious of the new feature, participants were asked if they noted the mobile app's version number, and if they had seen and read the application's information messages about the cell phone distraction. To analyze the data, we examined specific driving behaviors for both groups and compared the cognitive state with electroencephalography (EEG) recordings before and after Desjardins Insurance mobile app update.

3 Results

With regards to driving behavior, participants who were aware of the update of the mobile app had fewer accidents during Phase 2 (mean = 0.8) than in Phase 1 (mean = 1.3) compared to those who were not aware of the update. This latter group had more accidents in Phase 2 (mean = 2.05) than in Phase 1 (mean = 1.15). We did a logistic regression with mixed effect model (2-tailed p-value). The difference (aware vs unaware) in phase 2 vs phase 1 is significant ($r = -1{,}95$, $p < 0.05$) (Fig. 1). Furthermore, sudden braking and sudden accelerations trend in the same direction as the number of accidents during simulated driving, with a lower average number during the second phase than the first phase (average number of sudden braking = 6.05 vs 5.3; sudden accelerations = 5.9 vs 4.6). However, this difference is non-significant.

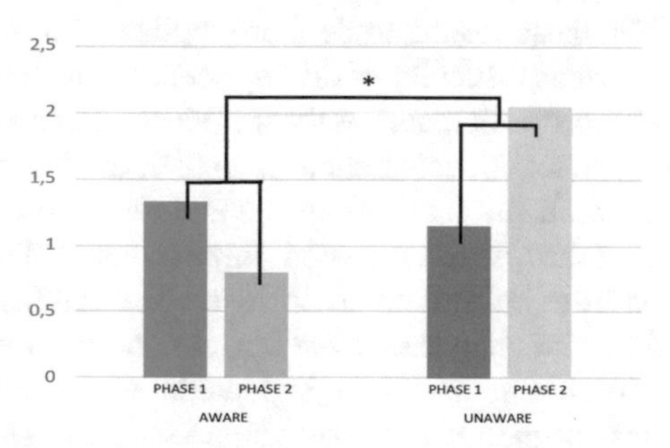

Fig. 1. Average number of accidents in the driving simulation per phase and awareness of the update (* statistically significant result $p < 0.05$)

For EEG, we performed a mixed effect linear regression model analysis. The results show that the difference (before-after) for α (alpha) and θ (theta) frequency bands is greater for participants who were aware of the Desjardins Insurance mobile app update ($p < 0.05$). Conversely, we observed the opposite for the β (beta) frequency band where the difference (before-after) is smaller for participants who were aware of the update ($p < 0.1$). In this context, interpreting increases in α and θ may signify that the driver is more aware and alert with a heightened level of vigilance. As for β frequency activity, a decrease may signify a more relaxed mental state [6, 7]. When level of difficulty is controlled for within the analysis, participants who were aware of the update of Desjardins Insurance mobile app had higher α and θ and lower β than those who were not aware of the update. The cognitive state of drivers who are aware of the mobile app update suggests that they were both more vigilant and potentially less stressed, as expressed by driving behavior metrics and positive outcomes.

Taken together, these results suggest that the awareness of the update of mobile app may have led to the creation and internalization of better driving habits during a 3-month period and thereafter, an improvement in behavioral outcomes in the simulated driving environment, in terms of a reduced number of accidents, sudden braking and sudden acceleration events.

4 Discussion - Conclusion

Our aim was to understand the impact of a new distraction feature of an insurance telematics mobile app on the policyholders' driving behaviors. The policyholders' behaviors and cognitive states in a driving simulator were analyzed to ascertain the effect of an app update aimed at assessing driver distractibility due to mobile phone usage.

Our results show that in this case when participants were made aware of the introduction of the new distraction feature in the mobile app, their driving behaviors changed positively in terms of driving safety profile, that is, reduced risk-taking while driving. Participants in the aware group drove more safely in the driving simulator with a reduction in the number of recorded accidents, sudden acceleration and sudden braking events. With regard to cognitive state the EEG results suggest that participants in the aware group appear from these results to be more vigilant and in better control of the vehicle throughout the various difficulty of driving scenario. It would appear from these results that a simple notice to participants of the introduction of a new feature appears to have had a far-reaching impact, inducing changes in their driving habits. These changes potentially reflect an internalization of the update and its personal implications on the part of participants during the test-retest period (three months). There is some evidence to support this assertion from information security studies and the saliency of training, which showed that behavioral change can last upwards of two weeks or more after an intervention [8]. However, in this case there is potentially an additional confound concerned with financial loss or gain which may help explain our results, all participants receive insurance from the supplier of the mobile app, and in this instance users of the app receive better (lower) premiums dependent on driving behaviors that are judged "safer" than others.

Therefore, we posit that the awareness of upcoming new features appears to change driver behavior, providing a positive impact on metrics of driving safety. We suggest that there is additional value in promoting upcoming updates to telematics mobile applications, which may encourage safer driving and thus reduce dangerous road incidents. The auto insurance sector could propose some personal advice for "risky" drivers or provide reports to policyholders outlining their traffic violations to encourage safer behaviors or changes in driving habits. However, this may have ethical considerations concerning data aggregation, usage and privacy.

Given the study was limited to 14 participants, we cannot exclude that other factors such as personality could play a role in explaining our results and future research should use additional measures to exclude confounding factors. Also, in our sample, participants had not been using the mobile application for the same duration before phase 1. Future research should target new users who are not familiar with the mobile application to investigate if there are also changes in driving behavior for new users.

To conclude, the introduction of a mobile phone distraction feature in a car insurer telematics application appears to have a positive impact on policyholders' driving habits. Participants, who were made aware of the mobile app update, appear to have internalized new safer driving habits over a period of 3 months.

References

1. Handel, P., et al.: Insurance telematics: opportunities and challenges with the smartphone solution. IEEE Intell. Transp. Syst. Mag. **6**(4), 57–70 (2014)
2. Ayuso, M., Guillen, M., Nielsen, J.P.: Improving automobile insurance ratemaking using telematics: incorporating mileage and driver behaviour data. Transportation **46**(3), 735–752 (2018)
3. Zhao, Y.: Telematics: safe and fun driving. IEEE Intell. Syst. **17**(1), 10–14 (2002)

4. Léger, P.M., Sénecal, S., Courtemanche, F., de Guinea, A.O., Titah, R., Fredette, M., Labonte-LeMoyne, É.: Precision is in the eye of the beholder: application of eye fixation-related potentials to information systems research. Association for Information Systems (2014)
5. Léger, P.M., Courtemanche, F., Fredette, M., Sénécal, S.: A cloud-based lab management and analytics software for triangulated human-centered research. In: Information Systems and Neuroscience, pp. 93–99. Springer, Cham (2019)
6. Campagne, A., Pebayle, T., Muzet, A.: Correlation between driving errors and vigilance level: influence of the driver's age. Physiol. Behav. **80**(4), 515–524 (2004)
7. Yang, L., Ma, R., Zhang, M., Guan, W., Jiang, S.: Driving behavior recognition using EEG data from a simulated car-following experiment. Accid. Anal. Prev. **116**(2018), 30–40 (2018)
8. Zafar, H., Randolph, A., Gupta, S., Hollingsworth, C.: Traditional SETA no more: investigating the intersection between cybersecurity and cognitive neuroscience. In Proceedings of the 52nd Hawaii International Conference on System Sciences (2018)

Understanding the Mental Recovery Effect of Cyberloafing: Attention Replenishment and Task-Set Inertia

Hemin Jiang(✉)

University of Science and Technology of China, Hefei, People's Republic of China
hmjiang@ustc.edu.cn

Abstract. Many employees justify their cyberloafing (i.e., non-work-related Internet use during work time) behavior as a mental break. However, there is little empirical research to examine the mental recovery effect of cyberloafing. This study aims to design a lab experiment to investigate the impact of cyberloafing on employee mental fatigue and task productivity. The study also aims to compare cyberloafing with a traditional means of mental breaks (i.e., walking outside for a while) in alleviating mental fatigue and improving productivity. The expected findings of this study are (1) cyberloafing can help employees reduce mental fatigue to some extent by replenishing their attentional resources; however, (2) compared with walking outside for a while, the mental recovery effect of cyberloafing may not be so good because it may take employees more time and effort to switch their attention from cyberloafing (than from walking outside) back to the work task. Neuroscience tools will be employed to support the expected findings above.

Keywords: Cyberloafing · Mental recovery · Task-set inertia · Task productivity · EEG · Eye tracker

1 Introduction

IT devices that are connected to the Internet, such as desktops, laptops, tablets, and smartphones, are playing a pivotal role in organizations. These IT resources have greatly facilitated employees to perform job tasks. At the same time, it also becomes increasingly common for employees to engage in cyberloafing, which is defined as employees' Internet usage for non-work-related purposes during work time [1, 2]. Examples of cyberloafing include surfing news sites, non-work-related emailing, visiting social network sites (SNS), booking personal travel, online stock-trading, shopping, gaming, and chatting, to name just a few.

Cyberloafing is very common in contemporary organizations. Recent studies suggest that employees spend 1–2 hours per workday on cyberloafing, accounting for up to 30% of their work time [3, 4]. Compared with other traditional non-work-related activities in the workplace, such as long lunch breaks and socializing with coworkers, cyberloafing

F. D. Davis et al. (Eds.): NeuroIS 2020, LNISO 43, pp. 204–210, 2020.
https://doi.org/10.1007/978-3-030-60073-0_24

does not require employees to be physically absent from the office and is thus often not as visible as traditional non-work-related behaviors [2, 5], this partially explains why cyberloafing is currently the main form of non-work-related behavior in the workplace [6]. Given the substantial amount of time employees spend on cyberloafing, many employers consider cyberloafing as a threat to employee productivity [3, 5].

Many employees claim that they engage in cyberloafing because they would like to have a mental break during work, and the mental break can increase productivity. However, the mental recovery effect of cyberloafing has not been fully examined. First, the underlying mechanism of the mental recovery effect of cyberloafing has not been fully discussed and empirically examined. Although some studies suggest that cyberloafing can replenish employees' attention so that they can better concentrate on the subsequent task [7–9], other studies suggest that cyberloafing can be a constant distraction to employees because it is easy for employees to switch their attention from work tasks to cyberloafing, but it is relatively difficult to switch attention from cyberloafing back to work tasks [10]. Such an attention switching cost can dampen the mental recovery effect of cyberloafing. Unfortunately, previous studies have paid little attention to employees' attention switching process in studying cyberloafing behavior. Second, it is not known from previous studies whether the mental recovery effect of cyberloafing is better or worse than traditional means of mental breaks. Comparing different means of mental breaks is important because such a comparison can help organizations and employees adopt appropriate interventions to regulate cyberloafing behavior.

To fill these research gaps, we are planning to design a lab experiment to understand and examine the neuro mechanism of the mental recovery effect of cyberloafing. We will study employees' attention switching process when they start and end their cyberloafing behavior as a mental break, and examine how cyberloafing can replenish participants' attention and how difficult it is to switch attention from cyberloafing back to work tasks. We will also compare the effect of cyberloafing with a traditional means of mental breaks (i.e., walking outside for a while) on participants' mental fatigue and subsequent task productivity. We will employ neuroscience tools to capture participants' attention switching process during the experiment.

The rest of this paper is organized as follows. In the next section, we discuss the theoretical background and present the research hypotheses. We then discuss the experiment design as well as data collection and analysis methodology. We conclude by discussing the potential theoretical and practical implications of our paper.

2 Theoretical Background and Research Hypotheses

According to attention restoration theory [11], individuals' attentional resources are limited at a given time. Therefore, individuals' attentional resources will be depleted and individuals are subject to mental fatigue after a certain period of performing cognition intensive tasks. As a result, individuals' concentration declines over time as attentional resources are consumed [11]. For example, previous studies have found that individuals begin to lose concentration after 5 to 15 min [12]. When individuals feel mental fatigue, activities that do not need to consume attentional resources can help replenish and restore individuals' attentional resources and improve concentration [13]. For example, compared with job tasks, most cyberloafing activities (such as surfing general news websites, visiting social networking sites, playing online games) require few cognitive and attentional resources, and therefore, can be a useful means for individuals to replenish their attention resources. The replenished attentional resources allow individuals to better concentrate on the subsequent task, which can improve task productivity. On the one hand, high concentration on a task facilitates individuals to select information and skills that are most relevant to the focal task. On the other hand, high concentration on a task can also help individuals to inhibit other stimuli or distractions that are not relevant to the focal task. Both selecting relevant information and skills and inhibiting irrelevant distractions are beneficial to individuals' performance of the focal task, and thus improve the task productivity [14]. Therefore, we propose the following hypotheses:

H1: Cyberloafing as a mental break can improve individuals' concentration on the subsequent task.
H2: Cyberloafing as a mental break can improve individuals' productivity of the subsequent task.

Although activities that do not consume cognitive resources can help individuals restore their attentional resources, the mental recovery effect can be attenuated by the difficulty of switching attention from the break activities back to work tasks. Switching attention from one task/activity to another tends to be difficult, and subsequent task performance easily suffers [15]. The reason for the attention switching difficulty can be explained by the concept of task-set inertia. Specifically, individuals always have a task-set in mind when performing a task [16], referring to the organization of cognitive processes and mental representations that enable the person to act in accordance with task requirements. Because an individual's cognitive resources are limited at a certain time [17, 18], they are hardly able to focus simultaneously on multiple tasks. Therefore, task-set reconfiguration is needed when switching between tasks [19]. However, the task set reconfiguration is sometimes not easy due to the task-set inertia [20]. As a result, a task-switching cost occurs in the form of extra time and effort that individuals need to complete the switched task [21, 22]. The task switching cost can be substantial, even if the tasks are rather simple [16].

We believe a good mental break should facilitate individuals to restore their attentional resources and, at the same time, produce low "task switching cost" when individuals switch back to work tasks. Therefore, it is important to compare the mental recovery effect of cyberloafing with that of traditional means of mental breaks. In this paper, we choose walking outside of the office as the traditional means of mental break to compare with cyberloafing, because walking outside is relatively less constrained physically and environmentally, and thus it is available for most, if not all, employees that work in offices. We compare the two means of mental break from the perspective of task-set inertia.

Generally speaking, the higher the test-set inertia, the more difficult of the task-set reconfiguration and attention switching. According to previous literature, two factors that affect task-set inertia can be important in the context of mental break: break activity arousing level and external stimuli to distinguish the work task and break activity. First, task-set inertia is higher when attention is switched from a task with high arousing level than when attention is switched from a task with a low arousing level. Arousal refers to a continuum of states of energization with physiological, subjective and behavioral expressions [23]. For example, compared with surfing general news, playing online games is more arousing because it involves real-time interactions among multiple players (or between the player and the computer). As a result, it is more difficult for employees' to switch their attention from playing computer games (relative to surfing news) back to work tasks [24]. Similarly, compared with walking outside of the office, most cyberloafing behaviors are more emotionally pleasurable and are thus more arousing. In this sense, the task-set inertia of cyberloafing is higher than that of walking outside for a while.

Second, task-set inertia is relatively high when switching between two tasks without external stimuli to differentiate from each other [25]. For example, both cyberloafing and many work tasks involve using computers and the Internet. Therefore, the external stimuli to distinguish from each other are inconspicuous. By contrast, walking outside of the office and come back from outside to office involve some physical movements (e.g., standing from the office chair, walking out, sitting back on the office chair), which can be a relatively strong stimuli to differentiate the two tasks (i.e., walking out and wok tasks). Therefore, from the perspective of external stimuli, the task-set inertia of cyberloafing is higher than that of walking outside for a while.

Taken together, task-set inertia of cyberloafing is higher than that of walking outside for a while, and therefore, switching attention from cyberloafing back to work tasks can be more difficult than switching attention from walking outside back to work tasks. The high task-set inertia is likely to result in two consequences. First, as mentioned earlier, it may take long time for individuals to focus on the work task after cyberloafing. Second, individuals have to devote effort to overcome the task-set inertia, which may consume part of individuals' replenished cognitive resources, leaving relatively less cognitive resources for the subsequent work task. As a result, the concentration duration will be shortened. Naturally, both the long time required for task switching and the short concentration duration on the subsequent task will negatively affect the subsequent task productivity. Therefore, we propose the following three hypotheses:

H3: Switching attention from cyberloafing to the subsequent task may take longer time than switching attention from walking outside back to the subsequent task.
H4: Individuals' concentration duration will be longer after walking outside as a mental break, compared with the concentration duration after cyberloafing as a mental break.
H5: The subsequent task productivity of individuals who walk outside for a while as a mental break is better than the task productivity of individuals who engage in cyberloafing as a mental break.

3 Experiment Design and Hypotheses Testing

We will conduct a laboratory experiment to test the hypotheses above. Power analysis would be conducted to determine the minimum numbers of participants. We would invite university students as the experiment participants. The experiment procedure is as follows.

Participants will be randomly divided into three groups, including one control group and two experimental groups. All participants will be asked to perform task1 (see details in the next paragraph) so that they are subject to mental fatigue after task1. Participants in the three groups are then placed in different conditions of mental breaks. Specifically, participants in the control group will have no break and immediately start performing task2 (which is a similar task to task1, and see also the details in the next paragraph) after the completion of task1. Participants in the experiment group1 will surf online news for 10 min as a mental break before starting to perform task2. Participants in the experiment group2 will walk outside of the room for 10 min as a mental break before starting to perform task2. We will not inform participants about the break schedule in advance so that they will not have an expectation of such a mental break. The expectation can confound the mental recovery effect.

Task1 and task2 are two 20 min tasks that require individuals' attention and concentration. We plan to use Toulouse–Piéron attention test (TP test) as task1. TP test requires participants to find the targeted graphic symbols among a larger group of very similar symbols. Previous studies suggest that participants are likely to feel mental fatigue after performing the TP test for 20 min [9]. We plan to choose a psychomotor vigilance task (PVT) as the task2. PVT is a reaction-timed task that measures the speed with which subjects respond to a visual stimulus. PVT is widely used in literature to study individuals' attention and concentration [26]. PVT is a simple, reliable and highly sensitive task for measuring attentional and performance deficits due to fatigue [27]. Therefore, PVT is a suitable task for our study.

4 Construct Measurements and Data Analysis

The key constructs in the experiment include participants' concentration and task productivity. We will employ two neuroscience tools to measure participants' attentional state during the experiment, including eye tracker and electroencephalogram (EEG). In terms of eye tracker, previous studies suggest that pupil diameter is positively related to individuals' attentional state, such that the larger of individuals' pupil diameter indicates higher individuals' concentration [28, 29], so individuals' concentration can be measured by their pupil diameter. In terms of EEG, previous studies suggest that different frequency bands of EEG signals can be an indicator of individuals' attentional state [30]. Therefore, participants' attentional state (and concentration) can also be measured by the frequency bands of EEG signals. In other words, we will combine the data of eye tracker and EEG to measure participants' concentration. Participants' task productivity will be measured via two dimensions, namely, task accuracy rate and response time. We will also control some variables that may affect participants' cognitive control abilities such as big five trait taxonomy [31]. Mix-ANOVA will be conducted to compare the differences of the three groups with respect to participants' attentional state (including attention switching time and concentration sustaining duration) and task performance.

References

1. Liberman, B., et al.: Employee job attitudes and organizational characteristics as predictors of cyberloafing. Comput. Hum. Behav. **27**(6), 2192–2199 (2011)

2. Khansa, L., et al.: To cyberloaf or not to cyberloaf: the impact of the announcement of formal organizational controls. J. Manage. Inf. Syst. **34**(1), 141–176 (2017)
3. Agarwal, U.: Impact of supervisors' perceived communication style on subordinate's psychological capital and cyberloafing. Australas. J. Inf. Syst. **23**, 1–27 (2019)
4. Askew, K.L., et al.: Disentangling how coworkers and supervisors influence employee cyber-loafing: what normative information are employees attending to? J. Leadersh. Organ. Stud. **26**, 154805181881309 (2018)
5. Wagner, D.T., et al.: Lost sleep and cyberloafing: evidence from the laboratory and a daylight saving time quasi-experiment. J. Appl. Psychol. **97**(5), 1068 (2012)
6. Ivarsson, L., Larsson, P.: Personal internet usage at work: a source of recovery. J. Workplace Rights **16**(1), 1–19 (2011)
7. Coker, B.L.: Freedom to surf: the positive effects of workplace Internet leisure browsing. New Technol. Work Employ. **26**(3), 238–247 (2011)
8. Coker, B.L.S.: Workplace internet leisure browsing. Hum. Perform. **26**(2), 114–125 (2013)
9. Santos, A.S., Ferreira, A.I., da Costa Ferreira, P.: The impact of cyberloafing and physical exercise on performance: a quasi-experimental study on the consonant and dissonant effects of breaks at work. Cogn. Technol. Work **22**, 1–15 (2019)
10. Jiang, H.: Employee personal Internet usage in the workplace. Jyväskylä Stud. Comput. **257**, 117 (2016)
11. Kaplan, S.: The restorative benefits of nature: toward an integrative framework. J. Environ. Psychol. **15**, 169–182 (1995)
12. Warm, J.S., Parasuraman, R., Matthews, G.: Vigilance requires hard mental work and is stressful. Hum. Factors **50**(3), 433–441 (2008)
13. Berto, R.: Exposure to restorative environments helps restore attentional capacity. J. Environ. Psychol. **25**(3), 249–259 (2005)
14. Kaplan, S., Berman, M.G.: Directed attention as a common resource for executive functioning and self-regulation. Perspect. Psychol. Sci. **5**(1), 43–57 (2010)
15. Leroy, S.: Why is it so hard to do my work? The challenge of attention residue when switching between work tasks. Organ. Behav. Hum. Decis. Process. **109**(2), 168–181 (2009)
16. Kiesel, A., et al.: Control and interference in task switching—a review. Psychol. Bull. **136**(5), 849 (2010)
17. Wickens, C.D.: The effects of divided attention on information processing in manual tracking. J. Exp. Psychol. Hum. Percept. Perform. **2**(1), 1 (1976)
18. Eriksen, C.W., James, J.D.S.: Visual attention within and around the field of focal attention: a zoom lens model. Percept. Psychophys. **40**(4), 225–240 (1986)
19. Rogers, R.D., Monsell, S.: Costs of a predictable switch between simple cognitive tasks. J. Exp. Psychol. Gen. **124**(2), 207 (1995)
20. Alport, A., Styles, E.A., Hsieh, S.: 17 shifting intentional set: exploring the dynamic control of tasks (1994)
21. Wylie, G., Allport, A.: Task switching and the measurement of "switch costs". Psychol. Res. **63**(3–4), 212–233 (2000)
22. Hsieh, S., Liu, L.-C.: The nature of switch cost: task set configuration or carry-over effect? Cogn. Brain. Res. **22**(2), 165–175 (2005)
23. Matthews, G., Amelang, M.: Extraversion, arousal theory and performance: a study of individual differences in the EEG. Pers. Individ. Differ. **14**(2), 347–363 (1993)
24. Reinecke, L.: Games at work: the recreational use of computer games during working hours. CyberPsychol. Behav. **12**(4), 461–465 (2009)
25. Pashler, H., Johnston, J.C., Ruthruff, E.: Attention and performance. Annu. Rev. Psychol. **52**(1), 629–651 (2001)
26. Ariga, A., Lleras, A.: Brief and rare mental "breaks" keep you focused: deactivation and reactivation of task goals preempt vigilance decrements. Cognition **118**(3), 439–443 (2011)

27. Lim, J., et al.: Imaging brain fatigue from sustained mental workload: an ASL perfusion study of the time-on-task effect. Neuroimage **49**(4), 3426–3435 (2010)
28. Unsworth, N., Robison, M.K.: Pupillary correlates of lapses of sustained attention. Cogn. Affect. Behav. Neurosci. **16**(4), 601–615 (2016)
29. Hopstaken, J.F., et al.: A multifaceted investigation of the link between mental fatigue and task disengagement. Psychophysiology **52**(3), 305–315 (2015)
30. Okogbaa, O.G., Shell, R.L., Filipusic, D.: On the investigation of the neurophysiological correlates of knowledge worker mental fatigue using the EEG signal. Appl. Ergon. **25**(6), 355–365 (1994)
31. Ophir, E., Nass, C., Wagner, A.D.: Cognitive control in media multitaskers. Proc. Natl. Acad. Sci. **106**(37), 15583–15587 (2009)

How to Measure Customers' Emotional Experience? A Short Review of Current Methods and a Call for Neurophysiological Approaches

Anna Hermes[1(✉)] and René Riedl[1,2]

[1] Department of Business Informatics–Information Engineering, Johannes Kepler University Linz, Linz, Austria
anna.hermes@jku.at
[2] University of Applied Sciences Upper Austria, Steyr, Austria
rene.riedl@fh-steyr.at

Abstract. In the digital age, retailers compete through various sales channels, both online and offline, with the effect that the customers' experiences have increasingly gained attention in the omnichannel era. Specifically, customer emotions have become an important topic, because they affect attitudes towards products and services as well as purchase decisions. While the phenomenon of customer experience is widely researched, surprisingly, to the best of our knowledge, no peer-reviewed journal publication exists that has studied the phenomenon from a NeuroIS angle. Against this background, we conducted a short literature review to obtain an overview of NeuroIS methods used to study customer behavior in a shopping and retailing context. Further, we outline a brief research agenda, thereby addressing the possible use of NeuroIS approaches in the context of customers' emotional experiences in retail.

Keywords: Customer experience · NeuroIS · Retail · Emotions

1 Introduction

Due to newly available channels to shop, including the internet, augmented reality, and smartphone apps, there is a need for retailers to provide a so-called omnichannel experience through the integration of various shopping channels [1–3]. Further, a flawless customer experience (hereafter CX) is said to be one of the keys for successful future retailers, hence, it is crucial for retailers to learn about the drivers of such an experience [4]. Hence, to better manage CX, it is critical for retailers to develop an in-depth understanding of CX to increase satisfaction and to optimize retail efforts [4, 5].

In the early 1980s, Holbrook and Hirschman [6] made the point that buying decisions are not only based on logical thinking. Since then, the knowledge and the interest in the personal determinants of CX have grown steadily [5]. More recently Lemon and Verhoef

F. D. Davis et al. (Eds.): NeuroIS 2020, LNISO 43, pp. 211–219, 2020.
https://doi.org/10.1007/978-3-030-60073-0_25

[4], for example, defined CX as "a multidimensional construct focusing on a customer's cognitive, emotional, behavioral, sensorial, and social responses to a firm's offerings during the customer's entire purchase journey" (p. 71). Hence, CX is multi-dimensional, yet, the literature on the conceptualization of CX is both diverse and fragmented [7].

Another issue frequently raised by researchers is the lack of a robust measurement of CX in combination with an explicit call for neuroscientific approaches in CX and emotion research [4, 8]. A recent literature review analyzed 45 retail CX research papers and, after examining 22 empirical studies in detail, concluded that emotions play a particular role for CX, and, that the research domain is dominated by self-report measures [7].

Yet, since self-report measures might not be able to properly measure emotions [9], the purpose of the present study is to review the literature to gain a better understanding of NeuroIS methods used to measure a customer's emotional retail experience. We hope that the findings of this research inspire new CX research under consideration of NeuroIS approaches.

2 Customer Experience and Emotions

Palmer [10] points out that CX can either be seen as a noun, highlighting the *outcome*, the retrospective, of the participation in an event or activity, or it can be seen as a *verb*, highlighting the feelings and involvement during the actual event or activity. Building on a recent literature review on the nature of CX [7], it can be concluded that there is no clear understanding of what an experience is and how it is best measured. In their conceptualizations, CX researchers included a vast majority of factors from retailer-related factors (e.g., atmospheric store experience, product presentation, and social experience) as well as various psychological factors (e.g., cognition). Yet, emotion-related concepts were the most studied factors in this context [7]. Just to name a few examples, researchers included the hedonic [11], the affective [12, 13], or an overall emotional experience [14]. Note: We are aware of the different meanings of the terms *emotion, affect, feeling* ([9], see also [15]), yet most of the reviewed literature used the terms interchangeably.

Bosnjak, Glasic and Tuten [16], for example, conducted a CX study in the context of online shopping, and they concluded that the affective involvement was a significant determinant of the willingness to shop online, while cognitive involvement was not. To learn more about the particular emotions researched, we extend the previously mentioned literature review by Hermes and Riedl [7] and reviewed the 22 empirical studies with a pure focus on emotions. Please refer to Table 1 for a list of studies and the studied emotions. We used a seminal classification of positive and negative emotions by Laros and Steenkamp [17] to search for and classify emotions during this review. We only looked at emotions mentioned as a construct to operationalize CX. First, we reviewed emotions listed in the research framework under CX subconstructs like *feel, hedonic, emotional* or *affective experience*. Yet, when there was no such specific construct, we reviewed all constructs operationalizing CX. In case the framework did not provide enough information, we expanded the search to, for example, survey designs (e.g., [18, 19]). We further made small word adjustments to classify the emotions in the schemata of Laros and Steenkamp [17] (e.g., *ease of navigation* [20] classified as *at ease* [17]; *Peace of mind* [21] as *Peaceful* [17]; *playfulness* [22] as *playful* [17]). Three papers did not include or specify specific emotions [23–25].

Table 1. List of emotion words used to operationalize CX in retail

Authors	Context	Emotion word by [17]	Positive (+)/Negative (−) emotions	Count
[20, 26–30]	Online, In-Store, Virtual	Enjoyment	+	6
[11, 12, 31–34]	In-Store, Online, Smart	Excitement	+	6
[11, 13, 14, 26, 32]	In-Store, Online, Smart	Entertainment	+	5
[12, 13, 26, 31]	In-Store, Online	Pleasure	+	4
[12, 20, 26]	Online	At ease	+	3
[28, 29, 31]	Online, Virtual, In-Store	Curiosity	+	3
[26, 34]	Online, Smart	Arousal	+	2
[12, 34]	Online, Smart	Calm	+	2
[11, 35]	In-Store, Smart	Comfortable	+	2
[18, 33]	In-Store, Mobile	Frustration	−	2
[21, 33]	In-Store	Peaceful	+	2
[19, 22]	Online, Mobile	Playful	+	2
[12, 34]	Online, Smart	Relaxed	+	2
[12]	Online	Annoyance	−	1
[18]	Mobile	Confused	−	1
[12]	Online	Content	+	1
[18]	Mobile	Disappointed	−	1
[20]	Online	Distrust	−	1
[18]	Mobile	Optimism	+	1
[18]	Mobile	Relief	+	1
[18]	Mobile	Satisfaction	+	1

Twenty-one emotions were revealed, as indicated in the third column in Table 1, of which sixteen were positive and five were negative (fourth column). The most researched emotions were enjoyment, excitement, entertainment, pleasure, at ease, curiosity, arousal, calm, comfortable, frustration, peaceful, playful, and relaxed. Thus, we conclude that positive as well as negative emotions play a central role in CX research, yet, researchers mostly consider positive emotions.

3 Measurement of Experiences in Consumer Behavior

To learn more about possible studies using NeuroIS tools when researching retail customer behavior in the context of CX, we conducted a literature review. We used the general keywords "customer experience" AND "emotion*" AND "neuro*". To find specific keywords for commonly used NeuroIS tools we consulted a recent NeuroIS review paper [36]. The most used NeuroIS tools in this review were: eye-tracking, electroencephalography (EGG), functional magnetic resonance imaging (fMRI), as well as the measurement of heart rate, blood pressure, skin conductance (electrodermal activity, EDA), and hormones. Additionally, we included facial electromyography due to its importance in emotion research [37]. Hence, we used the search queries "customer experience" AND "emotion*" AND "eye*/heart*/facial*/EGG/electroencephalography/fMRI/functional magnetic resonance imaging/blood pressure/skin/hormones/EDA/electrodermal" in the database *Web of Science*. To strengthen the focus on specific NeuroIS papers, we also searched for papers with the word "emotion*" in the title in the existing NeuroIS Retreat proceedings (2009–2019). We identified a total of 62 papers from which 56 were eliminated because they did not use NeuroIS tools, or did neither focus on consumer/shopping behavior nor on CX. Hence, we identified five studies and one literature review (see [38] which reviews the use of EDA measurement in customer emotions research).

Observing customer's liking and wanting when choosing one of two products, Ahn and Picard [39] used facial valence and self-reports to predict beverage preferences and market success. In another study, Guerreiro, Rita and Trigueiros [40] use eye-tracking, skin conductance, and self-reports. They conclude that attention and arousal were important markers when predicting product choices. The researchers also call for the use of physiological measurements in the domain of customer behavior. Kindermann and Schreiner [41] used the implicit association test (IAT) and measurements of pupil dilation, eye blinks, and skin response to study brand perception. Further, Popa et al. [42] study arousal and eight other emotions through facial impressions in the context of emotions toward products. Wang et al. [43] use eye-tracking to study the emotional shopping effects of human images as visual stimuli on B2C websites (with product types as moderator). Altogether, as shown in Table 2, only a very limited number of studies have used facial measurements, eye-tracking, and skin conductance to measure emotions in a brand and shopping-related context. Other methods, related to both brain activity measurement and autonomic nervous system activity measurement (for an overview, see Chapter 3 in [37]), have not been used so far.

Table 2. Customer behavior in retail research papers applying NeuroIS methods

Paper	Organism responses	Outcome	Methods
[39]	Affective liking value, affective wanting value, cognitive liking value, cognitive wanting value	Beverage Preference, Product market success	Facial valence, self-report
[40]	Visual attention, pleasure, emotional arousal	Consumer (product) choice	Eye-tracking, skin conductance, self-report
[41]	Arousal, valence	Brand perception	IAT, Eye-tracking, Galvanic skin response, self-report
[42]	Arousal, valence, admiration, like, happiness, dislike, disagreement, disgust, and neutral	Product appreciation	Facial expressions, self-report
[43]	Image appeal, enjoyment, perceived social presence	Attitude towards online shopping site	Eye-tracking, self-report

4 Discussion, Implications, and Future Research

With the increasing number of shopping channels, CX grows in importance for retailers. While emotions play a crucial role in the CX literature, the field is still dominated by survey research [7]. Our literature review revealed only a few studies applying NeuroIS tools (e.g., eye-tracking, skin conductance), we also found that research mostly focused on product, brand, or specific website experiences and not on omnichannel experiences as a whole. Hence, we call to deepen the research in this domain including the comparison of multiple channels along the whole customer journey.

Emotions are crucial to CX, yet, as demonstrated in this short review, the use of NeuroIS tools to examine CX is still rare. This is in line with findings from other researchers (e.g., dominance of surveys and questionnaires in the CX literature [7] as well as in IS research in general [44]). However, this finding is surprising considering that some researchers argue that emotions cannot be communicated properly [9]. Additionally, various researchers also made an explicit call to widen emotions research using NeuroIS methods [15, 36, 37, 45–47].

The NeuroIS community, as well as other scientific communities, should expand CX research to physiological measurements such as heart rate or heart rate variability, skin conductance, and other measures related to autonomic nervous system activity measurement [15, 36, 37, 45–47]. It follows that we propose the following research question: How can we utilize NeuroIS methods to collect unbiased data on emotional CX? As highlighted in Chapter 2 of this paper, a wide range of emotions play a crucial part in the individual CX process. Methods that allow to identify or cluster emotional CX responses into positive and negative states are, hence, especially important for CX research. In particular, we recommend to use NeuroIS tools like a combination of facial muscular movement and eye-tracking with pupil dilation to learn more about the roles of different emotional states during a customer's experience in digital contexts [37]. Further, using Walla and Koller's [48] Startle Reflex Modulation (SRM, eye-blink amplitudes) as well as skin conductance measurements could reveal valuable information on the role of stress and arousal during CX [37].

Another finding from our research was that none of the reviewed papers had considered the effects of psychological factors like personality on emotional CX. This is surprising since we know from other customer behavior studies in the retail domain (e.g., the Big Five and its influence on the willingness to shop online [16], or the influence of the Big Five traits on internet use, e-selling and e-buying [49]) that personality is an important factor predicting channel choice and shopping behavior, likely mediated by customer emotions. Thus, our causal logic that we suggest to empirically examine in future studies is: Personality (e.g., neuroticism) → Emotional Response (e.g., distrust or stress) → Shopping behavior (e.g., rejection of an offer). Hence, we propose the following research question: Does, and if so how emotional CX (e.g. distrust or stress) mediate the relationship between stable psychological factors (e.g. personality) and shopping behavior?

Despite the fact that we believe that our work provides value to the NeuroIS community, we emphasize limitations. The first limitation of our work concerns the process of the literature review. Our keyword list might not have involved all possibly relevant keywords and in not all possible constellations. Another limitation is that only one database (Web of Science) was examined. It follows that future research should extend the keyword list and also consider additional databases.

Acknowledgments. This study has been conducted within the training network project PERFORM funded by the European Union's Horizon 2020 research and innovation program under the Marie Skłodowska-Curie grant agreement No. 765395. Note: This research reflects only the authors' views. The Agency is not responsible for any use that may be made of the information it contains.

References

1. Stobart, J., Howard, V.: The Routledge Companion to the History of Retailing. Routledge, New York (2019)
2. Hilken, T., Heller, J., Chylinski, M., Keeling, D.I., Mahr, D., de Ruyter, K.: Making omnichannel an augmented reality: the current and future state of the art. J. Res. Interact. Mark. **12**, 509–523 (2018)
3. von Briel, F.: The future of omnichannel retail: a four-stage Delphi study. Technol. Forecast. Soc. Chang. **132**, 217–229 (2018)
4. Lemon, K.N., Verhoef, P.C.: Understanding customer experience throughout the customer journey. J. Mark. **80**, 69–96 (2016)
5. Puccinelli, N.M., Goodstein, R.C., Grewal, D., Price, R., Raghubir, P., Stewart, D.: Customer experience management in retailing: understanding the buying process. J. Retail. **85**, 15–30 (2009)
6. Holbrook, M.B., Hirschman, E.C.: The experiential aspects of consumption: consumer fantasies, feelings, and fun. J. Consum. Res. **9**, 132–140 (1982)

7. Hermes, A., Riedl, R.: The nature of customer experience and its determinants in the retail context: literature review. In: WI2020 Zentrale Tracks, pp. 1738–1749. GITO Verlag (2020)
8. Phelps, E.A.: Emotion and cognition: insights from studies of the human amygdala. Annu. Rev. Psychol. **57**, 27–53 (2006)
9. Walla, P.: Affective processing guides behavior and emotions communicate feelings: towards a guideline for the NeuroIS community. In: Lecture Notes in Information Systems and Organisation - Information Systems and Neuroscience - Gmunden Retreat on NeuroIS (2018)
10. Palmer, A.: Customer experience management: a critical review of an emerging idea. J. Serv. Mark. **24**, 196–208 (2010)
11. Foroudi, P., Gupta, S., Sivarajah, U., Broderick, A.: Investigating the effects of smart technology on customer dynamics and customer experience. Comput. Hum. Behav. **80**, 271–282 (2018)
12. Martin, J., Mortimer, G., Andrews, L.: Re-examining online customer experience to include purchase frequency and perceived risk. J. Retail. Consum. Serv. **25**, 81–95 (2015)
13. Foroudi, P., Jin, Z., Gupta, S., Melewar, T.C., Foroudi, M.M.: Influence of innovation capability and customer experience on reputation and loyalty. J. Bus. Res. **69**, 4882–4889 (2016)
14. Gentile, C., Spiller, N., Noci, G.: How to sustain the customer experience: an overview of experience components that co-create value with the customer. Eur. Manag. J. **25**, 395–410 (2007)
15. vom Brocke, J., Hevner, A., Léger, P.M., Walla, P., Riedl, R.: Advancing a NeuroIS research agenda with four areas of societal contributions. Eur. J. Inf. Syst. **29**, 9–24 (2020)
16. Bosnjak, M., Galesic, M., Tuten, T.: Personality determinants of online shopping: explaining online purchase intentions using a hierarchical approach. J. Bus. Res. **60**, 597–605 (2007)
17. Laros, F.J.M., Steenkamp, J.B.E.M.: Emotions in consumer behavior: a hierarchical approach. J. Bus. Res. **58**, 1437–1445 (2005)
18. McLean, G., Al-Nabhani, K., Wilson, A.: Developing a mobile applications customer experience model (MACE)- implications for retailers. J. Bus. Res. **85**, 325–336 (2018)
19. Shobeiri, S., Mazaheri, E., Laroche, M.: Creating the right customer experience online: the influence of culture. J. Mark. Commun. **24**, 270–290 (2018)
20. Pandey, S., Chawla, D.: Online customer experience (OCE) in clothing e-retail. Int. J. Retail Distrib. Manag. **46**, 323–346 (2018)
21. Deshwal, P.: Customer experience quality and demographic variables (age, gender, education level, and family income) in retail stores. Int. J. Retail Distrib. Manag. **44**, 940–955 (2016)
22. Dacko, S.G.: Enabling smart retail settings via mobile augmented reality shopping apps. Technol. Forecast. Soc. Chang. **124**, 243–256 (2017)
23. Blázquez, M.: Fashion shopping in multichannel retail: the role of technology in enhancing the customer experience. Int. J. Electron. Commer. **18**, 97–116 (2014)
24. Boyer, K.K., Hult, G.T.M.: Customer behavioral intentions for online purchases: an examination of fulfillment method and customer experience level. J. Oper. Manag. **24**, 124–147 (2006)
25. Lin, Z., Bennett, D.: Examining retail customer experience and the moderation effect of loyalty programmes. Int. J. Retail Distrib. Manag. **42**, 929–947 (2014)
26. Krasonikolakis, I., Vrechopoulos, A., Pouloudi, A., Dimitriadis, S.: Store layout effects on consumer behavior in 3D online stores. Eur. J. Mark. **52**, 1223–1256 (2018)
27. Roy, S.K., Balaji, M.S., Sadeque, S., Nguyen, B., Melewar, T.C.: Constituents and consequences of smart customer experience in retailing. Technol. Forecast. Soc. Chang. **124**, 257–270 (2017)
28. Visinescu, L.L., Sidorova, A., Jones, M.C., Prybutok, V.R.: The influence of website dimensionality on customer experiences, perceptions and behavioral intentions: an exploration of 2D vs. 3D web design. Inf. Manag. **52**, 1–17 (2015)

29. Piyathasanan, B., Mathies, C., Wetzels, M., Patterson, P.G., de Ruyter, K.: A hierarchical model of virtual experience and its influences on the perceived value and loyalty of customers. Int. J. Electron. Commer. **19**, 126–158 (2015)
30. Evanschitzky, H., Emrich, O., Sangtani, V., Ackfeldt, A.-L., Reynolds, K.E., Arnold, M.J.: Hedonic shopping motivations in collectivistic and individualistic consumer cultures. Int. J. Res. Mark. **31**, 335–338 (2014)
31. Terblanche, N.S.: Revisiting the supermarket in-store customer shopping experience. J. Retail. Consum. Serv. **40**, 48–59 (2018)
32. Insley, V., Nunan, D.: Gamification and the online retail experience. Int. J. Retail Distrib. Manag. **42**, 340–351 (2014)
33. Lucia-Palacios, L., Pérez-López, R., Polo-Redondo, Y.: Cognitive, affective and behavioural responses in mall experience: a qualitative approach. Int. J. Retail Distrib. Manag. **44**, 4–21 (2016)
34. Poncin, I., Garnier, M., Ben Mimoun, M.S., Leclercq, T.: Smart technologies and shopping experience: are gamification interfaces effective? The case of the Smartstore. Technol. Forecast. Soc. Chang. **124**, 320–331 (2017)
35. Srivastava, M., Kaul, D.: Exploring the link between customer experience-loyalty-consumer spend. J. Retail. Consum. Serv. **31**, 277–286 (2016)
36. Riedl, R., Fischer, T., Léger, P.-M., Davis, F.D.: A decade of NeuroIS research: progress, challenges, and future directions. Data Base Adv. Inf. Syst. **51**, 13–54 (2020)
37. Riedl, R., Léger, P.-M.: Fundamentals of NeuroIS – Information Systems and the Brain. Springer, Berlin, Heidelberg (2016)
38. Caruelle, D., Gustafsson, A., Shams, P., Lervik-Olsen, L.: The use of electrodermal activity (EDA) measurement to understand consumer emotions – a literature review and a call for action. J. Bus. Res. **104**, 146–160 (2019)
39. Ahn, H., Picard, R.W.: Measuring affective-cognitive experience and predicting market success. IEEE Trans. Affect. Comput. **5**, 173–186 (2014)
40. Guerreiro, J., Rita, P., Trigueiros, D.: Attention, emotions and cause-related marketing effectiveness. Eur. J. Mark. **49**, 1728–1750 (2015)
41. Kindermann, H., Schreiner, M.: IAT measurement method to evaluate emotional aspects of brand perception – a pilot study. In: Lecture Notes in Information Systems and Organisation - Information Systems and Neuroscience - Gmunden Retreat on NeuroIS (2017)
42. Popa, M.C., Rothkrantz, L.J.M., Wiggers, P., Shan, C.: Assessment of facial expressions in product appreciation. Neural Netw. World **27**, 197–213 (2017)
43. Wang, Q., Yang, Y., Wang, Q., Ma, Q.: The effect of human image in B2C website design: an eye-tracking study. Enterp. Inf. Syst. **8**, 582–605 (2014)
44. Riedl, R., Rueckel, D.: Historical development of research methods in the information systems discipline. In: Proceedings of the Americas Conference on Information Systems (AMCIS) (2011)
45. Dimoka, A., Banker, R.D., Benbasat, I., Davis, F.D., Dennis, A.R., Gefen, D., Gupta, A., Ischebeck, A., Henning, P.H., Pavlou, P.A., Müller-Putz, G., Riedl, R., vom Brocke, J., Weber, B.: On the use of neurophysiological tools in IS research: developing a research agenda for NeuroIS. MIS Q. **36**, 679–702 (2012)
46. Riedl, R., Banker, R.D., Benbasat, I., Davis, F.D., Dennis, A.R., Dimoka, A., Gefen, D., Gupta, A., Ischebeck, A., Kenning, P., Müller-Putz, G., Pavlou, P.A., Straub, D.W., vom Brocke, J., Weber, B.: On the foundations of NeuroIS: reflections on the Gmunden retreat 2009. Commun. Assoc. Inf. Syst. **27**, 243–264 (2010)
47. Riedl, R.: Zum Erkenntnispotenzial der kognitiven Neurowissenschaften für die Wirtschaftsinformatik: Überlegungen anhand exemplarischer Anwendungen. NeuroPsychoEconomics **4**, 32–44 (2009)

48. Walla, P., Koller, M.: Emotion is not what you think it is: startle reflex modulation (SRM) as a measure of affective processing in neuroIS. In: Lecture Notes in Information Systems and Organisation - Information Systems and Neuroscience - Gmunden Retreat on NeuroIS (2015)
49. McElroy, H.: Townsend, DeMarie: dispositional factors in internet use: personality versus cognitive style. MIS Q. **31**, 809–820 (2007)

Exploring Gender Differences on eCommerce Websites: A Behavioral and Neural Approach Utilizing fNIRS

Anika Nissen[1(✉)] and Caspar Krampe[2]

[1] University of Duisburg-Essen, Essen, Germany
anika.nissen@icb.uni-due.de
[2] Wageningen University & Research, Wageningen, Netherlands
caspar.krampe@wur.nl

Abstract. Whether males and females evaluate ecommerce websites differently has long been discussed and has resulted in inconsistent research findings. While some studies identified gender differences in the evaluation of websites, other studies indicate that these differences are inexistent. To shed light on these *hypothetical* gender differences on ecommerce website perceptions, a behavioral and functional near-infrared spectroscopy (fNIRS) experiment in which participants had to use and evaluate three different ecommerce websites was conducted. While the questionnaire-based behavioral results showed no significant differences between gender, neural gender differences could be discovered. In particular, well rated websites resulted in increased neural activity for men in brain regions of the dlPFC and vlPFC in the left hemisphere, while the lower evaluated websites resulted in an increased neural activity in brain regions of the vmPFC for men in the right hemisphere. Consequently, the results suggest that men seem to require higher neural activity for the emotional appraisal of, and decision making on ecommerce websites.

Keywords: Gender differences · fNIRS · eCommerce · Online shopping · Neural measurements · Behavioral measurements

1 Introduction

Gender differences in the perception and processing of real-world objects have been investigated across different research fields, demonstrating that gender differences are predominately driven by *biological*, *cognitive*, *behavioral*, or *social factors* [1–3]. In particular, men and women tend to process and perceive information differently [4, 5]. Aspects that seem to be particularly relevant for the field of ecommerce, given the fact that the way information is perceived determines how easily it is accessible and how trustworthy it appears [6, 7]. Research demonstrated that both of these aspects can be influenced through website design [8–10]. Consequently, different design choices can impact women and men in different ways. Against this background, given the anticipated

F. D. Davis et al. (Eds.): NeuroIS 2020, LNISO 43, pp. 220–232, 2020.
https://doi.org/10.1007/978-3-030-60073-0_26

gender-based differences in the perception of website designs, it seems to be crucial to also consider gender (defined as the biological sex and/or perceived gender role) as a critical influencing factor on the perception of a website (design). Previous research that investigated gender effects in an ecommerce context indicated that men focus more on the utility and usefulness of websites [9, 11, 12]. Whereas women tend to focus to a greater extent on consistent button use and the aesthetic design of the website [11, 13–15]. Nevertheless, there are also contradicting research findings, which did not find any significant differences related to website evaluations of men and women [16, 17]. Moreover, whether the indicated gender differences are caused by a websites' design, or whether they occur due to *biological*, *cognitive* or *societal* effects, has not been considered by now on a neural level.

Hence, the given research work aims to shed light on (possible) gender-related differences in the perception of ecommerce websites, while focusing on website design. Whereas prior studies frequently utilized self-reported measurements, which might not be sensitive enough to detect subtle differences in the processing and evaluation of websites [18], the current study applies the neuroimaging method functional near-infrared spectroscopy (fNIRS) in order to collect neural – for the user often unconscious – data. In the frame of NeuroIS research, the application of (neuro)physiological measurements has proven to be a reliable and additional data source, which provides profound insights of biological/neural processes of human beings [18, 19]. Especially the neuroimaging method of fNIRS seems to offer unique advantages, as it is a lightweight, portable, and user friendly method, which allows to measure users in real-life scenarios such as sitting in front of the computer screen or whilst operating a machine, due to the fact that fNIRS is mostly robust to movement artefacts [20–24].

Consequently, this paper addresses the following research question, which is answered by means of fNIRS: *How differently do men and women process ecommerce websites?* The remainder of the paper is structured as follows. First, relevant literature considering gender effects related to the perception of websites is reviewed. Second, the experimental design and the data analysis to investigate the users behavioral and neural reaction to the three predefined websites, employing fNIRS in a real-world setting, is explained. Finally, the results are discussed, and conclusions are drawn.

2 Gender Differences Related Literature

Given the tremendous amount of research concerning gender differences, the following literature review focuses on two overarching topics. The first topic reviews the general, ***neural*** gender differences, which can be observed cross-disciplinary. The second topic focuses on ***ecommerce-specific*** gender differences.

2.1 Neural Gender Differences

On a neural level, males tend to show generally higher oxygenated hemoglobin (HbO) – which is a proxy for increased neural activity – in brain regions of the prefrontal cortex (PFC) in comparison to females in several cognitive activities, excluding emotional processing [25–27]. In line with previous research [28], a possible explanation for this might be the fact that males use more rational, cognitive-focused strategies when evaluating a situation or an entity. Moreover, previous studies indicated that the female brain requires less neural activity for the same behavioral performance than men [27]. Also in other contexts, such as learning a second language, women tend to perform better than men – both on a behavioral and neural level [29]. This shows that there seem to be some general differences in the emotional and cognitive processing between males and females [26, 30–33].

These gender differences can also be found in the lateralization of the neural activity. Thereby, men's neural activity pattern of the PFC seems to be more left-lateralized [34]. Moreover, previous research findings indicated that apparently, neural activity is influenced by genes and sex-hormones [35, 36], which might explain the 'lateralization' phenomenon and gender differences. A prominent example in this regard might be the hormone estrogen. Estrogen is a hormone, which is important for sexual and reproductive development. Given the fact that estrogen is mostly found in women, it is expected that it modulates the neural activity in the dorsolateral prefrontal cortex (dlPFC), a brain region that is, among other functions, related to affective processing [37]. Consequently, it is thought to impact women to a greater extent than men, and thus, it is influencing the women's emotional processing. Consequently, based on the influence of the different hormones the neural processing might be different between men and women, a fact that also counts for the perception of ecommerce websites.

2.2 Gender Differences on eCommerce Websites

In this research work, the reviewed literature revealed four overarching areas for which self-reported gender differences apply: (i) *visual design and aesthetics*, (ii) *perceived trustworthiness*, (iii) *perceived usefulness* (PU) and *website navigation*, and finally, (iv) *perceived ease of use* (PEOU).

(i) Visual design and aesthetics are the first cues users perceive when visiting a website. Before any content is processed, the website's colors, shape, font styles, and pictures are perceived almost immediately [38, 39]. These components influence the perceived visual complexity of the interface, which have a direct impact on the users' evaluations and their associated behavior on the website. While the objective complexity (e.g. the amount of website cues) might be the same for several websites, the individual perceived complexity of users might be different, i.e. depending on their cultural background [40]. Nevertheless, differences in the perceived complexity of a website might also appear between gender, given the fact that women tend to prefer media-rich websites in comparison to men [15]. Interestingly, also the overall visual design of websites seems to be evaluated as more important for women than for men [9]. Consequently, the importance of the visual appeal and its association to the overall satisfaction of a website has been evaluated as more important for women in comparison to men across several studies [13, 41–44].

(ii) The influence of the *perceived trustworthiness* of a website seems to be consistent across most studies [45]. Thereby, women tend to be more skeptical, whether they can trust an ecommerce website or not. A fact, which seems to be particularly relevant for the first-time usage of a website. Besides, research results indicated that men are generally more willing to make online purchases (PUI) than women [46], which demonstrated that

security and privacy aspects seem to be more important for women than for men [41]. Preliminary studies, exploring the neural processes involved in the formation of trust in online environments, demonstrated that trust can be linked to (*1*) cognitive processes within the prefrontal cortex, and (*2*) emotional processes within the limbic system [6]. Furthermore, neural gender differences related to the perception of the trustworthiness of an internet offer identified that women significantly activate more brain regions than men [7]. This is crucial, given that the perceived trustworthiness of online shops influences also other concepts such as PU and PEOU and finally, the intention to make a purchase decision [47].

(iii) Perceived usefulness and *website navigation* are both functional aspects of websites, which are closer related to task performance than to emotional experience. Several studies have shown that for men PU seems to be more important than for women [3, 12, 44, 48, 49]. Findings that are supported by the fact that men tend to focus more on the content displayed and pay less attention to the visual appearance or the website design [50, 51]. With regard to *website navigation*, opposite results have been found, as in one study, men evaluated the ease of navigation as a more important construct than women [41], while in another study, women evaluated the ease of navigation as more important than men [11]. However, as men and women tend to have different navigational styles, their preferences might be severely dependent on the stimuli presented [52]. Furthermore, women seem to be influenced by norms and standards, which result in more extreme reactions to i.e. inconsistent button use [11, 48, 49].

(iv) Finally, the gender related differences of the *ease of use* are still unexplored [9, 53]. Some early studies focusing on PEOU and potential neural correlates in the prefrontal cortex even suggest that this construct might not have specific neural processing when investigated alone [54]. Further research that investigated the impact of PEOU on purchase intentions also found, that PU and trust have a greater impact on purchase intentions [55]. This lays the assumption near, that PEOU might not be observable in the brain or as an impact factor on PUI, when considered as only construct.

Therefore, in order to shed light on (possible) gender differences in the perception and evaluation of ecommerce websites, the given research work explores the constructs aesthetics (AEST), PU, PEOU, and PUI with self-reported evaluations, while also investigating the neural brain activity of the prefrontal cortex (PFC), during the perception and actual use of websites, while utilizing the neuroimaging method fNIRS. This is done because gender differences might only count for perception processes but cannot be directly observed by the intended or actually shown behavior or verbalized by the users [56]. The additional data source of neural data and the consideration of actual website usage in a realistic situation will, consequently, advance the understanding of the users' perception of websites and expose potential gender differences.

3 Study Design and Results

3.1 Method

The method of fNIRS has been applied in the field of neuroscience as well as cognate disciplines [18, 21, 23, 57]. Likewise, fNIRS has been shown to be a feasible method for measuring usability and user experiences of graphical user interfaces [20, 24, 58, 59].

Furthermore, fNIRS has proven to provide a reliable source to investigate cognitive and emotional processes. In comparison to functional magnetic resonance imaging (fMRI), fNIRS offers a lightweight and portable method, which can be applied in real-life contexts, while measuring the same biological processes of cortical brain regions. Further, when compared to electroencephalography (EEG), which offers the same mobility and even a better temporal resolution, fNIRS tends to be more robust to movement artefacts, when showing natural behavior (e.g. talking or body movements) [21, 58, 60].

The fNIRS device applied in this research work is an 8 source/7 detector continuous-wave headband with 22 channels from NIRX (montage depicted in Fig. 1). The sampling rate is 7.81 Hz and the average distance between sources and detectors was set to 30 mm. The wavelengths of the infrared light are 760 nm and 850 nm, with the associated differential pathlength factor (dpf) set to 7.25 for the 760 nm and to 8.38 for the 850 nm, which is in accordance to commonly reported values [61–63]. To avoid data biases due to experiment equipment interference, several variables were controlled for such as room luminance, head movements, and the relocation of hair to guarantee skin contact of the sources and detectors. The headband was calibrated individually for every participant through which data quality was assured.

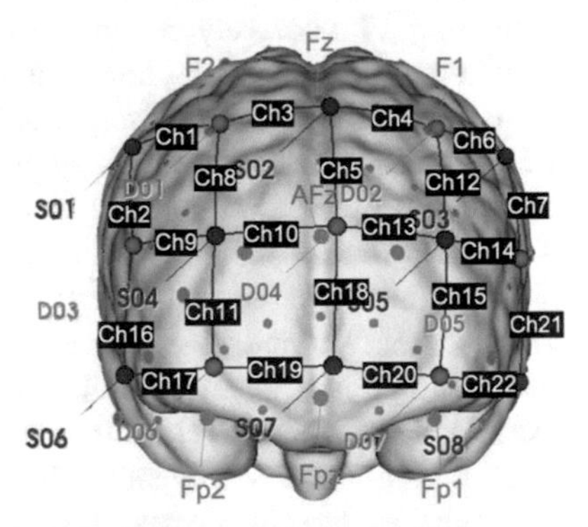

Fig. 1. fNIRS montage on PFC including optodes (red and blue), numbered channels (black) and EEG10–20 reference points (orange)

3.2 Sample, Stimuli and Data Pre-processing

For this study, a **sample** of $N = 20$ participants (40% female) were recruited from the local university, with 73% being students and 27% being employed. The here employed sample size thus represents the average employed in neuroscientific studies in general [64], and in studies using fNIRS in particular [65]. For the data analysis, the sample was divided into two groups. The female group consisted of $n = 8$ participants and the male group consisted of $n = 12$ participants. Regarding their age, no significant differences ($p > .05$) were found between females ($M = 25.63$) and males ($M = 26.50$) participants.

For the research scenario of this study, the recruited participants were asked to use three included websites on an iPad Pro 12″. On each website, they were asked to search for digital cameras; and after a while select their favored digital camera and place it in the shopping basket. With this, the task was completed, and participants were asked to answer a short website evaluation questionnaire. Afterwards, participants were asked to proceed with the next website. Having completed this procedure for all three websites, participants were asked to complete a final questionnaire, integrating demographical and control questions. Three ecommerce websites with comparable assortment, acting as **stimuli** material in the experiment (depicted in Fig. 2), were selected in a pre-study ($N = 150$ participants), in which the website's AEST, PU, PEOU, and PUI were measured. The three websites were selected based on their evaluations, integrating the highest significant differences related to the mentioned constructs; resulting in a 'high', 'medium' and 'low' evaluated website (with $p < .05$, uncorrected). The pre-study revealed that CU was rated 'highest' in all constructs, DIG was rated 'medium', and PN was rated 'lowest' in all constructs.

Fig. 2. Stimuli of the experiment

The raw fNIRS data was **pre-processed** by applying a band-pass filter with a low cut-off frequency of 0.01 Hz and a high cut-off frequency of 0.2 Hz, which allows to filter out noises such as heart rate and respiration [66–69]. Afterwards, the hemodynamic states of oxygenated hemoglobin (HbO) are computed using the modified Beer-Lambert law [70, 71]. Finally, task relevant changes in hemoglobin concentrations are calculated, using a general linear model (GLM) with the canonical hemodynamic response function (hfr) as a baseline. The included, self-reported data was analyzed using ANOVAs with post-hoc Tukey tests.

4 Results

To evaluate the TAM constructs (PU, PEOU, and PUI), previously applied questionnaires (taken from [72]) as well as the short version of the VisAWI scale to evaluate website aesthetics [73] have been used in this study. Each item was evaluated on a 5-point Likert scale, ranging from 1 'totally disagree' to 5 'totally agree'. To test the scales reliability, we calculated Cronbach's alpha (CA), which resulted in excellent reliability with CA for AEST being .934, for PU being .906, and for PEOU being .913, as well as very good reliability for PUI being .851 [74].

For the self-reported data, two one-way ANOVAs have been conducted - one, in which the three websites were compared to each other for the whole sample (within-group), and one in which the two gender groups were compared for each website (between-group). The results indicate significant differences for each construct between the three included websites for the within-group comparisons: $F_{PU}(2, 60) = 11.982$, $F_{PEOU}(2, 60) = 17.583$, $F_{PUI}(2, 60) = 16.926$, $F_{AEST}(2, 60) = 29.687$ $p < .001$. The website PN was evaluated significantly lower than the websites DIG and CU; and CU was evaluated highest on all constructs. Between gender, no significant differences were found in any of the included constructs for the websites between-groups: $F_{PU}(1, 61) = .074$, $F_{PEOU}(1, 61) = .063$, $F_{PUI}(1, 61) = .275$, $F_{AEST}(1, 61) = .739$ $p > .05$. However, a pattern was identified, signifying that male participants consistently evaluated CU highest on all constructs, while also rating PN lowest on all scales. Next to the questionnaire-based behavioral (intention) results, the neural activity of the PFC has been explored by utilizing fNIRS. In this study, the HbO levels have been used as an indicator for the neural activity of participants and contrasts between-groups for each website have been calculated. More precisely, to investigate the gender related neural activity patterns, the average HbO levels of the female group were contrasted to the average HbO levels of the male group. For the interpretation of the neural results, the HbO level acts as a proxy for (increased or decreased) neural activity.

The neural results for the between-group contrasts are displayed in Fig. 3. For the 'highest' rated website CU, significantly higher HbO levels were found for the male group in the left ventrolateral prefrontal cortex (vlPFC) and the left inferior dorsolateral prefrontal cortex (dlPFC) ($p < .01$). Furthermore, for the 'lowest' rated website PN the right superior ventromedial prefrontal cortex (vmPFC) indicated an increased neural activity for the male group ($p < .01$), in comparison to the female group. Finally, the 'moderately' rated website DIG showed an increased neural activity in the left superior dlPFC for the male group ($p < .01$).

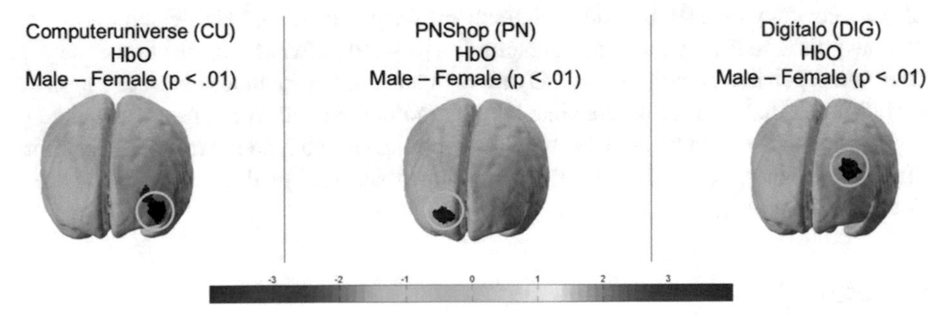

Fig. 3. T-statistic HbO results for between-group contrasts male > female

5 Discussion

The results indicate an increased neural activity in the *left vlPFC* and *inferior dlPFC* for the website, which was rated highest in all the questionnaire constructs for male participants. In general the *vlPFC* has been linked to semantic processing [75], controlling hand and eye movements [76, 77] as well as processing emotional stimuli due its connectivity to the amygdala [78]. Whereas the *right vlPFC* has also been related to the processing of negative emotional stimuli [78]. In line with these results and in accordance with the hemisphere specific hypothesis (HSH) [79], the increased neural activity only in the left hemisphere in male users might indicate that men perceive the website (unconsciously) as more pleasing than female users. A hypothesis, which is supported by the questionnaire-based behavioral (intention) results, indicating that men evaluate the website slightly higher than women. Therefore, it can be assumed that this finding is not due to biological differences, but also due to (unconscious) perceptional and evaluation differences. Further, the increased activation in the left vlPFC might also point to men requiring more oxygen to process text elements on websites. However, in this case the left vlPFC would have also been found on the lowest rated website PN. Consequently, and as we did not find significant increases in this area for the other websites, the prior interpretation relating to the HSH can be regarded as most plausible.

Furthermore, increased neural activity in the *left dlPFC* was found for the 'highest' as well as the 'medium' rated website for male users. The *dlPFC* has been frequently found to be activated in decision-making processes [80–83]. In this context, the *dlPFC* seems to play a major role in processing sensory information, whilst making a decision and planning further actions that have to be taken [84]. Moreover, an increased neural activity of the dlPFC was also shown when participants were confronted with their favorite product [57]. Consequently, based on the fact that men indicate a higher neural activity in the *dlPFC* when making a ecommerce related decision, it can be suggested that men require more neural activity to process information in comparison to women [25–27], which might indicate that men are more attracted by the website design. An aspect that might predominantly be driven by PU and PEOU, as some of the prior described studies found that PU is more important to men than to women.

Finally, the inferior *dmPFC,* which was found activated for men on the lowest rated website has also been linked to decision processes [80–83] and has been found to be active in the (re-)appraisal of emotional stimuli [85, 86]. While drawing back to the HSH, dmPFC increases for men were only found in the right hemisphere, which may imply that men require more neural activity to process aversive stimuli (in this case unaesthetic websites). This is also supported in the self-reported scales, as men consistently rated PN lower in all constructs compared to women.

Next to the mentioned effects, the neural results provide additional evidence that there are (unconscious) differences between men and women. These differences might be purely influenced by hormones such as estrogen. However, they might also relate to unconscious cognitive and emotional processes. This becomes especially evident for the lowest rated website, as men tend to, if at all, show increases primarily on the left hemisphere and not on the right. This is further supported as the significant neural findings represent the not significant, but observable findings of the questionnaire data.

6 Conclusion

Numerous *conclusions* and *implications* can be drawn from our preliminary research findings. Thereby, websites evaluated as aesthetically pleasing, useful and easy to use showed increased neural activity in the left hemisphere of the users' brain, whereas less aesthetically and as not useful evaluated websites result in increased neural activity in the right hemisphere of the users' brain. These results further support the HSH. On a general level, the results show that men require consistently more HbO to process the included websites for both, the emotional appraisals and the decision-making processes. A fact that supports recent gender related research findings, which pointed out that men require generally more HbO levels for the same performance [27, 29]. Against this background, the preliminary results support the suggestion that females like complex websites more than men and that men tend to favor clean and simple designs. A point which became ultimately evident in the activation of the right dmPFC for men on the lowest rated website, which was also the most complex website in terms of the number of content elements.

Consequently, from a practical point of view, website designers are advised to also consider their target groups' gender when designing a website, offering different complexity and media-richness levels. Whether this also counts for other software products requires further investigations. Nevertheless, the current research work provides a first step to address different user groups based on their neural, gender related information processing capabilities. Further research should therefore take also other website formats such as social networks or purely informational websites as well as individual characteristics such as age, culture, or experience levels into account.

While being a preliminary study, this paper has some *limitations*, which need to be taken into consideration in future research work. First, only three different ecommerce websites, focusing on electronic devices, were included. Thus, our results might not be generalizable to other website designs or software products. Consequently, future research might expand the research results of this study and investigate gender differences for other products and websites. Second, only German participants were included in the sample. This might be a limitation, given that cultural effects seem to have an impact on informational processing, too [8]. Third, the reported information in this study relies only on the HbO signal. To also integrate the deoxygenated hemoglobin signals can, consequently, add further information and insights to gender differences, whilst validating the research findings of the HbO signals. Finally, the here presented and discussed functions of the activated PFC areas represent only some PFC functions related to decision making processes. Future research could therefore also focus on a more holistic review of typical functions specifically for the dlPFC, vlPFC, and dmPFC to add additional insights and interpretations to the here observed neural patterns.

References

1. Costa, P.T., Terracciano, A., McCrae, R.R.: Gender differences in personality traits across cultures: robust and surprising findings. J. Pers. Soc. Psychol. **81**, 322–331 (2001)
2. Putrevu, S.: Exploring the origins and information processing differences between men and women: implications for advertisers. Acad. Mark. Sci. Rev. **10**, 1–16 (2003)
3. Sun, Y., Lim, K.H., Jiang, C., Peng, J.Z., Chen, X.: Do males and females think in the same way? An empirical investigation on the gender differences in Web advertising evaluation. Comput. Human Behav. **26**, 1614–1624 (2010)
4. Gefen, D., Nitza, G., Paravastu, N.: Vive la différence: the cross-culture differences within us. Int. J. e-Collab. **3**, 1–15 (2007)
5. Gefen, D., Ridings, C.M.: If you spoke as she does, sir, instead of the way you do. ACM SIGMIS Database **36**, 78–92 (2005)

6. Dimoka, A.: What does the brain tell us about trust and distrust? Evidence from a functional neuroimaging study. MIS Q. **34**, 373–396 (2010)
7. Riedl, R., Hubert, M., Kenning, P.: Are there neural gender differences in online trust? An fMRI study on the perceived trustworthiness of eBay offers. MIS Q. **34**, 397–428 (2010)
8. Cui, T., Wang, X., Teo, H.H.: Effects of cultural cognitive styles on users' evaluation of website complexity. In: The International Conference on Information Systems (ICIS), Orlando, pp. 1–17 (2012)
9. (Fone) Pengnate, S., Sarathy, R.: An experimental investigation of the influence of website emotional design features on trust in unfamiliar online vendors. Comput. Human Behav. **67**, 49–60 (2017)
10. Cyr, D., Head, M., Larios, H.: Colour appeal in website design within and across cultures: a multi-method evaluation. Int. J. Hum. Comput. Stud. **68**, 1–21 (2010)
11. Ramakrishnan, T., Prybutok, V., Peak, D.A.: The moderating effect of gender on academic website impression. Comput. Human Behav. **35**, 315–319 (2014)
12. Zhang, X., Prybutok, V.: TAM : The moderating effect of gender on online shopping. J. Int. Inf. Manag. **12**, 99–118 (2003)
13. Cyr, D., Bonanni, C.: Gender and website design in e-business. Int. J. Electron. Bus. **3**, 565 (2006)
14. Mahzari, A., Ahmadzadeh, M.: Finding gender preferences in e-commerce website design by an experimental approach. Context **5**, 35–40 (2013)
15. Simon, S.J., Peppas, S.C.: Attitudes towards product website design: a study of the effects of gender. J. Mark. Commun. **11**, 129–144 (2005)
16. Afshardost, M., Farahmandin, S., Sadiq Wshaghi, S.M.: Linking trust, perceived website quality, privacy protection, gender and online purchase intentions. IOSR J. Bus. Manag. **13**, 63–72 (2013)
17. Al-Maghrabi, T., Dennis, C.: The driving factors of continuance online shopping: gender differences in behaviour - the case of Saudi Arabia. In: European and Mediterranean Conference on Information Systems 2009, Izmir, pp. 1–19 (2009)
18. Riedl, R., Davis, F.D., Banker, R., Kenning, P.H.: Neuroscience in Information Systems Research. Springer, Cham (2017)
19. Dimoka, A., Banker, R.D., Benbasat, I., Davis, F.D., Dennis, A.R., Gefen, D., Gupta, A., Ischebeck, A., Kenning, P., Müller-Putz, G., Pavlou, P.A., Riedl, R., vom Brocke, J., Weber, B.: On the use of neurophysiological tools in is research: developing a research agenda for NeuroIS. MIS Q. **36**, 679–702 (2012)
20. Gefen, D., Ayaz, H., Onaral, B.: Applying functional near infrared (fNIR) spectroscopy to enhance MIS research. AIS Trans. Hum. Comput. Interact. **6**, 55–73 (2014)
21. Kim, H.Y., Seo, K., Jeon, H.J., Lee, U., Lee, H.: Application of functional near-infrared spectroscopy to the study of brain function in humans and animal models. Mol. Cells **40**, 523–532 (2017)
22. Kopton, I.M., Kenning, P.: Near-infrared spectroscopy (NIRS) as a new tool for neuroeconomic research. Front. Hum. Neurosci. **8**, 1–13 (2014)
23. Krampe, C., Gier, N., Kenning, P.: The application of mobile fNIRS in marketing research – detecting the 'first-choice-brand' effect. Front. Hum. Neurosci. **12**, 433 (2018)
24. Nissen, A., Krampe, C., Kenning, P., Schütte, R.: Utilizing mobile fNIRS to investigate neural correlates of the TAM in eCommerce. In: International Conference on Information Systems (ICIS), Munich, pp. 1–9 (2019)
25. Kalia, V., Vishwanath, K., Knauft, K., Von Der Vellen, B., Luebbe, A., Williams, A.: Acute stress attenuates cognitive flexibility in males only: an fNIRS examination. Front. Psychol. **9**, 1–15 (2018)

26. Cinciute, S., Daktariunas, A., Ruksenas, O.: Hemodynamic effects of sex and handedness on the Wisconsin Card Sorting Test: the contradiction between neuroimaging and behavioural results. PeerJ. **6**, e5890 (2018)
27. Li, T., Luo, Q., Gong, H.: Gender-specific hemodynamics in prefrontal cortex during a verbal working memory task by near-infrared spectroscopy. Behav. Brain Res. **209**, 148–153 (2010)
28. Mak, A.K., Hu, Z.G., Zhang, J.X., Xiao, Z., Lee, T.M.: Sex-related differences in neural activity during emotion regulation. Neuropsychologia **47**, 2900–2908 (2009)
29. Sugiura, L., Hata, M., Matsuba-Kurita, H., Uga, M., Tsuzuki, D., Dan, I., Hagiwara, H., Homae, F.: Explicit Performance in girls and implicit processing in boys: a simultaneous fNIRS–ERP study on second language syntactic learning in young adolescents. Front. Hum. Neurosci. **12**, 1–19 (2018)
30. Bidula, S.P., Przybylski, L., Pawlak, M.A., Króliczak, G.: Unique neural characteristics of atypical lateralization of language in healthy individuals. Front. Neurosci. **11**, 1–21 (2017)
31. Hill, A.C., Laird, A.R., Robinson, J.L.: Gender differences in working memory networks: a BrainMap meta-analysis. Biol. Psychol. **102**, 18–29 (2014)
32. Zilles, D., Lewandowski, M., Vieker, H., Henseler, I., Diekhof, E., Melcher, T., Keil, M., Gruber, O.: Gender differences in verbal and visuospatial working memory performance and networks. Neuropsychobiology **73**, 52–63 (2016)
33. McCarthy, M., Arnold, A., Ball, G., Blaustein, J., De Vries, G.: Sex differences in the brain: the not so inconvenient truth. J. Neurosci. **32**, 2241–2247 (2012)
34. Chuang, C.C., Sun, C.W.: Gender-related effects of prefrontal cortex connectivity: a resting-state functional optical tomography study. Biomed. Opt. Express **5**, 2503 (2014)
35. Schmitz, J., Lor, S., Klose, R., Güntürkün, O., Ocklenburg, S.: The functional genetics of handedness and language lateralization: insights from gene ontology, pathway and disease association analyses. Front. Psychol. **8**, 1–12 (2017)
36. Hausmann, M.: Why sex hormones matter for neuroscience: a very short review on sex, sex hormones, and functional brain asymmetries. J. Neurosci. Res. **95**, 40–49 (2017)
37. Amin, Z., Epperson, C.N., Constable, R.T., Canli, T.: Effects of estrogen variation on neural correlates of emotional response inhibition. Neuroimage **32**, 457–464 (2006)
38. Lindgaard, G., Fernandes, G., Dudek, C., Brown, J.: Attention web designers: you have 50 milliseconds to make a good first impression! Behav. Inf. Technol. **25**, 115–126 (2006)
39. Tractinsky, N., Cokhavi, A., Kirschenbaum, M., Sharfi, T.: Evaluating the consistency of immediate aesthetic perceptions of web pages. Int. J. Hum. Comput. Stud. **64**, 1071–1083 (2006)
40. Cui, X., Baker, J.M., Liu, N., Reiss, A.L.: Sensitivity of fNIRS measurement to head motion: an applied use of smartphones in the lab. J. Neurosci. Methods **245**, 37–43 (2015)
41. Cyr, D., Head, M.: Website design in an international context: the role of gender in masculine versus feminine oriented countries. Comput. Human Behav. **29**, 1358–1367 (2013)
42. Enoch, Y., Soker, Z.: Age, gender, ethnicity and the digital divide: university students' use of web-based instruction. Open Learn. J. Open Distance e-Learn. **21**, 99–110 (2006)
43. Zhou, G., Xu, J.: Adoption of educational technology: how does gender matter? Int. J. Teach. Learn. High. Educ. **19**, 140–153 (2007)
44. Terzis, V., Economides, A.A.: Computer based assessment: gender differences in perceptions and acceptance. Comput. Human Behav. **27**, 2108–2122 (2011)
45. Tuch, A.N., Bargas-Avila, J.A., Opwis, K.: Symmetry and aesthetics in website design: it's a man's business. Comput. Hum. Behav. **26**, 1831–1837 (2010)
46. Rodgers, S., Harris, M.A.: Gender and e-commerce: an exploratory study. J. Advert. Res. **43**, 322–329 (2003)
47. Gefen, D., Karahanna, E., Straub, D.W.: Trust and TAM in online shopping: an integrated model. MIS Q. **27**, 51–90 (2003)

48. Venkatesh, V., Morris, M.G., Davis, G.B., Davis, F.D.: User acceptance of information technology: toward a unified view. MIS Q. **27**, 425–478 (2003)
49. Hwang, Y.: The moderating effects of gender on e-commerce systems adoption factors: an empirical investigation. Comput. Hum. Behav. **26**, 1753–1760 (2010)
50. Burke, R.R.: Technology and the customer interface: what consumers want in the physical and virtual store. J. Acad. Mark. Sci. **30**, 411–432 (2002)
51. Richard, M.O., Chebat, J.C., Yang, Z., Putrevu, S.: A proposed model of online consumer behavior: assessing the role of gender. J. Bus. Res. **63**, 926–934 (2010)
52. Stenstrom, E., Stenstrom, P., Saad, G., Cheikhrouhou, S.: Online hunting and gathering: an evolutionary perspective on sex differences in website preferences and navigation. IEEE Trans. Prof. Commun. **51**, 155–168 (2008)
53. Venkatesh, V., Morris, M.G.: Why don't men ever stop to ask for directions? Gender, social influence, and their role in technology acceptance and usage behavior. MIS Q. **24**, 115 (2000)
54. Dumont, L., Larochelle-Brunet, F., Théoret, H., Riedl, R., Sénécal, S., Léger, P.M.: Non-invasive brain stimulation in information systems research: a proof-of-concept study. PLoS ONE **13**, 1–16 (2018)
55. Amin, M., Rezaei, S., Tavana, F.S.: Gender differences and consumer ' s repurchase intention: the impact of trust propensity, usefulness and ease of use for implication of innovative online retail. Int. J. Innov. Learn. **17**, 217–233 (2015)
56. Gefen, D., Straub, D.W.: Gender differences in the perception and use of e-mail: an extension to the technology acceptance model. MIS Q. **21**, 389 (2006)
57. Deppe, M., Schwindt, W., Kugel, H., Plaßmann, H., Kenning, P.: Nonlinear responses within the medial prefrontal cortex reveal when specific implicit information influences economic decision making. J. Neuroimaging **15**, 171–182 (2005)
58. Hill, A.P., Bohil, C.J.: Applications of optical neuroimaging in usability research. Ergon. Des. **24**, 4–9 (2016)
59. Pollmann, S., Eštočinová, J., Sommer, S., Chelazzi, L., Zinke, W.: Neural structures involved in visual search guidance by reward-enhanced contextual cueing of the target location. Neuroimage **124**, 887–897 (2016)
60. Irani, F., Platek, S.M., Bunce, S., Ruocco, A.C., Chute, D.: Functional near infrared spectroscopy (fNIRS): an emerging neuroimaging technology with important applications for the study of brain disorders. Clin. Neuropsychol. **21**, 9–37 (2007)
61. Essenpreis, M., Elwell, C.E., Cope, M., van der Zee, P., Arridge, S.R., Delpy, D.T.: Spectral dependence of temporal point spread functions in human tissues. Appl. Opt. **32**, 418 (1993)
62. Kohl, M., Nolte, C., Heekeren, H.R., Horst, S., Scholz, U., Obrig, H., Villringer, A.: Determination of the wavelength dependence of the differential pathlength factor from near-infrared pulse signals. Phys. Med. Biol. **43**, 1771–1782 (1998)
63. Zhao, H., Tanikawa, Y., Gao, F., Onodera, Y., Sassaroli, A., Tanaka, K., Yamada, Y.: Maps of optical differential pathlength factor of human adult forehead, somatosensory motor and occipital regions at multi-wavelengths in NIR. Phys. Med. Biol. **47**, 306 (2002)
64. Riedl, R., Banker, R.D., Benbasat, I., Davis, F.D., Dennis, A.R., Dimoka, A., Gefen, D., Gupta, A., Ischebeck, A., Kenning, P., Müller-Putz, G., Pavlou, P.A., Straub, D.W., vom Brocke, J., Weber, B.: On the foundations of NeuroIS: reflections on the Gmunden retreat 2009. Commun. Assoc. Inf. Syst. **27**, 243–264 (2010)
65. Vassena, E., Gerrits, R., Demanet, J., Verguts, T., Siugzdaite, R.: Anticipation of a mentally effortful task recruits Dorsolateral Prefrontal Cortex: an fNIRS validation study. Neuropsychologia **123**, 106–115 (2019)
66. Pinti, P., Scholkmann, F., Hamilton, A., Burgess, P., Tachtsidis, I.: Current status and issues regarding pre-processing of fNIRS neuroimaging data: an investigation of diverse signal filtering methods within a general linear model framework. Front. Hum. Neurosci. **12**, 1–21 (2019)

67. Trambaiolli, L.R., Biazoli, C.E., Cravo, A.M., Sato, J.R.: Predicting affective valence using cortical hemodynamic signals. Sci. Rep. **8**, 1–12 (2018)
68. Wang, L., Li, S., Zhou, X., Theeuwes, J.: Stimuli that signal the availability of reward break into attentional focus. Vis. Res. **144**, 20–28 (2018)
69. Zhang, D., Zhou, Y., Hou, X., Cui, Y., Zhou, C.: Discrimination of emotional prosodies in human neonates: a pilot fNIRS study. Neurosci. Lett. **658**, 62–66 (2017)
70. Delpy, D.T., Cope, M., van der Zee, P., Arridge, S., Wray, S., Wyatt, J.: Estimation of optical pathlength through tissue from direct time of flight measurement. Phys. Med. Biol. **33**, 1433–1442 (1988)
71. Kocsis, L., Herman, P., Eke, A.: The modified Beer-Lambert law revisited. Phys. Med. Biol. **51**, N91–N98 (2006)
72. Dimoka, A., Davis, F.D.: Where does TAM reside in the brain? The neural mechanisms underlying technology adoption. In: ICIS, pp. 1–18 (2008)
73. Moshagen, M., Thielsch, M.T.: Facets of visual aesthetics. Int. J. Hum. Comput. Stud. **68**, 689–709 (2010)
74. Blanz, M.: Forschungsmethoden der Statistik für die Soziale Arbeit: Grundlagen und Anwendungen. Kohlhammer, Stuttgart (2015)
75. Snyder, H.R., Banich, M.T., Munakata, Y.: Choosing our words: retrieval and selection processes recruit shared neural substrates in left ventrolateral prefrontal cortex. J. Cogn. Neurosci. **23**, 3470–3482 (2011)
76. Leung, H.C., Cai, W.: Common and differential ventrolateral prefrontal activity during inhibition of hand and eye movements. J. Neurosci. **27**, 9893–9900 (2007)
77. Heinen, S.J., Rowland, J., Lee, B.T., Wade, A.R.: An oculomotor decision process revealed by functional magnetic resonance imaging. J. Neurosci. **26**, 13515–13522 (2006)
78. Wager, T.D., Davidson, M.L., Hughes, B.L., Lindquist, M.A., Ochsner, K.N.: Prefrontal-subcortical pathways mediating successful emotion regulation. Neuron **59**, 1037–1050 (2008)
79. Killgore, W.D.S., Yurgelun-Todd, D.A.: The right-hemisphere and valence hypotheses: could they both be right (and sometimes left)? Soc. Cogn. Affect. Neurosci. **2**, 240–250 (2007)
80. Hutcherson, C.A., Plassmann, H., Gross, J.J., Rangel, A.: Cognitive regulation during decision making shifts behavioral control between ventromedial and dorsolateral prefrontal value systems. J. Neurosci. **32**, 13543–13554 (2012)
81. Chen, M.Y., Jimura, K., White, C.N., Todd Maddox, W., Poldrack, R.A.: Multiple brain networks contribute to the acquisition of bias in perceptual decision-making. Front. Neurosci. **9**, 1–13 (2015)
82. Greening, S.G., Finger, E.C., Mitchell, D.G.V.: Parsing decision making processes in prefrontal cortex: response inhibition, overcoming learned avoidance, and reversal learning. Neuroimage **54**, 1432–1441 (2011)
83. Mitchell, D.G.V., Luo, Q., Avny, S.B., Kasprzycki, T., Gupta, K., Chen, G., Finger, E.C., Blair, R.J.R.: Adapting to dynamic stimulus-response values: differential contributions of inferior frontal, dorsomedial, and dorsolateral regions of prefrontal cortex to decision making. J. Neurosci. **29**, 10827–10834 (2009)
84. Heekeren, H.R., Marrett, S., Ruff, D.A., Bandettini, P.A., Ungerleider, L.G.: Involvement of human left dorsolateral prefrontal cortex in perceptual decision making is independent of response modality. Proc. Natl. Acad. Sci. U. S. A. **103**, 10023–10028 (2006)
85. Buhle, J.T., Silvers, J.A., Wage, T.D., Lopez, R., Onyemekwu, C., Kober, H., Webe, J., Ochsner, K.N.: Cognitive reappraisal of emotion: a meta-analysis of human neuroimaging studies. Cereb. Cortex. **24**, 2981–2990 (2014)
86. Terasawa, Y., Fukushima, H., Umeda, S.: How does interoceptive awareness interact with the subjective experience of emotion? an fMRI study. Hum. Brain Mapp. 34, 598–612 (2013)

Physiological Measurement in the Research Field of Electronic Performance Monitoring: Review and a Call for NeuroIS Studies

Thomas Kalischko[1(✉)] and René Riedl[1,2]

[1] University of Applied Sciences Upper Austria, Steyr, Austria
{thomas.kalischko,rene.riedl}@fh-steyr.at
[2] Johannes Kepler University, Linz, Austria
rene.riedl@jku.at

Abstract. Electronic Performance Monitoring (EPM) refers to the computerized collection, storage, analysis, and reporting of information in the work context. Based on a literature review, we argue that the use of physiological measurement methods in the research field of electronic performance monitoring (EPM) should be considered more frequently in future studies. Analyses of the extant literature revealed that pulse rate, cheek-skin-temperature, blood pressure, and inter-heartbeat-latency measurements have been the only physiological measurement methods used to investigate EPM the outcomes stress and arousal, and that these few methods have been used in a very limited number of studies only. Most studies focused on retrospective measurement methods, predominantly survey. As the consequences of EPM application are known to be significantly related to employee reactions, including those related to the nervous system, application of physiological measurement methods promises to deliver novel research findings.

Keywords: Blood pressure · Brain · Computer monitoring · Electronic Performance Monitoring (EPM) · Heart rate · Physiological measurement

1 Introduction

Since the original proposal of "Electronic Work Monitoring", which describes the "computerized collection, storage, analysis and reporting of information" [1] about employees within an organization, many changes in digital technology have substantiated the relevance of this topic, both in science and practice. Nowadays the term "Electronic Performance Monitoring" (EPM) is more common within the literature and media. The ubiquitous use of technology at the workplace enables managers to monitor many aspects related to an organization and employee behavior [2]. Using new data sources and tools, the market research firm Gartner expects that about 80% of all companies worldwide will be monitoring employees in 2020 (gartner.com). The general development of technology has improved at a very high pace, thereby enabling new ways of monitoring. Smartphones and wearables (e.g., smartwatches) in order to track physiological parameters, such as heart rate or blood pressure, have become ubiquitous devices [3]. Many

F. D. Davis et al. (Eds.): NeuroIS 2020, LNISO 43, pp. 233–243, 2020.
https://doi.org/10.1007/978-3-030-60073-0_27

newspapers around the world reported cases where EPM leads to unacceptable employee privacy invasion and morally questionable monitoring methods. *The New York Times* [4] reports that Amazon is developing a wristband, which tracks every single movement of its warehouse workers in order to control the productivity level and to identify possible inefficient work behaviors. *Business Insider* [5] reported on measures to monitor employees working from home during the coronavirus pandemic, including being photographed every five minutes to check whether they are working or not. *BBC* [6] disclosed that Barclays, a British bank, installed software to track how much time their employees spend at their desk. Such developments towards EPM appear particularly bizarre in light of the European general data protection regulation (GDPR). Against the background of the increasing importance of EPM in organizations, we reviewed the scientific literature on EPM. Specifically, we were interested in the adoption of neurophysiological measurement for the study of EPM outcomes. We chose this focus on outcomes as application of EPM was expected to be significantly related to employee reactions (e.g., stress), including those related to the nervous system.

In Sect. 2, we briefly outline our review method. What follows is, in Sect. 3, a description of outcomes in EPM research. In Sect. 4, we discuss the few studies that used neurophysiological measurements. In Sect. 5, we provide our conclusion.

2 Review Methodology

In order to identify EPM studies, we conducted a keyword search using Scopus, Web of Science, and Google Scholar. Keywords included "electronic performance monitoring", "electronic monitoring", "computer monitoring", "workplace monitoring", "workplace surveillance", and "EPM". We only focused on peer-reviewed journal articles within the fields of management, business, human resource, computer science, social sciences, and psychology. We did not limit our search to a specific time period in order to get a full overview about the topic. Both authors of the paper then read article abstracts to check for relevance. We excluded articles that did not specifically focus on EPM, such as monitoring in medicine or public monitoring. In total, we investigated 165 articles and our analyses resulted in 124 articles with a clear focus on EPM. Based on analysis of the full content of these articles, we could only identify three papers that included physiological measurement methods in their data collection instrumentation. More information about the review methodology can be obtained from the authors upon request.

3 Outcomes in EPM Research

Based on our review, we identified six major outcome variables. We discuss the variables in the following.

Stress – The effect of EPM on stress is described by many authors. The ubiquitous technology and its pervasiveness can lead to technostress and consequences such as fatigue, burnout, or even depression [7–9]. Most authors who investigated the impact of EPM on employee stress reveal adverse effects. Thus, most studies show that application of EPM in an organization increases employee stress (e.g., [10]). More specifically, physiological stress [11] and psychological stress [12–15] is reported in the literature. Yet, some studies also report only weak, or even no, correlation between EPM use and stress [16–18]. Because (techno)stress is a construct with an outstanding reference to human physiology [8, 19], investigating the relationship between EPM and stress should see more studies with physiological measurements in the future. Table 1 groups EPM studies on stress based on self-report versus physiological measurement. As can be seen, self-reports dominate by far.

Table 1. Referenced measurement methods of stress.

Stress	Authors
Survey/Questionnaire	Aiello et al. 1995, Bartels et al. 2012, Carayon 1994, Davidson et al. 2000, DiTecco et al. 1992, Galletta et al. 1995, Hawk 1994, Henderson et al. 1998, Huston et al. 1993, Kolb et al. 1996, Mallo et al. 2007, Nebeker et al. 1993, Rogers et al. 1990, Smith et al. 1992, Sarpong et al. 2014, Sprigg et al. 2006, Varca 2006, Visser et al. 2008, Westin 1992
Physiological	Galletta et al. 1995, Henderson et al. 1998, Huston et al. 1993

Motivation – In order to accomplish objectives, motivation is an important concept within the work environment [20]. The results of EPM and its effect on motivation are, despite the relative lack of studies, rather inconsistent. Individual EPM seems to have a positive effect on motivation if compared to group monitoring [16, 21, 22]. Nevertheless, there are also studies that show no effect [23, 24] or even a negative effect [25]. EPM seems to have a positive impact on job motivation when simple and repetitive tasks have to be executed by the employee. Yet, the limited number of available studies does not allow for definitive conclusions. Table 2 summarizes our results. We could not identify studies with physiological measurement in this domain.

Table 2. Referenced measurement methods of motivation.

Motivation	Authors
Survey/Questionnaire	Aiello et al. 1995, Arnaud et al. 2013, Bartels et al. 2012, Gichuhi et al. 2016, O'Donnell et al. 2013, Rietzschel et al. 2014
Physiological	No studies

Job Satisfaction – A crucial factor from a business perspective and a major determinant of organizational performance is job satisfaction (e.g., [26]). Receiving positive feedback about monitoring itself is a crucial part, since this affects job satisfaction in a positive way. Importantly, perceived inappropriateness of EPM methods leads to negative intrinsic and extrinsic job satisfaction [27]. Moreover, there are studies that found a decrease in job satisfaction with increasing monitoring intensity [16, 24]. Privacy concerns also lead to a satisfaction decrease [28]. Employees' possibility to turn off EPM systems may positively affect job satisfaction [29]. In addition, it was found that providing employees information about monitoring in advance and informing them that quality aspects get monitored, rather than their behavior, are two factors that may increase job satisfaction [30]. Table 3 summarizes our results. We could not identify studies with physiological measurement in this domain.

Table 3. Referenced measurement methods of jobs satisfaction.

Job satisfaction	Authors
Survey/Questionnaire	Bartels et al. 2012, Chalykoff et al. 1989, Douthitt et al. 2001, Holman et al. 2002, Jeske et al. 2014/15, Nebeker et al. 1993, McNall et al. 2011, Rietzschel et al. 2014, Stanton et al. 1996/2002, Wells et al. 2007, Zweig et al. 2007
Physiological	No studies

Trust – Trust between employees and their supervisors, as well as employee trust into the organization, is critical for organizational success (e.g., [31]). EPM may have a negative impact on employees' trust towards the organization [30, 32, 33]. However, there are also studies indicating that there may be a positive impact on trust when employees receive notice on EPM system implementation in advance and its purpose [30, 34, 35]. What follows is that negative effects of EPM on trust are reported frequently, but it might not be the mere use and invasion of monitoring technology that leads to a decrease of trust, but more the lack of transparent communication why monitoring is used. Table 4 summarizes our results. We could not identify studies with physiological measurement in this domain.

Table 4. Referenced measurement methods of trust.

Trust	Authors
Survey/Questionnaire	Alder et al. 2006, Alge et al. 2004, Carpenter et al. 2016, Holland et al. 2015, Hovorka-Mead et al. 2002, McNall et al. 2009, Stanton et al. 2003, Westin 1992, Workman 2009
Interview	Stanton et al. 2003
Physiological	No studies

Commitment – There are three different types of commitment: organizational commitment [36], organizational citizenship behaviour (OCB), and counterproductive work behaviour [33]. Organizational commitment is defined as a behavioural attitude, where employees' identify themselves with the organization [36]; OCB is defined as positive employee behaviour that is beneficial to the organization [33], and counterproductive work behaviour refers to where employees' want to harm the organization and its stakeholders [33]. EPM has a negative impact on organizational commitment when the extent of monitoring is too strong [37, 38]. The results on OCB are mixed. There are results for a positive impact of EPM on OCB [39], negative results [25, 40], as well as no impact [33, 41]. For counterproductive work behaviour, only a positive correlation is reported [33, 42–44]. Table 5 summarizes our results. We could not identify studies with physiological measurement in this domain.

Table 5. Referenced measurement methods of commitment.

Commitment	Authors
Survey/Questionnaire	Bhave 2014, Chang et al. 2015, Vries et al. 2015, Greenberg et al. 1999, Jensen et al. 2012, Jeske et al. 2015, Martin et al. 2016, Niehoff et al. 1993, O'Donnell et al. 2013, Spitzmüller et al. 2006, Visser et al. 2008, Wellen et al. 2009, Yost et al. 2019
Interview	Sherif et al. 2017
Physiological	No studies

Performance – Taking Zajonc's [45] social facilitation theory into account, the sheer presence of EPM should affect performance positively. This does not seem to hold true for regular work of employees or students [46], as well as for simple tasks [47]. Yet, there are also studies that indicate a positive performance effect [48–51]. Irving et al. [10] report an increase in office productivity on easy short-circle tasks. Similar results are reported in other studies [18, 52]. In addition to positive and negative impact studies, we also found examinations, which report no significant effect of EPM on performance at all [51, 53, 54]. Table 6 groups EPM studies on stress based on self-report versus physiological measurement. As can be seen, self-reports dominate by far.

Table 6. Referenced measurement methods of performance.

Performance	Authors
Survey/Questionnaire	Aiello et al. 1993, Al-Rjoub et al. 2009, Bartels et al. 2012, Becker et al. 2014, Brewer 1995, Brewer et al. 1998, Davidson et al. 2000, Douthitt et al. 2001, Griffith 1993, Henderson et al. 1998, Huston et al. 1993, Irving et al. 1986, Kolb et al. 1996, Larson et al. 1990, Mallo et al. 2007, Nebeker et al. 1993, O'Donnell et al. 2013, Stanton et al. 2003
Physiological	Henderson 1998, Huston et al. 1993

4 Physiological Measurement in EPM Research

A majority of the reviewed studies investigated EPM based on survey (e.g. [16, 21]) and only a few authors quantified the impact of EPM based on physiological measurement.

We only found three publications with a focus on EPM using physiological measurement methods. Considering that our review covers a total of 124 papers, we observe a rate of 2.4% of papers with physiological measurement in the EPM literature. Table 7 summarizes the three papers. We briefly discuss the papers in the following.

Table 7. Referenced physiological measurement methods.

Author	Year	Published	Sample size	Measurement method	Investigated outcome
Huston et al. [17]	1993	HICSS[a]	18[d]	Earlobe pulse meter; cheek skin temperature probe	Performance [~], Stress [~]
Galletta et al. [55]	1995	AMIT[b]	18	Pulse rate; skin temperature	Stress [~]
Henderson et al. [56]	1998	IJHCS[c]	32	Blood pressure; inter-heartbeat latency	Performance [~], Arousal/Stress [-]

[a]Proceedings of the Annual Hawaii International Conference on System Sciences
[b]Accounting, Management and Information Technologies
[c]International Journal of Human-Computer Studies
[d]Full sample size was 44. Yet, physiological measures were only used from 18 participants.

Huston et al. [17] performed two separate studies. In the first study, 18 participants had to fill out several STAI (Spielberger State-Trait Anxiety Index) forms and an earlobe pulse meter sensor as well as a cheek skin temperature probe were attached to each participant. Regarding physiological measurement, the authors write: "[T]he physiological measures used in the first experiment were discarded because of the extremely low variation in the measures […] Also, because the physiological measures of experiment one did not yield significant results and, because the subjects may have been affected by the physical attachment of these devices, it was helpful to assess the performance of a group that was placed in a more natural work environment without the possible interference of the physiological devices […] promising physiological measures should be considered once again. Options for such measurement would include skin conductance; cortisol secretions in saliva, blood, or urine; or combinations of skin temperature, pulse rates, and skin conductance as measured by a polygraph. Selection of measures is highly dependent on their obtrusiveness, ease of capture, and sensitivity. Promising measures might be galvanic skin response and salivary cortisol; researchers who hope to make use of these measures should investigate ways to diminish their obtrusiveness" (pp. 571–573). What can we conclude based on this study by Huston et al. [17]? Despite the fact that the authors report low inter-individual variation and potential obtrusiveness issues, they made a call for application of neurophysiological measurement in future studies.

Galletta et al. [55] also conducted two separate studies with 18 participants to investigate the relationship between EPM and stress. Participants were randomly assigned to either a computer-monitored or a human-monitored group. During the experiment, the participants had to complete a competitive and a non-competitive task. Pulse rate and skin temperature measurements were used. They also used the STAI form to collect retrospective stress measurements. The pulse rates decreased over time, because participants probably became more comfortable and lost their initial laboratory anxiety from task to task. The authors concluded that pulse rate only revealed small differences that were not significant, they write: "Pulse rates. For virtually all subjects, pulse rates slowed as they performed each task. The task was not strenuous, therefore it is not surprising to find this decline. The subjects probably became more comfortable over time as they lost their initial laboratory anxiety and became accustomed to the tasks. What was important, therefore, was the within subjects perspective afforded by this study; our analysis concerned the differential rate at which pulse rates slowed for each group. By the final composition task, the mean pulse rate for each group slowed to about 95% of what it was for the practice task. Both ANOVA and repeated measures ANOVA revealed that the small differences were not significant for any of the five pulse measures" (p. 170).

Skin temperature measurements yielded similar results. The authors indicate: "Skin temperatures. The skin temperature of all subjects appeared to increase along with the room temperature, which drifted upwards as the day progressed. In fact, there was a striking temperature pattern for all subjects; by the end of the last task, temperature had increased nearly one degree. ANOVA and repeated measures ANOVA again confirmed that none of the fractional differences in any of the five measures was significant" (p. 170). Thus, this study failed to discover any significant physiological differences between the computer-monitored and a human-monitored group.

Henderson et al. [56] conducted another study, based on 32 participants, in order to investigate the impact of EPM on performance and stress. They used inter-heartbeat-latency and blood pressure measurement. The computer task required the participants to enter mock clinical case notes under various conditions. In condition 1, subjects were required to enter the case notes while keystroke data were collected. Condition 2 was divided into three discrete stages. In stage 1, the security baseline condition, participants were informed that a keystroke security monitoring system had been instituted, but no security challenges occurred. In stage 2, the security challenge condition, a several security challenges occurred. In stage 3, the performance monitoring condition, subjects were informed that their data entry speed was monitored and they were placed on a response-cost schedule for poor performance. Results indicate that monitoring systems caused "altered arousal states in the form of increased heart rate and blood pressure" (p. 143) and "Electronic monitoring systems, whether they be security or performance, would seem to have the potential to be stress evoking. This was evidenced by the decreased inter-heartbeat latency observed during the security baseline condition. The altered cardiovascular state was elicited just by the knowledge that such a monitoring system was in place" (p. 154). Moreover, it is reported that the hypothesized improvement in task performance within the performance monitoring condition could not be observed.

5 Conclusion

Current studies in the EPM literature mainly use retrospective measures (predominantly survey). We only found three papers that used physiological measurement methods.

Those papers investigated the impact of EPM on stress and arousal. What follows is that physiological measurement has not been applied so far in the investigation of EPM consequences on motivation, job satisfaction, trust, and commitment. Another finding of our study is that the methods that were used so far only refer to measurement of autonomic nervous system activity, but not to brain activity measurement (e.g., fMRI, EEG, fNIRS) [57]. Based on these findings, we argue that future studies on the consequences of EPM use should consider application of physiological measurement. Importantly, physiological measurements should be used as complements to self-reports and other methods (such as behaviour observation), and not as substitute. Because important EPM outcomes such as motivation, satisfaction, and trust have physiological correlates (see, for example, a review by Riedl and Javor [58], on the neurobiological correlates of trust, as well as literature cited in Riedl and Léger [57], on other correlates in the cognitive neuroscience literature), we foresee high potential of neuroscience approaches in the research field of EPM.

Acknowledgement. This research was funded by the Austrian Science Fund (FWF) as part of the project "Technostress in organizations" (project number: P 30865) at the University of Applied Sciences Upper Austria.

References

1. U.S. Congress, Office of Technology Assessment, The Electronic Supervisor: New Technology, New Tensions, OTA-CIT-333 (1987)
2. Tredinnick, L., Laybats, C.: Workplace surveillance. Bus. Inf. Review. **36**, 50–52 (2019)
3. Sherif, K., Al-Hitmi, M.: The moderating role of competition and paradoxical leadership on perceptions of fairness towards IoT monitoring. In: AMCIS 2017 - America's Conference on Information Systems: A Tradition of Innovation (2017)
4. Yeginsu, C.: If Workers Slack Off, the Wristband Will Know. (And Amazon Has a Patent for It). https://www.nytimes.com/2018/02/01/technology/amazon-wristband-tracking-privacy.html
5. Holmes, A.: Employees at home are being photographed every 5 minutes by an always-on video service to ensure they're actually working — and the service is seeing a rapid expansion since the coronavirus outbreak. https://www.businessinsider.com/work-from-home-sneek-webcam-picture-5-minutes-monitor-video-2020-3?r=DE&IR=T
6. Barclays scraps 'Big Brother' staff tracking system. https://www.bbc.com/news/business-51570401
7. Ayyagari, R., Grover, V., Purvis, R.: Technostress: technological antecedents and implications. MIS Q. **35**(4), 831–858 (2011)
8. Riedl, R.: On the biology of technostress: literature review and research agenda. SIGMIS Database **44**(1), 18–55 (2013)
9. Tarafdar, M., Cooper, C.L., Stich, J.F.: The technostress trifecta - techno eustress, techno distress and design. Theoretical directions and an agenda for research. Inf. Syst. J. **29**, 6–42 (2017)
10. Irving, R.H., Higgins, C.A., Safayeni, F.R.: Computerized performance monitoring systems: use and abuse. Commun. ACM **29**, 794–801 (1986)
11. Amick, B.C., Smith, M.J.: Stress, computer-based work monitoring and measurement systems: a conceptual overview. Appl. Ergon. **23**, 6–16 (1992)

12. DiTecco, D., Cwitco, G., Arsenault, A., André, M.: Operator stress and monitoring practices. Appl. Ergon. **23**, 29–34 (1992)
13. Carayon, P.: Effects of electronic performance monitoring on job design and worker stress: results of two studies. Int. J. Hum. Comput. Interact. **6**, 177–190 (1994)
14. Hawk, S.R.: The effects of computerized performance monitoring: an ethical perspective. J. Bus. Ethics **13**, 949–957 (1994)
15. Varca, P.E.: Telephone surveillance in call centers: prescriptions for reducing strain. Manag. Serv. Qual. **16**, 290–305 (2006)
16. Bartels, L.K., Nordstrom, C.R.: Examining big brother's purpose for using electronic performance monitoring. Perform. Improv. Q. **25**, 65–77 (2012)
17. Huston, T.L., Galletta, D.F., Huston, J.L.: The effects of computer monitoring on employee performance and stress: results of two experimental studies. In: Proceedings of the 26th Hawaii International Conference on System Sciences, vol. 4, pp. 568–574 (1993)
18. Nebeker, D.M., Tatum, B.C.: The effects of computer monitoring, standards, and rewards on work performance, job satisfaction, and stress. J. Appl. Soc. Psychol. **23**, 508–536 (1993)
19. Riedl, R., Kindermann, H., Auinger, A., Javor, A.: Technostress from a neurobiological perspective: system breakdown increases the stress hormone cortisol in computer users. Bus. Inf. Syst. Eng. **4**(2), 61–69 (2012)
20. Herzberg, F., Maunser, B., Snyderman, B.: The Motivation to Work. Wiley, New York (1959). Relations Industrielles 15, 275
21. Aiello, J.R., Kolb, K.J.: Electronic performance monitoring and social context: impact on productivity and stress. J. Appl. Psychol. **80**, 339–353 (1995)
22. Gichuhi, J.K., Ngari, J.M., Senaji, T.: Employees' response to electronic monitoring: the relationship between CCTV surveillance and employees' engagement. Int. J. Innov. Res. Dev. **5**, 141–150 (2016)
23. Arnaud, S., Chandon, J.: Will monitoring systems kill intrinsic motivation? An empirical study. Revue de gestion des ressources humaines **90**, 35 (2013)
24. Rietzschel, E.F., Slijkhuis, M., Van Yperen, N.W.: Close monitoring as a contextual stimulator: how need for structure affects the relation between close monitoring and work outcomes. Eur. J. Work. Organ. Psychol. **23**, 394–404 (2014)
25. O'Donnell, A.T., Ryan, M.K., Jetten, J.: The hidden costs of surveillance for performance and helping behavior. Group Process. Intergroup Relat. **16**, 246–256 (2013)
26. Bakotić, D.: Relationship between job satisfaction and organisational performance. Econ. Res. **29**, 118–130 (2016)
27. Alder, G.S., Ambrose, M.L.: An examination of the effect of computerized performance monitoring feedback on monitoring fairness, performance, and satisfaction. Organ. Behav. Hum. Decis. Process. **97**, 161–177 (2005)
28. Seppänen, M., Pajarre, E., Kuparinen, P.: The effects of performance monitoring technology on privacy and job autonomy. Int. J. Bus. Inf. Syst. **20**, 139–156 (2015)
29. Douthitt, E.A., Aiello, J.R.: The role of participation and control in the effects of computer monitoring on fairness perceptions, task satisfaction, and performance. J. Appl. Psychol. **86**, 867–874 (2001)
30. Stanton, J.M., Sarkar-Barney, S.T.M.: A detailed analysis of task performance with and without computer monitoring. Int. J. Hum. Comput. Interact. **16**, 345–366 (2003)
31. Mayer, R.C., Davis, J.H., Schoorman, F.D.: An integrative model of organizational trust. Acad. Manag. Rev. **20**, 709–734 (1995)
32. Holland, P.J., Cooper, B., Hecker, R.: Electronic monitoring and surveillance in the workplace. Pers. Rev. **44**, 161–175 (2015)
33. Jensen, J.M., Raver, J.L.: When self-management and surveillance collide. Group Organ. Manag. **37**, 308–346 (2012)

34. Alder, G.S., Noel, T.W., Ambrose, M.L.: Clarifying the effects of Internet monitoring on job attitudes: the mediating role of employee trust. Inf. Manag. **43**, 894–903 (2006)
35. Hovorka-Mead, A., Ross, W.H., Whipple, T., Renchin, M.B.: Watching the detectives: seasonal student employee reactions to electronic monitoring with and without advanced notification. Pers. Psychol. **55**, 329–362 (2002)
36. Mowday, R.T., Steers, R.M., Porter, L.W.: The measurement of organizational commitment. J. Vocat. Behav. **14**, 224–247 (1979)
37. Chang, S.E., Liu, A.Y., Lin, S.: Exploring privacy and trust for employee monitoring. Ind. Manag. Data Syst. **115**, 88–106 (2015)
38. Visser, W.A., Rothmann, S.: Exploring antecedents and consequences of burnout in a call centre: empirical research. J. Ind. Psychol. **34**, 79–87 (2008)
39. Bhave, D.P.: The invisible eye? Electronic performance monitoring and employee job performance. Pers. Psychol. **67**, 605–635 (2014)
40. Jeske, D., Santuzzi, A.M.: Monitoring what and how: psychological implications of electronic performance monitoring. New Technol. Work. Employ. **30**, 62–78 (2015)
41. Niehoff, B.P., Moorman, R.H.: The effects of computer monitoring, standards, and rewards on work performance, job satisfaction, and stress. J. Appl. Soc. Psychol. **23**, 508–536 (1993)
42. Greenberg, L., Barling, J.: Predicting employee aggression against coworkers, subordinates and supervisors: the roles of person behaviors and perceived workplace factors. J. Organ. Behav. **20**, 897–913 (1999)
43. Martin, A.J., Wellen, J.M., Grimmer, M.R.: An eye on your work: how empowerment affects the relationship between electronic surveillance and counterproductive work behaviours. Int. J. Hum. Resour. Manag. **27**, 2635–2651 (2016)
44. Wellen, J., Martin, A., Hanson, D.: The impact of electronic surveillance and workplace empowerment on work attitudes and behaviour. In: Industrial Organisational Psychology Conference, vol. 8, pp. 145–149 (2009)
45. Zajonc, R.B.: Social facilitation. Science **149**, 218–232 (1965)
46. Mallo, J., Nordstrom, C.R., Bartels, L.K., Traxler, A.: Electronic performance monitoring the effect of age and task difficulty. Perform. Improv. Q. **20**, 49–63 (2007)
47. Becker, T.E., Marique, G.: Observer effects without demand characteristics: an inductive investigation of video monitoring and performance. J. Bus. Psychol. **29**(4), 541–553 (2013)
48. Brewer, N.: The effects of monitoring individual and group performance on the distribution of effort across tasks. J. Appl. Soc. Psychol. **25**, 760–777 (1995)
49. Brewer, N., Ridgway, T.: Effects of supervisory monitoring on productivity and quality of performance. J. Exp. Psychol. **4**, 211 (1998)
50. Goomas, D.T., Ludwig, T.D.: Standardized goals and performance feedback aggregated beyond the work unit: optimizing the use of engineered labor standards and electronic performance monitoring. J. Appl. Soc. Psychol. **39**, 2425–2437 (2009)
51. Davidson, R., Henderson, R.: Electronic performance monitoring: a laboratory investigation of the influence of monitoring and difficulty on task performance, mood state, and self-reported stress levels. J. Appl. Soc. Psychol. **30**, 906–920 (2000)
52. Al-Rjoub, H., Zabian, A., Qawasmeh, S.: Electronic monitoring: the employees point of view. J. Soc. Sci. **4**, 189–195 (2008)
53. Griffith, T.L.: Monitoring and performance: a comparison of computer and supervisor monitoring. J. Appl. Soc. Psychol. **23**, 549–572 (1993)
54. Kolb, K.J., Aiello, J.R.: The effects of electronic performance monitoring on stress: locus of control as a moderator variable. Comput. Hum. Behav. **12**, 407–423 (1996)
55. Galletta, D., Grant, R.A.: Silicon supervisors and stress: merging new evidence from the field. Account. Manag. Inf. Technol. **5**, 163–183 (1995)

56. Henderson, R., Mahar, D., Saliba, A., Deane, F., Napier, R.: Electronic monitoring systems: an examination of physiological activity and task performance within a simulated keystroke security and electronic performance monitoring system. Int. J. Hum. Comput. Stud. **48**, 143–157 (1998)
57. Riedl, R., Léger, P.-M.: Fundamentals of NeuroIS – Information Systems and the Brain. Springer, Berlin (2016)
58. Riedl, R., Javor, A.: The biology of trust: Integrating evidence from genetics, endocrinology and functional brain imaging. J. Neurosci. Psychol. Econ. **5**, 63–91 (2012)

The Effect of Individual Coordination Ability on Cognitive-Load in Tacit Coordination Games

Dor Mizrahi[1(✉)], Ilan Laufer[1], and Inon Zuckerman[1,2]

[1] Department of Industrial Engineering and Management, Ariel University, Ariel, Israel
dor.mizrahi1@msmail.ariel.ac.il, {ilanl,inonzu}@ariel.ac.il
[2] Data Science and Artificial Intelligence Center, Ariel University, Ariel, Israel

Abstract. Tacit coordination games are coordination games in which communication between the players is not allowed or not possible. Some players manage to reason about the selections made by the co-player while others fail to do so and might turn to rely on guessing. The aim of this study is to examine whether good coordinators are associated with a higher or lower cognitive load relative to weaker coordinators. We aimed to answer this question by using an electrophysiological marker of cognitive load, i.e., Theta/Beta Ratio. Results show that good coordinators are associated with a higher cognitive load with respect to weaker coordinators.

Keywords: Tacit coordination games · EEG · Theta/Beta Ratio

1 Introduction

In this study we have applied a dual process account to human decision making in the context of a pure coordination game. In pure coordination games both players share common interests, and each player has an equal interest to successfully coordinate with the other player since coordinating on the same solution is beneficial for both [1].

Dual process cognitive framework posits that an interaction exists between intuitive automatic processes and more deliberate processes which are more controlled and reflective [2]. These two processes (intuitive and deliberate) are assumed to be involved in effective coordination [2] which requires reliance on common knowledge [3].

From the perspective of a dual process account, good coordinators might rely on a certain strategy [4–6] that may ease the coordination process and therefore reduce the associated cognitive load. Thus, the convergence on the same solution to achieve coordination might be entirely intuitive, involve heuristic choice strategies [7] and may therefore be regarded as a highly automatic process [8]. On the other hand, it might be that deliberate coordination relies on more cognitive resources and therefore entails a higher cognitive load when coordination is performed.

Thus, good coordinators manage to reason about the selections made by the co-player while weak coordinators fail to do so and might turn to rely on guessing. In tacit coordination games, the pure coordination game used here, players reach an agreement

F. D. Davis et al. (Eds.): NeuroIS 2020, LNISO 43, pp. 244–252, 2020.
https://doi.org/10.1007/978-3-030-60073-0_28

regarding a salient solution (i.e. a focal point) without any communication, (e.g. [1, 14, 20, 22]) considering only pay-irrelevant cues [9] such as spatial location. Therefore, tacit coordination games provide an excellent experimental environment for testing cooperative decision making in the context of dual process theory, since it is the most basic form of coordination and does not include any form of communication or conflicts of interest [9].

In this study we aimed to test whether there is a difference in cognitive load between good and poor coordinators in the context of a tacit coordination game by using an electrophysiological marker of cognitive load, i.e., Theta/Beta Ratio (TBR) [10–12]. TBR is known to decrease as cognitive load increases and vice-versa. Hence, in this study we utilize the TBR to find out whether good coordinators are associated with a lower TBR with respect to weaker coordinators.

2 Materials and Methods

The EEG Data acquisition process during the tacit coordination game session was recorded by a 16-channel g.USBAMP biosignal amplifier (g.tec, Austria) at a sampling frequency of 512 Hz. 16 active electrodes were used for collecting EEG signals from the scalp based on the international 10–20 system. Recording was done by the OpenVibe [13] recording software. Impedance of all electrodes was kept below the threshold of 5K [ohm] during all recording sessions.

In this study players were presented with a tacit coordination task in which they had to select a word from a given set of four words (in Hebrew) in order to coordinate with an unknown co-player [14]. This task consists of 12 different instances each with a different set of words. For example game board #1 displayed in Fig. 1 (B) contains the set {"Water", "Beer", "Wine", "Whisky"}. All the words belong to the same semantic category, however, there is at least one word that stands out from the rest of the set because it is different in some prominent feature, e.g., in the current example, a non-alcoholic beverage ("water") which stands out among other alcoholic beverages. The more salient is the outlier, the easier it is to converge on the same focal point [15, 16].

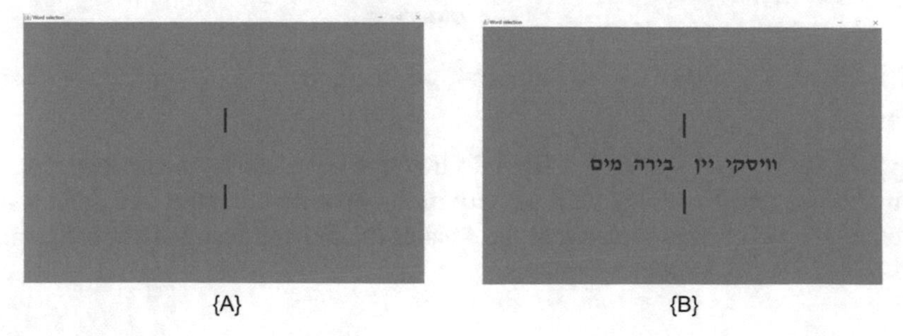

Fig. 1. (A) Stand by screen (B) Game board #1 ["Water", "Beer", "Wine", "Whisky"]

Figure 2 portrays the outline of the experiment. The list of four words were embedded within a sequence of standby screens each presented for *U(2,2.5)* sec. The slide presenting the list of words was presented for a maximal duration of 8 s and the next slide appeared following a button press. The order of the 12 games was randomized in each session. In each of the games the players were told that they have to coordinate with an unknown randomly selected co-player by choosing the same word from the given set of words. Participants were further informed that they will receive an amount of 100 points each in case of successful coordination and that otherwise they will get nothing. The participants were 10 students from Ariel University that were enrolled in one of the courses on campus (right-handed, mean age = ~26, SD = 4).

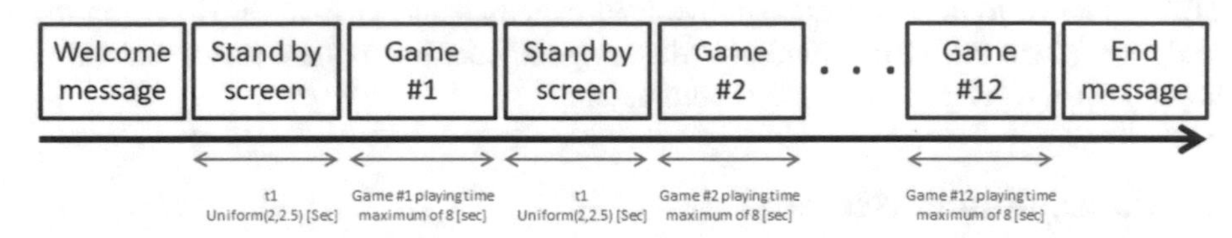

Fig. 2. Experimental paradigm with timeline

In this study the following measures were computed.

Individual Coordination Ability (iCA) – The iCA measure reflects the individual coordination ability of each player with respect to the other players in the group [5, 6, 17]. The iCA calculates the total number of games in which each player was able to coordinate their responses against the entire population. The iCA is formally defined as follows:

$$iCA(i) = \sum\nolimits_{j=1|(j \neq i)}^{N} \sum\nolimits_{k=1}^{t} \frac{CF(i,j,k)}{(N-1) * t} \tag{1}$$

Where *i* denote the ith participant, j denotes the index of the jth co-player, *N* denotes the total number of participants, and *t* denotes the number of games in the experiments. The CF (coordination function) is defined as follows:

$$CF(i,j,k) = \begin{cases} 1; & \textit{if players i and j chose the same label in game k} \\ 0; & \textit{otherwise} \end{cases} \tag{2}$$

The higher the player's iCA value (ranged in [0, 1]), the higher the coordination ability.

Theta Beta Ratio (TBR) – is known from the literature to reflect cognitive load in various cognitive tasks and to covary with activity in the executive control and default mode networks [10, 12]. It was shown that the smaller the TBR, the cognitive load is higher [10, 12] (see Sect. 3 for more details).

3 EEG Metrics for Assessing Cognitive Load

Cognitive load refers to the amount of working memory resources required to perform a particular task [18] and there are two basic approaches for estimating cognitive load from EEG data. The first approach relies on power spectrum analysis of continuous EEG that reveals the distribution of signal power over frequency. In this method the EEG signal is divided into different frequency bands (i.e. delta, theta, alpha, and beta) in order to detect the bands sensitive to variations in load as a function of task demands. To estimate cognitive load, power-based features are extracted such as the signal's average or maximum power, and the ratio between bands may also be calculated (e.g. [19–22]) as was done in the current study (i.e. the energy ratio between the theta and beta bands, the TBR measure, was computed). Power spectrum analysis was used in various studies associated with the information systems (IS) discipline [27].

The second approach involves measuring the neural signal complexity that has been associated with both memory ability [23] and cognitive load [21]. Common methods in this category include fractal dimension (e.g. [24]), multi-scale entropy (e.g. [25]), and detrended fluctuation analysis [26, 27]. Table 1 summarizes the above-mentioned EEG metrics.

Table 1. EEG metrics for assessing cognitive load

Cognitive load estimation technique	Metric 1	Metric 2	Metric 3
Power spectrum analysis	Accumulated band power ratio (e.g. accumulated TBR)	Maximal band power ratio (e.g. maximal TBR)	Average band power ratio (e.g. mean TBR)
Neural signal complexity	Fractal dimension	Multi-scale entropy	Detrended fluctuation analysis

Xie and Salvendy [28, 29] have differentiated between several main indices meant to quantify mental workload. These measures include instantaneous load (dynamic changes in load during task performance), peak load, average load, overall load and accumulated load. In the current study we have created a hybrid index as follows. For each epoch we have first calculated the accumulated cognitive load [12], by calculating the energy ratio between theta and beta bands for each participant on each single epoch. Then, we have averaged the ratio across all epochs of an individual player to obtain the average cognitive load.

4 Data Processing and Analysis

Based on the literature (e.g. [11, 30–32]), we focused on the following cluster of frontal and prefrontal electrodes (Fp1, F7, Fp2, F8, F3, and F4). The preprocessing pipeline (see Fig. 3) consisted of band-pass filtering [1, 32] Hz Subsequent by notch filtering of [50] Hz for an artifact removal following iCA. The preprocessing pipeline (see Fig. 3) consisted of band-pass filtering [1, 32] Hz an artifact removal following iCA. The data was re-referenced to the average reference and down sampled to 64 Hz following baseline correction. Data was analyzed on a 1-s epoch window from the onset of each game.

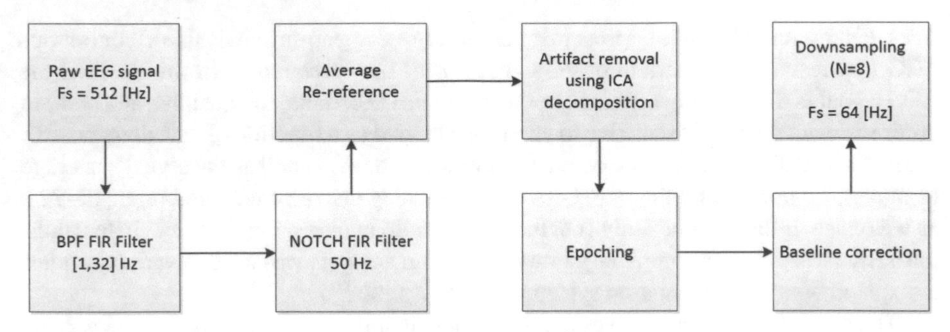

Fig. 3. Preprocess pipeline

To calculate the energy in the Theta and Beta bands, for each epoch, we have used the Discrete Wavelet Transform (DWT) [33, 34]. The DWT is based on a multiscale feature representation. Every scale represents a unique thickness of the EEG signal [35]. Each filtering step contains two digital filters, a high pass filter, $g(n)$, and a low pass filter $h(n)$. After each filter, a downsampler with factor 2 is used in order to adjust time resolution. In our case, we used a 3-level DWT, with the input signal having a sampling rate of 64 Hz. As can be seen in Fig. 4, this specific DWT scheme resulted in the coefficients of the four EEG main frequency bands (see red rectangles in Fig. 4).

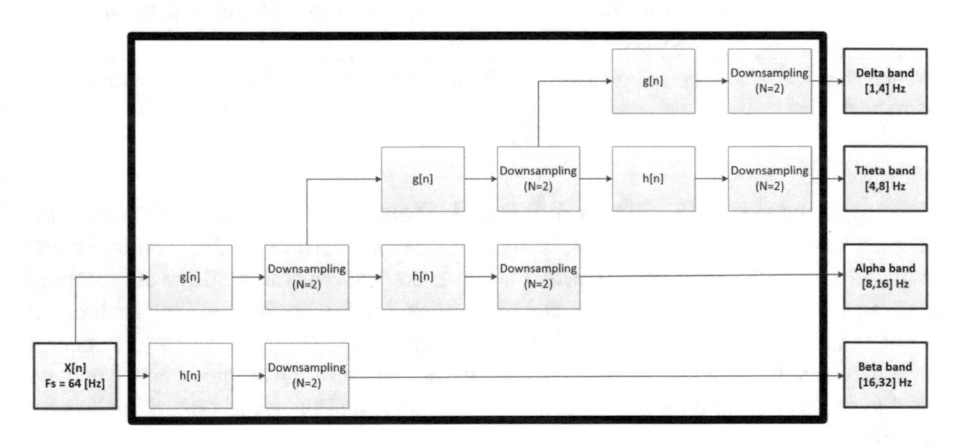

Fig. 4. 3 level DWT scheme

To calculate the cognitive load, which is expressed by the TBR, the DWT was applied on all the epochs to calculate the TBR. That is the ratio of the average energy in each one of the Theta and Beta bands (Theta/Beta) was calculated in each of the epochs.

To find out whether there is a direct and significant relationship between each player's individual coordination ability (iCA) and the cognitive load (TBR) we performed several computations. First, we calculated the iCA value of each player. Next, the average individual TBR was calculated for each of the six channels by weighting the 12 epochs. Finally, for each channel, the relationship was calculated between the mean individual TBR and the iCA using linear regression (Table 2).

Table 2. Results of modeling the relationship between ICA* and TBR

	Regression model	p-value	R-squared
Channel 1 – Fp1	TBR = 18.415 **– 46.802***iCA	p = 0.0147	0.5456
Channel 2 – F7	TBR = 17.886 **– 45.396***iCA	p = 0.0240	0.4901
Channel 5 – Fp2	TBR = 18.401 – **46.191***iCA	p = 0.0251	0.4859
Channel 6 – F8	TBR = 18.200 – **44.376***iCA	p = 0.0235	0.4934
Channel 9 – F3	TBR = 19.660 – **50.880***iCA	p = 0.0475	0.4062
Channel 13 – F4	TBR = 19.785 – **50.793***iCA	p = 0.0068	0.6207

(*in this table ICA denotes individual coordination ability)

Table 2 presents the regression model for each of the six channels. It can be seen that all regression lines turned out to be significant ($p < 0.05$). The highest coefficients in the regression model were obtained for F3 and F4, while the most significant p-value was obtained for F4. The negative coefficient denotes that there is a negative relationship between iCA and TBR. That is, the higher is the iCA, the smaller is the TBR. Hence, higher iCA is associated with a higher cognitive load. Figure 5 presents the regression line for the F4 channel. The regression line shows a clear negative relationship between iCA and TBR.

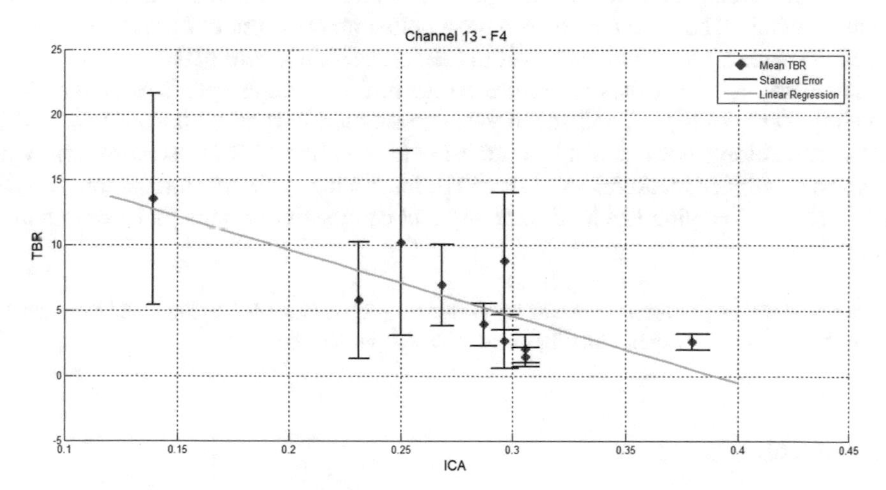

Fig. 5. The regression model of F4

5 Conclusions and Future Work

To the best of our knowledge, this is the first time in which a correlation is shown between individual coordination ability in tacit coordination games and electrophysiological marker of cognitive load. Specifically, we have demonstrated this relationship by using a prefrontal and frontal cluster of electrodes. This result corroborates previous research showing that complex cognitive tasks depend on prefrontal [11] and frontal [12] cortex activation. This result also strengthens the connections found between TBR fluctuations and executive control [11]. It might have been hypothesized that better coordinators lean on a certain behavioral strategy, and in turn utilize less cognitive resources than weaker coordinators. However, our results do not coincide with this assumption but rather demonstrate the opposite, namely that better coordinators use more cognitive resources in order to coordinate. Hence, our results also support the dual process theory as was utilized by Tversky and Kahneman to discuss human bounded rationality [36]. It appears that in order to coordinate, players rely on deliberation (System-2) mode of thinking which is more deliberate and analytical than intuition (System-1) which relies on fast heuristics and is therefore more automatic.

Apparently, the results of this study stand in contrast to previous findings showing significant negative correlations between glucose consumption and task performance [37–39]. These studies indicate that good performers of a complex task may use less brain circuits or less inefficient brain areas compared to poor performers and underscore the effects of practice. These results are also in agreement with behavioral findings showing that experts perform more efficiently than novices in decision-making contexts [40–42]. Taken together with the glucose metabolism studies, it may very well be the case that more efficient performers, as a result of expertise and practice, rely more efficiently on cognitive resources. However, in our study better coordinators may be associated with a higher cognitive load, not because of less efficiency in the functionality of brain circuits, but rather because they reason more deliberately (a system 2 process) about the selections made by the other co-player to reach successful coordination.

There are many avenues for future research. For example, previous studies have shown that culture [6, 43] and social value orientation [44] affect human behavior in tacit coordination games. It will be interesting to see if the TBR is also correlated with the abovementioned measures. Also, it will be interesting to observe fluctuations in TBR as a function of varying levels of difficulty and complexity of other tacit coordination games.

Acknowledgement. This research was supported by grant number RA1900000666 provided by the Data Science and Artificial Intelligence center at Ariel University.

References

1. McAdams, R.H.: Conventions and norms (Philosophical aspects). In: International Encyclopedia of Social & Behavioral Sciences (2001)
2. Belloc, M., Bilancini, E., Boncinelli, L., D'Alessandro, S.: Intuition and deliberation in the stag hunt game. Sci. Rep. **9**, 1–7 (2019)
3. De Freitas, J., Thomas, K., DeScioli, P., Pinker, S.: Common knowledge, coordination, and strategic mentalizing in human social life. Proc. Natl. Acad. Sci. U. S. A. **116**, 13751–13758 (2019)
4. Duffy, S., Smith, J.: Cognitive load in the multi-player prisoner's dilemma game: are there brains in games? J. Behav. Exp. Econ. **51**, 47–56 (2014)

5. Mizrahi, D., Laufer, I., Zuckerman, I.: Individual strategic profiles in tacit coordination games. J. Exp. Theor. Artif. Intell. 1–16 (2020)
6. Mizrahi, D., Laufer, I., Zuckerman, I.: Collectivism-individualism: strategic behavior in tacit coordination games. PLoS ONE **15**, e0226929 (2020)
7. Poulsen, A., Sonntag, A.: Focality is intuitive - experimental evidence on the effects of time pressure in coordination games (2019)
8. Krueger, J.I.: From social projection to social behaviour. Eur. Rev. Soc. Psychol. **18**, 1–35 (2008). https://doi.org/10.1080/10463280701284645
9. Sitzia, S., Zheng, J.: Group behaviour in tacit coordination games with focal points - an experimental investigation. Games Econ. Behav. **117**, 461–478 (2019)
10. van Son, D., de Rover, M., De Blasio, F.M., van der Does, W., Barry, R.J., Putman, P.: Electroencephalography theta/beta ratio covaries with mind wandering and functional connectivity in the executive control network. Ann. N. Y. Acad. Sci. **1452**, 52–64 (2019)
11. Gartner, M., Grimm, S., Bajbouj, M.: Frontal midline theta oscillations during mental arithmetic: effects of stress. Front. Behav. Neurosci. **9**, 1–8 (2015)
12. Bagyaraj, S., Ravindran, G., Shenbaga Devi, S.: Analysis of spectral features of EEG during four different cognitive tasks. Int. J. Eng. Technol. **6**, 725–734 (2014)
13. Renard, Y., Lotte, F., Gibert, G., Congedo, M., Maby, E., Delannoy, V., Bertrand, O., Lécuyer, A.: Openvibe: an open-source software platform to design, test, and use brain–computer interfaces in real and virtual environments. Presence Teleoperators Virtual Environ. **19**, 35–53 (2010)
14. Bardsley, N., Mehta, J., Starmer, C., Whitehead, K.: Explaining focal points: cognitive hierarchy theory versus team reasoning about the centre or contact. Econ. J. **120**, 40–79 (2009)
15. Schelling, T.C.: The Strategy of Conflict, Cambridge (1960)
16. Mehta, J., Starmer, C., Sugden, R.: Focal points in pure coordination games: an experimental investigation. Theory Decis. **36**, 163–185 (1994)
17. Mizrahi, D., Laufer, I., Zuckerman, I.: Modeling individual tacit coordination abilities. In: International Conference on Brain Informatics, pp. 29–38. Springer, Cham (2019)
18. Antonenko, P., Paas, F., Grabner, R., van Gog, T.: Using electroencephalography to measure cognitive load. Educ. Psychol. Rev. **22**, 425–438 (2010)
19. Zarjam, P., Epps, J., Chen, F.: Spectral EEG features for evaluating cognitive load. In: Proceedings of the Annual International Conference of the IEEE Engineering in Medicine and Biology Society, EMBS, pp. 3841–3844 (2011)
20. Kumar, N., Kumar, J.: Measurement of cognitive load in HCI systems using EEG power spectrum: an experimental study. Procedia Comput. Sci. **84**, 70–78 (2016)
21. Friedman, N., Fekete, T., Gal, K., Shriki, O.: EEG-based prediction of cognitive load in intelligence tests. Front. Hum. Neurosci. **13**, 191 (2019)
22. Müller-Putz, G.R., Riedl, R., Wriessnegger, S.C.: Electroencephalography (EEG) as a research tool in the information systems discipline: foundations, measurement, and applications. Commun. Assoc. Inf. Syst. **37**, 46 (2015)
23. Sheehan, T.C., Sreekumar, V., Inati, S.K., Zaghloul, K.A.: Signal complexity of human intracranial EEG tracks successful associative-memory formation across individuals. J. Neurosci. **38**, 1744–1755 (2018)
24. Stokić, M., Milovanović, D., Ljubisavljević, M.R., Nenadović, V., Čukić, M.: Memory load effect in auditory–verbal short-term memory task: EEG fractal and spectral analysis. Exp. Brain Res. **233**(10), 3023 (2015)
25. Escudero, J., Abásolo, D., Hornero, R., Espino, P., López, M.: Analysis of electroencephalograms in Alzheimer's disease patients with multiscale entropy. Physiol. Meas. **27**, 1091–1106 (2006)

26. Peng, C.K., Havlin, S., Stanley, H.E., Goldberger, A.L.: Quantification of scaling exponents and crossover phenomena in nonstationary heartbeat time series. Chaos **5**, 82–87 (1995). https://doi.org/10.1063/1.166141
27. Rubin, D., Fekete, T., Mujica-Parodi, L.R.: Optimizing complexity measures for fMRI data: algorithm, artifact, and sensitivity. PLoS ONE **8** (2013). https://doi.org/10.1371/journal.pone.0063448
28. Xie, B., Salvendy, G.: Review and reappraisal of modelling and predicting mental workload in single-and multi-task environments. Work Stress **14**, 74–99 (2010)
29. Xie, B., Salvendy, G.: Prediction of mental workload in single and multiple tasks environments. Int. J. Cogn. Ergon. **4**, 213–242 (2000). https://doi.org/10.1207/S15327566IJCE0403
30. De Vico Fallani, F., Nicosia, V., Sinatra, R., Astolfi, L., Cincotti, F., Mattia, D., Wilke, C., Doud, A., Latora, V., He, B., Babiloni, F.: Defecting or not defecting: how to "read" human behavior during cooperative games by EEG measurements. PLoS ONE **5**, e14187 (2010)
31. Boudewyn, M., Roberts, B.M., Mizrak, E., Ranganath, C., Carter, C.S.: Prefrontal transcranial direct current stimulation (tDCS) enhances behavioral and EEG markers of proactive control. Cogn. Neurosci. **10**, 57–65 (2019)
32. Moliadze, V., Sierau, L., Lyzhko, E., Stenner, T., Werchowski, M., Siniatchkin, M., Hartwigsen, G.: After-effects of 10 Hz tACS over the prefrontal cortex on phonological word decisions. Brain Stimul. **12**, 1464–1474 (2019)
33. Shensa, M.J.: The discrete wavelet transform: wedding the. A trous and mallat algorithms. IEEE Trans. signal Process. **40**, 2464–2482 (1992)
34. Jensen, A., la Cour-Harbo, A.: Ripples in Mathematics: The Discrete Wavelet Transform. Springer (2001)
35. Hazarika, N., Chen, J.Z., Tsoi, A.C., Sergejew, A.: Classification of EEG signals using the wavelet transform. Signal Process. **59**, 61–72 (1997)
36. Kahneman, D.: Thinking, Fast and Slow. Macmillan, London (2011)
37. Haier, R.J., Siegel, B.V., MacLachlan, A., Soderling, E., Lottenberg, S., Buchsbaum, M.S.: Regional glucose metabolic changes after learning a complex visuospatial/motor task: a positron emission tomographic study. Brain Res. **570**, 134–143 (1992)
38. Haier, R.J., LaFalase, J., Katz, M., Nuechterlein, K., Buchsbaum, M.S.: Brain efficiency and intelligence: inverse correlations between cerebral glucose metabolic rate and abstract reasoning. Manuscript submitted for publication (1992)
39. Hazletr, E.: Cortical glucose metabolic rate correlates of abstract reasoning and attention studied with positron emission tomography. Intelligence **12**, 199–217 (1988)
40. Leger, P.-M., René, R., vom Brocke, J.: Emotions and ERP information sourcing: the moderating role of expertise. Ind. Manag. Data Syst. **114**(3), 456–471 (2014)
41. Hong, J.C., Liu, M.C.: A study on thinking strategy between experts and novices of computer games. Comput. Hum. Behav. **19**, 245–258 (2003)
42. Hung, S.Y.: Expert versus novice use of the executive support systems: an empirical study. In: Proceedings of the 34th Annual Hawaii International Conference on System Sciences (2003). https://doi.org/10.1016/S0378-7206(02)00003-4
43. Cox, T.H., Lobel, S.A., Mcleod, P.L.: Effects of ethnic group cultural differences on cooperative and competitive behavior on a group task. Acad. Manag. J. **34**, 827–847 (1991)
44. Mizrahi, D., Laufer, I., Zuckerman, I., Zhang, T.: The effect of culture and social orientation on player's performances in tacit coordination games. In: Proceedings of the International Conference on Brain Informatics – BI 2018 Arlington, TX, USA, 7–9 December 2018, pp. 437–447 (2018)

Analysis of Contextual Effects of Advertising Banners

Sebastian Schöber(✉) and Harald Kindermann

University of Applied Sciences Upper Austria, Campus Steyr, Wels, Austria
Sebastian.schoeber@gmail.com, Harald.kindermann@fh-steyr.at

Abstract. In the light of the fact that advertising contacts can be affected by the surrounding contents, this paper focuses on answering the question to what extent the mood, triggered by an online news article in which a banner is placed, influences the brand decision. It seems that, the content of an article will not impact the effect of the banner and thus on brand decisions. A conducted experiment revealed that neither negative, neutral nor positive articles had a significant impact on the brand decision.

Due to the exploratory approach of the study and the fact that similar publications show different results, more research is needed to clarify the discrepancy.

Keyword: Banner advertising · Online news article · Mood · Brand decision

1 Introduction

Online advertising has been part of the standard bouquet of companies to spread advertising messages and generate clicks or purchases. Investments in this area are also increasing annually [1].

Because banner ads are embedded in websites, they provide a context to the advertising. Therefore, this work aims to investigate whether online news articles have an influence on brand decisions through the moods they trigger.

In order to evaluate the effectiveness of banner advertising, companies usually use click-through rates (CTR). However, it is also possible to measure constructs such as attention, recall, recognition, etc., as well as some long-term communication effects such as behavioral changes [2, 3]. However, research on how banners influence effectiveness is still at an early stage [4]. At the same time, a number of studies on this topic came to controversial empirical results [5].

We speak of a context effect when an independent context feature is significantly related to a dependent individual feature, even when other relevant individual and context features are controlled [6].

Weber/Fahr came to the conclusion that especially in media advertising the media context can influence the effect of the web contact [7]. In addition, Internet advertising is

F. D. Davis et al. (Eds.): NeuroIS 2020, LNISO 43, pp. 253–258, 2020.
https://doi.org/10.1007/978-3-030-60073-0_29

valued more favorably and results in a greater purchase intention of consumers compared to the advertised products when ads are embedded in contextually relevant websites [8].

When speaking about context effects, the term priming plays a role. This is a well-known phenomenon in psychology, which describes the influence of a preceding stimulus that has activated implicit memory content on the processing of stimuli. This activation based on previous experiences occurs often or in most cases unconsciously [9]. Psychological basis for these effects are the neuronal networks, which consist of many different, interconnected nodes. If one of these nodes is activated, an activation of the other connected nodes follows (spreading activation). At the same time, the concepts that are not connected to the triggered node are suppressed. Connections between the nodes can be more or less distinct. In addition, these components have different valence regarding positive or negative evaluation [10–12].

Schemer writes about the priming effect of media coverage that certain information activates certain cognitions in the cognitive network of people. These are then more readily available than non-activated concepts. This effect can also influence behavior [9].

If one relates to the behavior-influencing impacts of contextual effects to online banner advertising [9], it is reasonable to assume that the sympathy of brands that place banners depends, among other things, on the context in which the banner is placed.

Therefore, it is the aim of this thesis to find out to what extent the mood triggered by an online news article in which a banner is placed influences the brand decision.

In order to avoid misunderstandings the authors find it necessary to explain what is meant by the term “mood”. It is important to remark that the term mood may not synonymously be used with the terms “emotion”, “affective state” and “feeling”. In this paper, when using the term “mood”, the authors refer to a mental state, that is triggered by reading an online news article. By using a pretest, these articles were classified by the mood they were supposed to trigger: positive, neutral and negative. This was done through a simple self-assessment of the test subjects. According to the results of the pretest, we assumed that by reading the articles the respective moods would indeed be triggered in the subjects: That means for example, it was assumed that those subjects reading the article which was classified as positive mood triggering, were actually set into a positive mood.

2 Methodology

Empirical research was conducted to answer the research question. It was designed as an exploratory study just to find out any coherences. The experiment was conducted using an online questionnaire.

Initially, the test subjects were asked to assess different brands to the extent that they like them. After that, the subjects were given the actual task. The aim was to determine whether the type of content of a typical article in daily news will have a short term impact on a purchase decision or whether no influence can be demonstrated. For this purpose, the following experiment was carried out.

All subjects (n = 157) were randomly divided into four groups (three experimental groups and one control group). At the beginning, the test subjects of the experimental

UNIVERSITY OF APPLIED SCIENCES UPPER AUSTRIA

Kronen Zeitung

REZEPT DER WOCHE mit Peter Tichatschek

Nachrichten Oberösterreich Sport Adabei Digital Freizeit Auto Trends Videos Service

Wien NÖ / Bgld. Oberösterreich Steiermark Kärnten Salzburg Tirol / Vlbg.

Mattsee

Bauboom

Mattsees Bevölkerung wächst um 10% in drei Jahren

Zehn Prozent Bevölkerungswachstum in nur drei Jahren! Bald wird das Flachgauer Mattsee an die 3500 Einwohner haben, denn derzeit werden 100 neue Wohnungen gebaut - und es werden noch mehr. Aber kann ein Ort ein derart schnelles Wachstum vertragen? Laut Bürgermeister Rene Kuel im Moment schon noch.

Traumhaft liegt die Flachgauer Gemeinde Mattsee eingebettet zwischen den Trumerseen. Von den Hügeln ein herrlicher Blick auf die Gewässer, im Ort der direkte Seezugang mit dem Strandbad, die Häfen der beiden Segelclubs und eine große Parkanlage. Viele Menschen zieht das an.
Natürlich erkannten das auch große Bauunternehmen und eine Großbaustelle nach der anderen schoss aus dem Boden. Im Flachgau ist das zwar kein neues Phänomen, wenn aber die Bevölkerungszahl in nur drei Jahren um zehn Prozent - über 300 Bewohner - wächst ist es doch nicht alltäglich. 100 Wohnungen werden im Moment gebaut und zum Teil schon bezogen. „Auf die Bauaktivitäten haben wir keinen Einfluss, da die Flächen ja schon vor Jahren gewidmet wurden", erklärt Bürgermeister Rene Kuel. Sein Ort wird bald 3500 Einwohner haben. Zum Vergleich: Seit den 1960ern stieg die Zahl meist um 300-500 Bürger im Jahrzehnt.

„Im Moment ist das noch kein Problem, aber die Kinderbetreuung muss in den nächsten ein bis zwei Jahren wieder erweitert werden", sagt der Bürgermeister und ist sicher: „Noch so einen Sprung in kurzer Zeit würde der Ort nicht schaffen."Noch Thema: Ein neues Gemeindeamt im Ort

„Es werden noch weitere Projekt verwirklicht. Diese fallen aber kleiner aus", erklärt Kuel. Auch die Gemeinde selbst könnte noch aktiv werden: Ob das Amt auf den Dorfplatz übersiedeln wird, soll bis spätestens 2019 entschieden werden. „Das wäre eine zusätzliche Belebung des Ortskerns."

Felix Roittner

Aktuelle Schlagzeilen

Arsenal-Legende: „Özil spielt wie ein Geist" — FUSSBALL INTERNATIONAL

LIVE: Bestzeit! Max Franz setzt sich an die Spitze — WINTERSPORT

Das brauchen Sie für Ihre Detox-Kur

Videos

Nach Attacke auf Bus: Ägyptische Polizei tötet 40 „Terroristen"

Rückzug aus Syrien: Werden US-Truppen Kurden ihre Waffen überlassen?

Kurz stellt klar: „Werden die Digitalsteuer in Österreich einführen"

Top 3 (der letzten 72 Stunden)

GELESEN KOMMENTIERT

1 Brutaler Raubüberfall auf Ordensbrüder in Kirche — 157.585 mal gelesen

2 Sänger (29) ging bei Kreuzfahrt über Bord — 132.767 mal gelesen

3 Silbereisen sagte Trennung von Fischer voraus — 125.400 mal gelesen

Newsletter

Melden Sie sich hier mit Ihrer E-Mail-Adresse an, um täglich den "Krone"-Newsletter zu erhalten.

Ihre E-Mail Adresse — LOS

Mehr Salzburg

Das letzte Wort ist noch nicht gesprochen

Rabiater Mann bedrohte Trio mit einer Gaspistole

Vier Alkolenker in einer Nacht aus Verkehr gezogen

Schon wieder drei Einbrüche in der Mozartstadt

Totaler Triumph für Salzburg

Weit

Sebastian Schöber, Digital Business Management, University of Applied Sciences Upper Austria, Campus Steyr – 2019

30% ausgefüllt

Fig. 1. Exemplary screenshot of the neutral article

groups were asked to assess a typical newspaper article, to what extent this article was formulated objectively. As a reward for this task, a bottle of mineral water was promised. Depending on the random group allocation, the test subjects were presented with an article with a neutral (group 1), a negative (group 2) or a positive message (group 4), which they should evaluate. A banner from the brand "Römerquelle" was built into each article. In order to determine whether the respective content of the article influences the banner effect, the subjects were immediately afterwards offered to choose out of two similar mineral water brands (Römerquelle versus Vöslauer). The remaining control group did not have to assess an article. So, they could choose from the same two mineral water brands without a preceding specific task. Figure 1 shows exemplary the neutral article which the experimental group 1 had to read.

Chi-squared tests of independence were carried out to determine whether the content of the article influenced the choice of brand. The brand decision served as a dependent variable. Group membership was the independent variable. The Kronen Zeitung was used as the medium from which the articles were taken, as it is the newspaper with the highest print run in Austria [13]. The news articles thus originated from the website krone.at and were presented as screenshots in the questionnaire.

The articles were selected by means of a pretest. A total of nine articles were preselected – three negative, three neutral and three positive. For the pretest, these were not yet preperated with the banner used later in the experiment. The articles were presented to a total of 20 people for reading. Using a questionnaire, the subjects of the pretest were instructed to rate the different articles on a seven-part Likert scale ("−3" = negative, "0" = neutral, "+3" = positive) according to the mood they triggered (question: " What mood does the article trigger in you?"). The articles with the most extreme mean value of the negative, neutral or positive mood were selected and later used for the experiment. For the experiment as such, screenshots of these articles were used and included in the questionnaire. These were edited to eliminate disruptive factors as far as possible. The date and the region reference (krone.at has subpages for each region in Austria) removed and the side news bars unified.

The format chosen for the banner was Medium Rectangle, because this format integrates the ad into the editorial part of the website and thus increases the visibility for the users. In addition to the increased visibility, a higher credibility of the advertisement can be achieved [14].

3 Results

As described at the beginning, the aim is to find out whether the moods triggered by the individual articles influence the brand decision.

The data was evaluated using contingency tables and chi-square tests. Each brand selection decision was evaluated individually. Table 1 shows the results. In the table, which represents the evaluation by means of cross tables, the expected number indicates how many people should select the respective brand with complete independency. The more the "number" differs from the "expected number", the more the result was influenced by the article.

The Chi-square tests statistic of Chi-Square = 4,411 (df = 3) indicate non significant results (p = 0.220).

Table 1. Selection decisions Römerquelle vs. Vöslauer: Cross table

		Group			
		Nobanner	Negative article	Neutral article	Positive article
Vöslauer	Number	31,0	18,0	27,0	23,0
	Expected number	27,7	22,7	24,6	24,0
Römerquelle	Number	13,0	18,0	12,0	15,0
	Expected number	16,3	13,3	14,4	14,0

4 Discussion

Combining the results of the evaluation according to individual groups (negative, neutral, positive, no article), it can be clearly deduced that the mood triggered by an article with an integrated banner has no significant influence on the brand decision.

This result is insofar surprising, since most of the existing papers that investigated the same or very similar topics with similar research designs came to the conclusion that the media content in which the banners are embedded influences the attitude towards the brand [5, 15–17].

For example Kindermann, who has examined to what extent the banner blindness occurs, came to very different results. For this study, banners of three different soft drink brands (Pfanner, Bravo, Happy Day) were inserted into the lower right corner of an existing online newspaper site. Similar to the current study, the test persons in Kindermann's study were confronted with the screenshot of the website with one of the above mentioned banners or with the version without such a banner, depending on their group affiliation. Immediately after the confrontation with the websites including the banners and a distraction task, the test persons were asked to indicate their preferences for these brands. For this purpose they were then confronted with all three possible pairs of the three brands mentioned above. The results show, among other things, that an initially positive presentation of a brand has a positive effect on brand selection, and an initially negative presentation of a brand has a negative effect. A banner is therefore only effective if someone has mainly positive images of the brand in mind. In all other cases, the opposite is the case and the probability is high that such a banner will worsen future buying behavior [15].

In another paper, Kindermann came to the conclusion, among other things, that a negative banner brand could benefit from the positive brand by being placed on the side of a positive brand and was therefore perceived more positively than at the beginning of the experiment. Huang drew similar conclusions from the results of his research. Here it could be observed that a positive attitude of the test persons towards the website content had a positive influence on their attitude towards the advertising banner [5, 16].

In light of the big contrast of the results this paper conducted compared to other studies, more research has to be done in different settings, to come to a final conclusion, if and how context effects influence banner brands.

References

1. eMarketer. Werbeinvestitionen weltweit in den Jahren 2011 bis 2016 und Prognose bis 2021 (in Milliarden US-Dollar) [Internet]. 2017. https://de.statista.com/statistik/daten/studie/190508/umfrage/entwicklung-der-werbeinvestitionen-weltweit/. Accessed 15 Apr 2019
2. Yoo, C.Y.: Unconscious processing of Web advertising: effects on implicit memory, attitude toward the brand, and consideration set. J. Interact. Market. **22**(2), 2–18 (2008). https://doi.org/10.1002/dir.20110
3. Hamborg, K.-C., Bruns, M., Ollermann, F., Kaspar, K.: The effect of banner animation on fixation behavior and recall performance in search tasks. Comput. Hum. Behav. **28**(2), 576–582 (2012). https://doi.org/10.1016/j.chb.2011.11.003
4. Li, K., Huang, G., Bente, G.: The impacts of banner format and animation speed on banner effectiveness: evidence from eye movements. Comput. Hum. Behav. **54**, 522–530 (2016). https://doi.org/10.1016/j.chb.2015.08.056
5. Kindermann, H.: Priming and context effects of banner ads on consumer based brand equity: a pilot study. In: Nah, F.F.-H., Tan, C.-H., (eds.) HCI in Business, Government and Organizations. Supporting Business, 4th International Conference, HCIBGO 2017, Held as Part of HCI International 2017, Vancouver, BC, Canada, July 9–14, 2017, Proceedings, Part II. Lecture Notes in Computer Science, pp. 55–70. Springer International Publishing, Cham (2017)
6. Alpheis, H.: Kontextanalyse: Die Wirkung des sozialen Umfeldes, untersucht am Beispiel der Eingliederung von Ausländern. Wiesbaden: Deutscher Universitätsverlag, 326 p. ger (1988)
7. Weber, P., Fahr, A.: Werbekommunikation. In: Schweiger, W., Fahr, A. (eds.) Handbuch Medienwirkungsforschung, pp. 333–352. Springer VS, Wiesbaden (2013)
8. Jeong, Y., King, C.M.: Impacts of website context relevance on banner advertisement effectiveness. J. Promot. Manag. **16**(3), 247–264 (2010). https://doi.org/10.1080/10496490903281395
9. Myers, D.G.: Psychologie. Springer , Heidelberg (2014)
10. Schemer, C.: Priming, Framing, Stereotype. In: Schweiger, W., Fahr, A. (eds.) Handbuch Medienwirkungsforschung, pp. 153–169. Springer VS, Wiesbaden (2013)
11. Higgins, E.T.: Knowledge activation: Accessibility, applicability, and salience. In: Higgins, E.T., Kruglanski, A.W. (eds.) Social Psychology, Handbook of Basic Principles, pp. 133–168. U.S. Guilford Press, New York (1996)
12. Collins, A.M., Loftus, E.F.: A spreading-activation theory of semantic processing. Psychol. Rev. **82**(6), 407–428 (1975). https://doi.org/10.1037//0033-295X.82.6.407
13. Statista. Media-Analyse (Österreich): Durchschnittliche Anzahl der Leser von Tageszeitungen in Österreich von 2016 bis 2018 (in 1.000) [Internet] (n.d.). https://de.statista.com/statistik/daten/studie/307114/umfrage/tageszeitungen-in-oesterreich-nach-anzahl-der-leser/. Accessed 24 Apr 2019
14. Kreutzer, R.T.: Praxisorientiertes Online-Marketing. Springer Fachmedien Wiesbaden, Wiesbaden (2018)
15. Kindermann, H.: A short-term twofold impact on banner ads. In: Nah, F.F.-H., Tan, C.-H. (eds.) HCI in Business, Government, and Organizations: eCommerce and Innovation, Third International Conference, HCIBGO 2016, Held as Part of HCI International 2016, Toronto, Canada, July 17–22, 2016, Proceedings, Part I. Lecture Notes in Computer Science, pp. 417–26. Springer, Cham (2016)
16. Huang, S.: The impact of context on display ad effectiveness: automatic attitude activation and applicability. Electron. Commer. Res. Appl. **13**(5), 341–354 (2014). https://doi.org/10.1016/j.elerap.2014.06.006
17. Rieger, D., Bartz, F., Bente, G.: Reintegrating the ad: effects of context congruency banner advertising in hybrid media. J. Media Psychol. **27**(2), 64–77 (2015). https://doi.org/10.1027/1864-1105/a000131

Measuring Extraversion Using EEG Data

Hermann Baumgartl(✉), Samuel Bayerlein, and Ricardo Buettner

Aalen University, Aalen, Germany
hermann.baumgartl@hs-aalen.de

Abstract. Using a modern fine-graded machine learning approach we show that it is possible to distinguish extraverts from introverts on the basis of resting-state EEG data. We correctly identify extraverts with a prediction performance of 67% and achieve a balanced accuracy of 60.6%. Our results have theoretical and practical implications.

Keywords: Electroencephalography · Machine learning · Big-five · Personality traits · Extraversion

1 Introduction

For some time scholars have argued that personality influences behavior [1]. As a result, researchers in the field of psychology have intensively investigated the impact of personality on behavior and have shown that it, in fact, has a lot of influence on how people act, think and behave [2–5].

Extraversion describes social skills, talkative ability, and personal charm [6]. Furthermore, extraverted people have a higher need for self-presentation [3]. Currently, extraversion is assessed through questionnaires [7].

EEG is one of the most widely used tools in NeuroIS research [8, 9] and new studies show that extraversion has a positive correlation with alpha subbands in social interactions, showing the influence of personality traits on EEG data [10]. EEG measurements can hardly be manipulated by the patient, therefore, provide more unbiased results [11]. It is also more reliable than the time-consuming assessments using questionnaires. Therefore, we investigate the possibility of predicting personality traits based on EEG data using a modern fine-graded machine learning approach. We successfully developed a novel machine learning approach for the detection of extraversion from resting state EEG data.

2 Methodology

2.1 Data and Participants

The data we used in our research is part of the Leipzig Study for Mind-Body-Emotion Interactions (LEMON) Dataset [12]. This data is publicly available and consists of a 62-channel resting-state EEG experiment and the NEO-FFI personality assessment [12].

F. D. Davis et al. (Eds.): NeuroIS 2020, LNISO 43, pp. 259–265, 2020.
https://doi.org/10.1007/978-3-030-60073-0_30

The EEG data collected is a 16-min resting state record, consisting of 16 blocks of 60-s records, eight with eyes open and eight with eyes closed. The dataset comprised of 202 patients. Answers for the NEO-FFI range from 0 (strong denial) to 4 (strong approval) on a 5-point Likert scale. The extraversion score reached from 1.00 to 3.92 with a mean score of 2.41 and a standard deviation of ±0.52. The participants were grouped into introverts and extroverts by splitting at the mean score.

2.2 Preprocessing

We applied an anti-aliasing FIR filter before downsampling the data from 2,500 Hz to 250 Hz, applied a bandpass filter (0.5–50 Hz) and removed EEG artifacts using the FASTER automated ICA algorithm [13]. The tool used in our preprocessing, containing the automated ICA and other tools was provided by eegUtils v.0.5.0, R x64 3.6.1.

2.3 Machine Learning and Feature Extraction

To extract a rich feature space for machine learning, we used the fine-grained EEG spectrum [14] to get a 99-power band spectrum. The signal was transformed into a frequency signal via the fast Fourier transformation, using the eight blocks of eyes closed data Contrary to the traditional division into five bands (delta, theta, alpha, beta, gamma [15]) of unequal step size, the fine-grained sub-bands have an equal 0.5 Hz range. The fine grained EEG spectra approach was already successfully applied in the detection of different health conditions such as alcoholism [16], schizophrenia [17], sleep disorders [18] and epilepsy [19]. As for the machine learning algorithm, a Random Forest algorithm was used [20]. By using a Random Forest, we can also evaluate the frequency sub-bands with the most support for the correct classification of extraversion. In order to find these specific frequency bands and gain more insights into the underlying processes, we evaluated the variable importance of the Random Forest algorithm [21].

2.4 Validation Procedure

For training and evaluation purposes, the dataset was split into a 75% training partition and 25% evaluation partition. To ensure the validity of our model we used a holdout 10-fold cross-validation [22, 23] and evaluated on the holdout testing part. The dataset used to train contained 152 participants with 71 encoded as introverts and 81 encoded as extraverts. In the validation set, the distribution of the labels was 23 introverts and 27 extraverts.

3 Results

The results are shown in Table 1. The trained model has a mean balanced accuracy of 60.60%.

Table 1. Model performance based on the validation set

Performance indicator	Value	SD
Balanced accuracy	60.60%	1.7%
True positive rate	54.54%	6.5%
True negative rate	66.66%	6.0%
Positive predictive value	58.39%	2.4%
Negative predictive value	63.39%	1.9%
Prevalence	46.00%	–

4 Discussion

While former approaches using the traditional wide-ranged EEG bands (e.g. Korjus et al. [24]) were not able to predict extraversion from EEG data, we are the first to predict extraversion using the fine-grained EEG spectrum approach. The current baseline for extraversion detection is at 50% (e.g. random guess). By using the three most predictive EEG sub-bands (1.5–2 Hz; 2.5–3 Hz; 13.5–14 Hz) based on the Random Forest feature importance (all above the importance of 80) we achieved a balanced accuracy of 60.60%, showcasing the potential to predict personality traits solely from resting state EEG data. We could archive a predictive gain of 21.2% from the baseline model. As we can see in Table 1, we can identify extraverted people with a performance of 66.66%. With these results, we can provide a basis for the further development of an alternative measure for personality traits and a better classification of the extraversion trait.

5 Limitations and Future Work

While introverts show greater reactivity to sensory stimulation than extroverts, under neutral conditions such as resting-state EEG, there is little difference in baseline levels [25]. In our study, this could mean that stimuli induced differences between extroverts and introverts are not present in the data. Using EEG data including a stimulus could show these differences and might support a better separation between the two groups. Another limitation concerns the ground truth of the extraversion trait for the model training. Since our model is trained on the scores obtained through the NEO-FFI questionnaire, our model cannot outperform the questionnaire at the current stage and any potential bias in the extraversion scores are inherited by our model. Therefore, further research is required in order to assess the extraversion score more objectively and established a new ground truth for the model training.

In future work we evaluate the robustness of the approach by assessing the influence of individual states and mental concepts such as cognitive workload [26–28], concentration [29] and mindfulness [5, 30] in multi-agent-settings [31–34]. Furthermore, we will triangulate psychophysiological and physiological data (i.e., electroencephalographic data and spectra [35, 36], electrocardiographic data [37, 38], electrodermal activity [39], eye fixation [40, 41], eye pupil diameter [42–44], facial data [45]) to increase reliability.

In addition, we will evaluate technology acceptance [46–49] and trust in our machine learning-based personality trait prediction method [50–56]. Furthermore, we want to apply our identified resting-state EEG sub-bands to convolutional neural networks [57–60] and test for the influence of eyes closed vs. eyes open conditions on the prediction of personality traits.

References

1. Gale, A., Coles, M., Blaydon, J.: Extraversion-introversion and the EEG. Br. J. Psychol. **60**, 209–223 (1969)
2. Barrick, M.R., Mount, M.K., Judge, T.A.: Personality and performance at the beginning of the new millennium: what do we know and where do we go next? Int. J. Sel. Assess. **9**, 9–30 (2001)
3. Buettner, R.: Personality as a predictor of business social media usage: an empirical investigation of XING usage patterns. In: PACIS 2016 Proceedings, p. 163 (2016)
4. Buettner, R.: Predicting user behavior in electronic markets based on personality-mining in large online social networks. Electron. Markets **27**(3), 247–265 (2016)
5. Sauer, S., Buettner, R., Heidenreich, T., Lemke, J., Berg, C., Kurz, C.: Mindful machine learning. Eur. J. Psychol. Assess. **34**, 6–13 (2018)
6. Bai, S., Hao, B., Li, A., Yuan, S., Gao, R., Zhu, T.: Predicting big five personality traits of microblog users. In: IEEE/WIC/ACM WI-IAT 2013 Proceedings, pp. 501–508. IEEE (2013)
7. McCrae, R.R., Costa, P.T.: Empirical and theoretical status of the five-factor model of personality traits. In: Boyle, G.J. (ed.) Personality Theories and Models, pp. 273–294. SAGE, Los Angeles (2010)
8. Davis, F., Riedl, R., Hevner, A.: Towards a NeuroIS research methodology: intensifying the discussion on methods, tools, and measurement. JAIS **15**, I–XXXV (2014)
9. Riedl, R., Fischer, T., Léger, P.-M., Davis, F.: A Decade of NeuroIS research: progress, challenges, and future directions. data base for advances in information systems **51** (2020, in Press)
10. Roslan, N.S., Izhar, L.I., Faye, I., Amin, H.U., Mohamad Saad, M.N., Sivapalan, S., Abdul Karim, S.A., Abdul Rahman, M.: Neural correlates of eye contact in face-to-face verbal interaction: an EEG-based study of the extraversion personality trait. PLoS ONE **14**, e0219839 (2019)
11. Vom Brocke, J., Riedl, R., Léger, P.-M.: Application strategies for neuroscience in information systems design science research. J. Comput. Inf. Syst. **53**, 1–13 (2013)
12. Babayan, A., Erbey, M., Kumral, D., Reinelt, J.D., Reiter, A.M.F., Röbbig, J., Schaare, H.L., Uhlig, M., Anwander, A., Bazin, P.-L., et al.: A mind-brain-body dataset of MRI, EEG, cognition, emotion, and peripheral physiology in young and old adults. Sci. Data **6**, 180308 (2019)
13. Nolan, H., Whelan, R., Reilly, R.B.: FASTER: fully automated statistical thresholding for EEG artifact rejection. J. Neurosci. Methods **192**, 152–162 (2010)
14. Buettner, R., Rieg, T., Frick, J.: Machine learning based diagnosis of diseases using the unfolded EEG spectra: towards an intelligent software sensor. In: Davis, F.D., Riedl, R., Vom Brocke, J., Léger, P.-M., Randolph, A.B., Fischer, T. (eds.) Information Systems and Neuroscience. NeuroIS Retreat 2019, vol. 32, pp. 165–172. Springer, Cham (2019)
15. Müller-Putz, G.R., Riedl, R., Wriessnegger, S.C.: Electroencephalography (EEG) as a research tool in the information systems discipline: foundations, measurement, and applications. CAIS **37**, 46 (2015)

16. Rieg, T., Frick, J., Hitzler, M., Buettner, R.: High-performance detection of alcoholism by unfolding the amalgamated EEG spectra using the Random Forests method. In: HICSS-52 Proceedings, pp. 3769–3777 (2019)
17. Buettner, R., Beil, D., Scholtz, S., Djemai, A.: Development of a machine learning based algorithm to accurately detect schizophrenia based on one-minute EEG recordings. In: HICSS-53 Proceedings, pp. 3216–3225 (2020)
18. Buettner, R., Grimmeisen, A., Gotschlich, A.: High-performance diagnosis of sleep disorders: a novel, accurate and fast machine learning approach using electroencephalographic data. In: HICSS-53 Proceedings, pp. 3246–3255 (2020)
19. Buettner, R., Frick, J., Rieg, T.: High-performance detection of epilepsy in seizure-free EEG recordings: A novel machine learning approach using very specific epileptic EEG sub-bands. In: ICIS 2019 Proceedings, pp. 1–16 (2019)
20. Breiman, L.: Random Forests. Mach. Learn. **45**, 5–32 (2001)
21. Louppe, G., Wehenkel, L., Sutera, A., Geurts, P.: Understanding variable importances in forests of randomized trees. In: NIPS 2013 Proceedings, pp. 431–439. Curran Associates Inc. (2013)
22. Bengio, Y., Grandvalet, Y.: No unbiased estimator of the variance of k-fold cross-validation. J. Mach. Learn. Res. **5**, 1089–1105 (2004)
23. Fushiki, T.: Estimation of prediction error by using K-fold cross-validation. Stat. Comput. **21**, 137–146 (2011)
24. Korjus, K., Uusberg, A., Uusberg, H., Kuldkepp, N., Kreegipuu, K., Allik, J., Vicente, R., Aru, J.: Personality cannot be predicted from the power of resting state EEG. Front. Hum. Neurosci. **9**, 63 (2015)
25. Stelmack, R.M.: Biological bases of extraversion: psychophysiological evidence. J. Pers. **58**, 293–311 (1990)
26. Buettner, R.: The relationship between visual website complexity and a user's mental workload: a NeuroIS perspective. In: Davis, F.D., Riedl, R., Vom Brocke, J., Léger, P.-M., Randolph, A.B. (eds.) Information Systems and Neuroscience. Gmunden Retreat on NeuroIS 2016, pp. 107–113. Springer, Cham (2016)
27. Buettner, R.: A user's cognitive workload perspective in negotiation support systems: An eye-tracking experiment. In: PACIS 2016 Proceedings, p. 115 (2016)
28. Buettner, R., Scheuermann, I.F., Koot, C., Rössle, M., Timm, I.J.: Stationarity of a user's pupil size signal as a precondition of pupillary-based mental workload evaluation. In: Davis, F.D., Riedl, R., vom Brocke, J., Léger, P.-M., Randolph, A.B. (eds.) Information Systems and Neuroscience, vol. 25, pp. 195–200. Springer, Cham (2018)
29. Buettner, R., Baumgartl, H., Sauter, D.: Microsaccades as a predictor of a user's level of concentration. In: Davis, F.D., Riedl, R., Vom Brocke, J., Léger, P.-M., Randolph, A.B. (eds.) Information Systems and Neuroscience. NeuroIS Retreat 2018, vol. 29, pp. 173–177. Springer, Cham (2018)
30. Sauer, S., Lemke, J., Zinn, W., Buettner, R., Kohls, N.: Mindful in a random forest: assessing the validity of mindfulness items using random forests methods. Pers. Individ. Diff. **81**, 117–123 (2015)
31. Buettner, R.: A Classification structure for automated negotiations. In: IEEE/WIC/ACM WI-IAT 2006 Proceedings, pp. 523–530. IEEE (2006)
32. Buettner, R., Kirn, S.: Bargaining power in electronic negotiations: a bilateral negotiation mechanism. In: Psaila, G., Wagner, R. (eds.) EC-Web 2008, vol. 5183, pp. 92–101. Springer, Heidelberg (2008)
33. Buettner, R.: Cooperation in hunting and food-sharing: a two-player bio-inspired trust model. In: Altman, E., Carrera, I., El-Azouzi, R., Hart, E., Hayel, Y. (eds.) Bioinspired Models of Network, Information, and Computing Systems, vol. 39, pp. 1–10. Springer, Heidelberg (2010). https://doi.org/10.1007/978-3-642-12808-0_1

34. Landes, J., Buettner, R.: Argumentation–based negotiation? Negotiation–based argumentation! In: EC-Web 2012 Proceedings, 123, pp. 149–162. Springer (2012)
35. Buettner, R., Hirschmiller, M., Schlosser, K., Roessle, M., Fernandes, M., Timm, I.J.: High-performance exclusion of schizophrenia using a novel machine learning method on EEG data. In: IEEE Healthcom 2019 Proceedings, pp. 1–6. IEEE (2019)
36. Buettner, R., Fuhrmann, J., Kolb, L.: Towards high-performance differentiation between narcolepsy and idiopathic hypersomnia in 10 minute EEG recordings using a novel machine learning approach. In: IEEE Healthcom 2019 Proceedings, pp. 1–7. IEEE (2019)
37. Buettner, R., Bachus, L., Konzmann, L., Prohaska, S.: Asking both the user's heart and its owner: empirical evidence for substance dualism. In: Davis, F.D., Riedl, R., Vom Brocke, J., Léger, P.-M., Randolph, A.B. (eds.) Information Systems and Neuroscience. NeuroIS Retreat 2018, vol. 29, pp. 251–257. Springer, Cham (2018)
38. Buettner, R., Schunter, M.: Efficient machine learning based detection of heart disease. In: IEEE Healthcom 2019 Proceedings, pp. 1–6. IEEE (2019)
39. Eckhardt, A., Maier, C., Buettner, R.: The influence of pressure to perform and experience on changing perceptions and user performance: a multi-method experimental analysis. In: ICIS 2012 Proceedings (2012)
40. Buettner, R.: Cognitive workload of humans using artificial intelligence systems: towards objective measurement applying eye-tracking technology. In: Timm, I.J., Thimm, M. (eds.) KI 2013: Advances in Artificial Intelligence, vol. 8077, pp. 37–48. Springer, Heidelberg (2013). https://doi.org/10.1007/978-3-642-40942-4_4
41. Eckhardt, A., Maier, C., Hsieh, J.J., Chuk, T., Chan, A.B., Hsiao, J.H., Buettner, R.: Objective measures of IS usage behavior under conditions of experience and pressure using eye fixation data. In: ICIS 2013 Proceedings (2013)
42. Buettner, R.: Social inclusion in e-participation and e-government solutions: a systematic laboratory-experimental approach using objective psychophysiological measures. In: EGOV/ePart 2013 Proceedings, pp. 260–261. Gesellschaft für Informatik e.V, Bonn (2013)
43. Buettner, R., Daxenberger, B., Eckhardt, A., Maier, C.: Cognitive workload induced by information systems: introducing an objective way of measuring based on pupillary diameter responses. In: Pre-ICIS HCI/MIS 2013 Proceeding, Paper 20 (2013)
44. Buettner, R., Sauer, S., Maier, C., Eckhardt, A.: Towards ex ante prediction of user performance. a novel NeuroIS methodology based on real-time measurement of mental effort. In: HICSS-48 Proceedings, pp. 533–542. IEEE (2015)
45. Buettner, R.: Robust user identification based on facial action units unaffected by users' emotions. In: HICSS-51 Proceedings, pp. 265–273 (2018)
46. Buettner, R., Daxenberger, B., Woesle, C.: User acceptance in different electronic negotiation systems - a comparative approach. In: ICEBE 2013 Proceedings, pp. 1–8. IEEE (2013)
47. Buettner, R.: Towards a new personal information technology acceptance model: conceptualization and empirical evidence from a bring your own device dataset. In: AMCIS 2015 Proceedings (2015)
48. Buettner, R.: Analyzing the problem of employee internal social network site avoidance: are users resistant due to their privacy concerns? In: HICSS-48 Proceedings, pp. 1819–1828. IEEE (2015)
49. Buettner, R.: Getting a job via career-oriented social networking markets. Electron. Markets **27**(3), 371–385 (2017)
50. Buettner, R.: The state of the art in automated negotiation models of the behavior and information perspective. ITSSA **1**, 351–356 (2006)
51. Buettner, R.: Electronic Negotiations of the Transactional Costs Perspective. In: IADIS 2007 Proceedings, pp. 99–105 (2007)
52. Buettner, R.: Imperfect information in electronic negotiations: an empirical study. In: IADIS 2007 Proceedings, pp. 116–121 (2007)

53. Landes, J., Buettner, R.: Job allocation in a temporary employment agency via multi-dimensional price VCG auctions using a multi-agent system. In: MICAI 2011 Proceedings, pp. 182–187. IEEE (2011)
54. Buettner, R., Landes, J.: Web service-based applications for electronic labor markets: a multi-dimensional price VCG auction with individual utilities. In: ICIW 2012 Proceedings, pp. 168–177 (2012)
55. Buettner, R.: A systematic literature review of crowdsourcing research from a human resource management perspective. In: HICSS-48 Proceedings, pp. 4609–4618. IEEE (2015)
56. Rodermund, S.C., Timm, I.J., Buettner, R.: Towards simulation-based preplanning for experimental analysis of nudging. In: WI-2020 Proceedings, pp. 1219–1233. GITO Verlag (2020)
57. Buettner, R., Baumgartl, H.: A highly effective deep learning based escape route recognition module for autonomous robots in crisis and emergency situations. In: HICSS-52 Proceedings, pp. 659–666 (2019)
58. Baumgartl, H., Buettner, R.: Development of a highly precise place recognition module for effective human-robot interactions in changing lighting and viewpoint conditions. In: HICSS-53 Proceedings, pp. 563–572 (2020)
59. Baumgartl, H., Tomas, J., Buettner, R., Merkel, M.: A deep learning-based model for defect detection in laser-powder bed fusion using in-situ thermographic monitoring. Progress Addit. Manufact. **5**(3), 277–285 (2020). https://doi.org/10.1007/s40964-019-00108-3
60. LeCun, Y., Bengio, Y., Hinton, G.: Deep learning. Nature **521**, 436–444 (2015)

AttentionBoard: A Quantified-Self Dashboard for Enhancing Attention Management with Eye-Tracking

Moritz Langner(✉), Peyman Toreini, and Alexander Maedche

Karlsruhe Institute of Technology (KIT), Institute of Information Systems and Marketing (IISM), Karlsruhe, Germany

Moritz.langner@student.kit.edu, {peyman.toreini,alexander.maedche}@kit.edu

Abstract. In the age of information, office workers process huge amounts of information and distribute their attention to several tasks in parallel. However, attention is a scarce resource and attentional breakdowns, such as missing important information, may occur while using information systems (IS). Currently, there is a lack of support to understand and improve attention management to avoid such breakdowns. In the meantime, self-tracking applications are becoming popular due to the increasing sensory capabilities of smart devices. These systems support their users in understanding and reflecting their behavior. In this research-in-progress paper, we suggest leveraging self-tracking concepts for attention management while working with ISs and describe the design of the NeuroIS-based system called "AttentionBoard". The goal of AttentionBoard is to help office workers in improving their attention management competencies. The system records attention allocation in real-time using eye-tracking and presents the aggregated data as metrics and visualizations on a dashboard. This paper presents the first step by motivating and introducing an initial design following the design science research (DSR) methodology.

Keywords: Attention · Eye-tracking · Quantified-self · Self-tracking · Design science research · NeuroIS

1 Introduction

With the emergence of sensor-enriched smartphones and smart wearables, tracking information about oneself like steps, climbed stairs, calories, etc. became very easy. Thus, we observe a massive increase in the usage of such self-tracking systems. In 2007, Gary Wolf and Kevin Kelly founded the blog quantifiedself.com that is focused on sharing the best practices of the self-tracking community [1]. The quantified-self community is built around the credo that self-knowledge is gained through data [1]. The research stream focusing on systems that support tracking and reflecting personal information to better understand behavior is called personal informatics [2]. Self-tracking, self-analytics, life

F. D. Davis et al. (Eds.): NeuroIS 2020, LNISO 43, pp. 266–275, 2020.
https://doi.org/10.1007/978-3-030-60073-0_31

logging or personal informatics are common paradigms used for regularly tracking information about oneself to create statistics and data visualizations and to analyze behaviors, habits or feelings [3]. Existing research puts a strong emphasis on technology [4] and tracking physical activities [3].

Nowadays, attention allocation is becoming more important because attention is distributed continuously partial and not focused [5]. Thus, it can be a promising new field of application for leveraging on self-tracking. Mobile operating systems, like iOS or Android, provide users the summary of their interaction with the system to increase the users' self-awareness of usage. These systems are working mainly based on analyzing touch interactions with the system and are not based on actual users' attention. In addition, self-tracking is mostly applied in private life rather than in the working environment [6]. In the age of information, office workers frequently need to shift their attention between various tasks and get used to have continuous partial attention. However, in various situations, like monitoring tasks while working with an information dashboard, focused attention is required. It has been seen that users of these systems have difficulties to manage their limited attentional resources while exploring dashboards [7]. Besides, the huge amount of available information creates a poverty of attention [8]. Following the NeuroIS paradigm, one can design advanced self-tracking systems that increase users' awareness about specific affective and cognitive states [9–11]. Therefore, office workers would benefit of a self-tracking feature that helps them to analyze and understand their attention management while using an IS like an information dashboard.

Based on the eye-mind hypothesis, attentional resources are dedicated to the location where users are fixating as it reflects their underlying cognitive process [12]. Moreover, eye-movements are indicating overt attention [13]. Thus, tracking eye-movements can provide more accurate information about attention allocation than interaction data. Eye-tracking also proved to be a reliable source of information of attention for neuro-adaptive systems in previous research [14–16]. With the help of eye-trackers, we are capable of designing a neuro-adaptive IS that is sensitive to the attention of the user. Therefore, we suggest using eye-tracking devices to track the attention of the user and design self-aware services for office workers while working with IS applications.

Dashboards are a common visualization technique of information to ease analysis in business and learning [17, 18]. Executive IS included the first dashboards that aggregated information from different sources to key performance indicators (KPIs) and visualized those on a single screen in a comprehensible way [17, 19]. Since then dashboards were used in various forms [19]. Following this paradigm, we suggest designing a dashboard called "AttentionBoard" to visualize the previous attention allocation of office workers based on eye-movement data as an attention management support. Although the NeuroIS community called for the usage of eye-tracking devices to design and evaluate innovative systems [20–22], to the best of our knowledge, personal informatics systems to support the users in tracking and reflecting attention are not investigated well so far [10]. As an example, in a previous study we investigated the usage of offline records as well as real-time sensing of eye-movement data to increase users' self-awareness about their attention [14]. However, this system only included the gaze duration of the user and presented it in a time format. Considering more eye-tracking metrics and proper visualizations are suggested as future work for such systems [14]. In this research, we focus on providing a comprehensive overview of users' attention in the form of a dashboard, AttentionBoard. Therefore, we answer the following question:

RQ: *How to design a quantify-self dashboard to enhance office worker's attention management?*

In this research project, we will follow the design science research (DSR) methodology and use eye-tracking as a NeuroIS tool [22]. This research-in-progress paper is focused on motivating the need for such a dashboard and on providing the first design ideas for AttentionBoard. Furthermore, the system architecture and an initial user interface (UI) idea are presented.

2 Methodology

In this DSR project, we follow the approach suggested by Kuechler and Vaishnavi [23]. The first step of the DSR project is identifying and understanding the problem. Previous studies in human-computer interaction (HCI) highlighted the need for designing different types of attention support systems [24–26]. Moreover, by investigating the basket of eight in the IS field, we uncovered that although researchers in the NeuroIS field highlight the need for designing neuro-adaptive systems [21] and providing live-biofeedback [10], there is a lack of research about systems that actually support users' attention management. Furthermore, our previous study showed the difficulty of users in managing their limited attentional resources while working with IS applications such as information dashboards [7]. In that study, we designed a lab experiment in which participants received a dashboard with six graphs with the same level of complexity. As a result, we identified that users are biased to the left part of the dashboard and ignored the information on the right side. Also, by giving them a second chance to improve their behavior, participants had difficulty in remembering their previous information processing and repeated the same behavior. Finally, the findings from this study highlight six meta-requirements for designing user-adaptive information dashboards that include providing attention feedback to support user's data exploration.

Besides, we conducted a focused group study with office workers that create or work with information dashboards, as an example of IS application, within a big energy company. In total 13 persons (4 females, 9 males) participated in this focus group. After a brief introduction to eye-tracking and its capability, we did a brainstorming to extract their opinion about eye-tracking and providing attention support features. The result of the brainstorming was a SWOT analysis where they classified their ideas. During this process, the participants reported the need for having a summary of their previous attention to conduct their task properly while working with IS applications.

In 2007, Gary Wolf and Kevin Kelly started their blog quantifiedself.com that became the first address for the quantified-self community and provides access to best practices of self-tracking. Since then quantified-self has been researched in different domains like technological, medical and social domain [27]. In IS and other domains, research is mainly focused on usage and adoption of self-tracking technology [4]. However, there is little research on how to support reflecting self-tracking information, like in a format of a dashboard. Also, to the best of our knowledge, there is no research investigating self-tracking features for attention management support.

To design such support, the system needs to track the attention of the user while interacting with it and providing live biofeedback by visualizing the previous attention allocation. Therefore, in this research-in-progress paper, we come up with a preliminary list of suggestions for designing an attention management support system called "AttentionBoard". The identified meta-requirements (MRs) as well as preliminary design principles (DPs) are explained in the following chapter. In future steps, we plan to develop and evaluate AttentionBoard in a large-scale study to identify the effects of it. In the last chapter, possible further applications of the AttentionBoard concept are provided.

3 Designing the AttentionBoard

3.1 Meta-Requirements

In this chapter, we describe the MRs of AttentionBoard. The goal of AttentionBoard is to support office workers in attention management when required. Therefore, it should provide information about the previous attention allocation while using an IS application, like an information dashboard. This information can be visualizations of the user's eye-movement data as an indicator for overt attention.

While using an IS application, like an information dashboard, the office workers should be able to decide whether they want support in attention management or not. The capability of having individualized attention management support based on the user's need and character was identified in our previous study [7]. Furthermore, privacy has become an important aspect to consider when introducing new software. To cover for privacy concerns, it is important that the feature can be turned off when the user doesn't want to feel observed. Therefore, if the office worker has concerns about using eye-trackers, the system should provide the office worker with the ability to control the attention tracking feature by turning it on and off **(MR1)**.

Attention is considered as a selective process that guides the perception of incoming sensory information [28]. It is known as an important component in processing information. Attention is divided into two different types, covert and overt attention [29]. Overt attention refers to the type of attention that users need to change their head or eye to process information. In contrast, covert attention is happening internally such as allocating attention to a topic while thinking about it. Both types are considered as limited resource and users cannot allocate attention to several stimuli at the same time. Overt attention is measurable with eye-tracking devices and covert attention is trackable with fMRI or EEG tools. In this study, we focus on the overt attention of the users to enhance attention management. Eye-movements are a good estimator for overt attention [13, 29]. Our previous study shows that tracking user's eye-movements in real-time and providing the summary of attention allocation supports them in having a better information processing performance [14]. Therefore, the system should track the office worker's overt attention and estimate it by recording eye-movements **(MR2)**.

The recorded raw eye-movement data points do not provide meaningful insights on attention allocation and thus need to be further aggregated. Identification of fixations and saccades plays an essential role in analyzing users attention as common metrics [30].

Therefore, the raw eye-movement data should be aggregated to fixations and saccades which are stored on a database (**MR3**). As office workers probably are not familiar with the used terminology in eye-tracking, such as fixation, saccade, etc., visualizations and understandable KPIs are necessary to increase the usefulness of the dashboard for unexperienced users. Researchers suggest heatmaps and scan paths as useful visualizations to present recorded eye-movement data [31]. Therefore, to ease the analysis of attention for the office worker, the recorded eye-movement data should be aggregated to comprehensible visualizations and dedicated KPIs (**MR4**).

The information gathered and visualized by the eye-tracking system should be available in an easy and comprehensible way to enhance the attention management capability of the office worker. Dashboards aggregate information on a single screen to support decision making [17, 18]. Therefore, the recorded eye-movement data and extracted attention should be visualized on a dashboard to provide a comprehensive overview (**MR5**).

3.2 Initial Design Principles

In the following, the design principles of the attention management support system, called AttentionBoard, will be described. To support the office worker when facing attention management problems, it is essential that AttentionBoard can be easily turned on. As this attention management support is not required all the time, the user should also be able to turn it off when not required. Additionally, we found out in the focus group that being able to turn off the attention tracking feature is important to increase acceptance by office workers to avoid privacy concerns. Moreover, there has to be a function to get updated visualizations and metrics of previous attention allocation. Therefore, the attention management support system has dedicated buttons to start and stop the attention tracking feature and to create the analysis in order to support attention management upon request (**DP1**).

As AttentionBoard aims to enhance office worker's attention management, we focus on overt attention while working with IS. When taking a closer look at the workplace, most office workers interact with IS applications via computers. Thus, the overt attention should be measured while working with these devices. As we want to focus on desktop applications, like information dashboards, the attention management support system uses a display mounted eye-tacker in order to record the overt attention of the user in real-time (**DP2**).

In dashboard design, it is common to rely on graphical visualization of data instead of pure metrics [19]. A good dashboard balances between visual complexity and information utility [17]. Furthermore, the usage of KPIs to aggregate data is a key concept in dashboard design [19]. Thus, the attention management support system provides the corresponding visualizations and KPIs of recorded eye-movement data on a dashboard in order to support the reflection and enhancement of attention management (**DP3**). Table 1 presents the summary of derived MRs and DPs for designing AttentionBoard.

Table 1. Meta requirements and design principles

Meta requirements	Design principles
MR1: If the office worker has concerns about using eye-trackers, the system should provide the office worker with the ability to control the attention tracking feature by turning it on and off	**DP1:** Provide the attention management support system with dedicated buttons to start and stop the attention tracking feature in order to support attention management upon request
MR2: The system should track the office worker's overt attention and estimate it by recording eye-movements	**DP2:** Provide the attention management support system with a display mounted eye-tacker in order to track the users eye-movement data and extract overt attention of the users
MR3: Raw eye-movement data should be aggregated to fixations and saccades which are stored on a database	**DP3:** Provide the attention management support system with corresponding visualizations and KPIs of recorded eye-movement data on a dashboard in order to support the reflection and enhancement of attention management
MR4: Recorded eye-movement data should be aggregated to comprehensible visualizations and KPIs	
MR5: The recorded eye-movement data and extracted attention should be visualized on a dashboard	

3.3 Development: System Architecture and Mockup

Figure 1 visualizes the proposed system architecture which instantiates the DPs. This system architecture consists of three subsystems: event handler, eye-tracking, dashboard. Providing such an architecture supports a further extension of the system and allows for future integration of new subsystems, like for interruptions or mind-wandering. To increase the adoption of AttentionBoard, we rely on Tobii 4C Eye-Tracker as an apparatus. We chose this eye-tracker since it is easy to carry, to setup on different systems and it is known as a low-cost eye-tracker useful for real-word use cases.

The event handler subsystem is the instantiation of DP1. On starting the attention management feature, this subsystem connects to the eye-tracker and starts the recording of eye-movement data. Furthermore, the event handler subsystem triggers the eye-tracking subsystem in order to store the recorded eye-movements on the database. Whenever the user stops the recording of eye-movements, the event-handler hands over this command to the eye-tracking subsystem to stop the recording of eye-movement data. Additionally, the analysis of eye-movement data is triggered, and the update dashboard function opens AttentionBoard in the browser. It is implemented in C# to ensure a native integration into Microsoft Windows.

The eye-tracking subsystem takes care of recording, storing, calculating and visualizing the eye-movement data. The eye-movement recorder and storage are the instantiation of DP2 and coded in C# due to the better integration of the Tobii 4C and the database

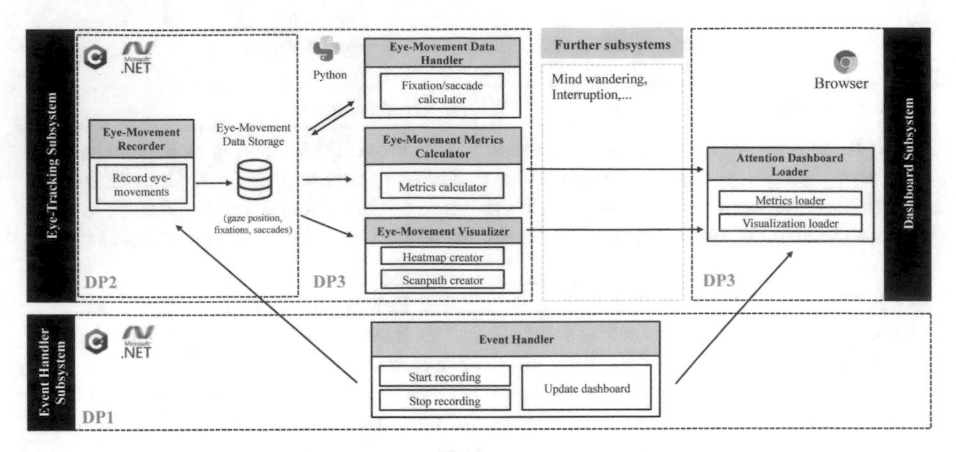

Fig. 1. System architecture of AttentionBoard

into Microsoft Windows. The eye-movement handler, calculator and visualizer are supporting the instantiation of DP3. We developed this handler with Python as we rely on the code of Pygaze Analyser to produce the heatmaps and scan paths [32]. Also, the eye-movement data handler calculates fixations and saccades out of the stored eye-movements and saves them on the database. The eye-movement metrics calculator loads the fixations and saccades as an input and calculates related measures such as number of fixations, fixation duration, saccade length, etc. Later, the eye-movement visualizer loads the calculated fixations and saccades and creates heatmaps and scan path visualizations. These visualizations are dynamic videos that show the temporal course of fixations and saccades.

The output of the eye-movement metrics calculator and visualizer are then handed over to the dashboard subsystem. The dashboard subsystem is the final instantiation of DP3. Up on user request, the dashboard system loads the generated visualizations (dynamic heatmap and scanpath) and metrics of attention allocation to the dashboard.

Figure 2 shows the proposed UI of the AttentionBoard and the corresponding DPs. At the top of AttentionBoard, there are dedicated areas for eye-movement visualizations like heatmap and scan path. These can be created for a specific time frame of the users' interaction with the IS application. Furthermore, an overview of usage is given in the form of an attention timeline. For deeper analysis, detailed KPIs of eye-movements, gaze and general usage are given (e.g. average fixation duration, number of fixations, average saccade length, system usage).

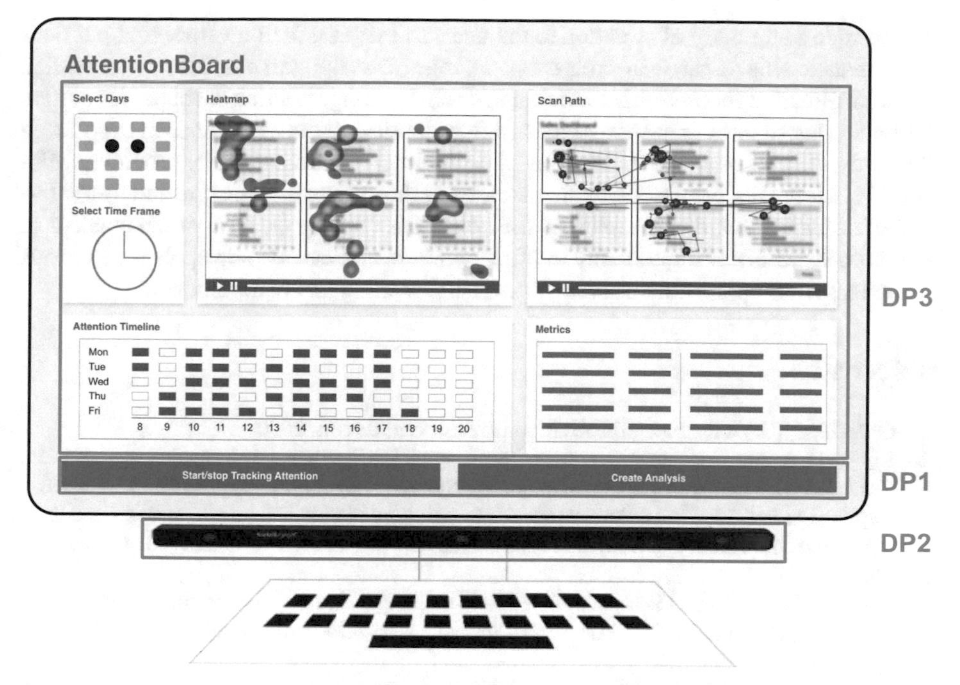

Fig. 2. Mockup of the AttentionBoard

4 Discussion and Next Steps

We presented AttentionBoard, a self-tracking system to support attention management of office workers. Following the DSR paradigm, we identified MRs by focusing on the users' difficulties in managing limited attentional resources which were identified in previous studies. Furthermore, we articulated three DPs that aim to increases the attention management performance of office workers while conducting complex tasks. Later, we described an initial system architecture and presented the first draft of the UI for AttentionBoard in a form of dashboard. This research-in-progress paper comes with several limitations and open issues. In the next step, we plan to instantiate AttentionBoard as an extension for users working with IS applications such as business intelligence dashboards. Later, we will test the designed AttentionBoard in a controlled lab experiment to evaluate whether having such a system for an IS application affects users' attention awareness and management. We assume that by increasing users' capability in managing their attention, they are able to process more information and avoid missing important information while working with IS applications.

5 Conclusion

In the age of information, attention is a limited resource and attentional breakdowns happen on a regular base while using IS applications. Advances in eye-tracking technology as a NeuroIS tool enabled measuring attention within the working environment and provide a summary of attention to the users to increase their awareness about their performance. This research-in-progress paper represents the start of a DSR project called AttentionBoard. It focuses on how to design a self-tracking attention dashboard for office workers using their eye-movement data. After identifying the need for such support, in this study, we presented the initial steps of designing AttentionBoard by providing MRs and DPs. Moreover, we described the first idea of the system architecture and user interface of AttentionBoard and explained the next steps. The application of the concept of AttentionBoard is not limited only to office workers and can also be applied in various contexts, like IS application training, e-learning, reading and entertainment.

References

1. Quantified Self, https://quantifiedself.com/. Accessed 03 Mar 2020
2. Li, I., Dey, A., Forlizzi, J.: A stage-based model of personal informatics systems. In: Conference on Human Factors in Computing Systems – Proceedings, pp. 557–566 (2010)
3. Lupton, D.: Self-tracking cultures: towards a sociology of personal informatics. In: Proceedings of the 26th Australian Computer-Human Interaction Conference, OzCHI 2014, pp. 77–86 (2014)
4. de Moya, J.F., Pallud, J., Scornavacca, E.: Self-tracking technologies adoption and utilization: a literature analysis. In: 25th Americas Conference on Information Systems, AMCIS 2019, pp. 1–10 (2019)
5. Bulling, A.: Pervasive attentive user interfaces. Computer (Long. Beach. Calif) **49**, 94–98 (2016)
6. Lupton, D.: The diverse domains of quantified selves: self-tracking modes and dataveillance. Econ. Soc. **45**, 101–122 (2016)
7. Toreini, P., Langner, M.: Designing user-adaptive information dashboards: considering limited attention and working memory. In: 27th European Conference on Information Systems (ECIS), pp. 0–16 (2019)
8. Simon, H.A.: Designing organizations for an information-rich world. In: Computers, Communications, and the Public Interest, vol. 72, p. 37 (1971)
9. Riedl, R., Léger, P.-M.: Fundamentals of NeuroIS. Springer, Heidelberg (2016)
10. Lux, E., Dorner, V., Knierim, M.T., Adam, M.T.P., Helming, S., Weinhardt, C.: Live biofeedback as a user interface design element: a review of the literature. Commun. Assoc. Inf. Syst. **43**, 257–296 (2018)
11. Adam, M.T.P., Gimpel, H., Maedche, A., Riedl, R.: Design blueprint for stress-sensitive adaptive enterprise systems. Bus. Inf. Syst. Eng. **59**(4), 277–291 (2016)
12. Just, M.A., Carpenter, P.A.: Eye fixations and cognitive processes. Cogn. Psychol. **8**, 441–480 (1976)
13. Kowler, E.: Eye movements: the past 25 years. Vision. Res. **51**, 1457–1483 (2011)
14. Toreini, P., Langner, M., Maedche, A.: Using eye-tracking for visual attention feedback. In: Davis, F., Riedl, R., vom Brocke, J., Léger, P.M., Randolph, A., Fischer, T. (eds.) Information Systems and Neuroscience. Lecture Notes in Information Systems and Organisation, pp. 261–270. Springer, Cham (2020)
15. Pfeiffer, J., Prosiegel, J., Meißner, M., Pfeiffer, T.: Identifying goal-oriented and explorative information search patterns. In: Proceedings of the Gmunden Retreat on NeuroIS 2014, pp. 1–3 (2014)
16. Pfeiffer, T., Pfeiffer, J., Meißner, M.: Mobile recommendation agents making online use of visual attention information at the point of sale. In: Proceedings of the Gmunden Retreat on NeuroIS 2013, p. 3 (2013)

17. Yigitbasioglu, O.M., Velcu, O.: A review of dashboards in performance management: implications for design and research. Int. J. Account. Inf. Syst. **13**, 41–59 (2012)
18. Verbert, K., Govaerts, S., Duval, E., Santos, J.L., Van Assche, F., Parra, G., Klerkx, J.: Learning dashboards: an overview and future research opportunities. Pers. Ubiquitous Comput. **18**, 1499–1514 (2014)
19. Few, S.: Information Dashboard Design: The Effective Visual Communication of Data. O'Reilly Media, Inc., Sebastopol (2006)
20. Riedl, R., Davis, F.D., Hevner, A.R.: Towards a neuroIS research methodology: intensifying the discussion on methods, tools, and measurement (2014)
21. Dimoka, A., Davis, F.D., Pavlou, P.A., Dennis, A.R.: On the use of neurophysiological tools in IS research: developing a research agenda for NeuroIS. MIS Q. **36**, 679–702 (2012)
22. vom Brocke, J., Riedl, R., Léger, P.-M.: Application strategies for neuroscience in information systems design science research. J. Comput. Inf. Syst. **53**, 1–13 (2013)
23. Kuechler, B., Vaishnavi, V.: Theory development in design science research: anatomy of a research project (2008)
24. Kern, D., Marshall, P., Schmidt, A.: Gazemarks: gaze-based visual placeholders to ease attention switching. In: Proceedings of SIGCHI Conference Human Factors Computing System, pp. 2093–2102 (2010)
25. Mariakakis, A., Goel, M., Aumi, M.T.I., Patel, S.N., Wobbrock, J.O.: SwitchBack: using focus and saccade tracking to guide users' attention for mobile task resumption. In: Proceedings of the 33rd Annual ACM Conference on Human Factors in Computing Systems, pp. 2953–2962. ACM, New York (2015)
26. Jo, J., Kim, B., Seo, J.: EyeBookmark: assisting recovery from interruption during reading. In: Proceedings of the 33rd Annual ACM Conference on Human Factors in Computing Systems - CHI 2015, pp. 2963–2966 (2015)
27. Pallud, J., De Moya, J.-F.: Quantified self: a literature review based on the funnel paradigm. In: Proceedings of the ECIS 2017, pp. 1678–1694 (2017)
28. Roda, C., Thomas, J.: Attention aware systems: theories, applications, and research agenda. Comput. Human Behav. **22**, 557–587 (2006)
29. Posner, M.I.: Orienting of attention. Q. J. Exp. Psychol. **32**, 3–25 (1980)
30. Salvucci, D.D., Goldberg, J.H.: Identifying fixations and saccades in eye-tracking protocols. Eye Tracking Res. Appl. **2000**, 71–78 (2000)
31. Duchowski, A.T.: Eye Tracking Methodology (2017)
32. Dalmaijer, E.S., Mathôt, S., Van Der, S.S.: PyGaze: an open-source, cross-platform toolbox for minimal-effort programming of eye-tracking experiments Edwin, pp. 1–16

Machine Learning-Based Diagnosis of Epilepsy in Clinical Routine: Lessons Learned from a Retrospective Pilot Study

Thilo Rieg(✉), Janek Frick, and Ricardo Buettner

Aalen University, Aalen, Germany
thilo.rieg@hs-aalen.de

Abstract. In this work-in-progress paper, we present preliminary results of a large pilot study for implementing a novel machine learning approach presented at HICSS 2019 [1] and ICIS 2019 [2] in a German hospital to detect epileptic episodes in EEG data. While the algorithm achieved a balanced accuracy of 75.6% on real clinical data we could gain valuable experience regarding the implementation barriers of machine learning algorithms in practice, which is discussed in this paper. These lessons learned have practical implications for future work.

Keywords: Electroencephalography · Random forests · Spectral analysis · Machine learning

1 Introduction

Research in the field of machine learning for disease detection is a currently trending scientific topic. The data gathered for studies are mostly collected under perfect laboratory conditions and on the basis of two clearly separable groups of test persons.

In order to test the performance of machine learning algorithms in everyday clinical practice, 21-channel EEG data were collected from the archive of the Ostalbklinikum Aalen. Electroencephalography is a relevant topic for NeuroIS research [3]. In contrast to the normal procedure using five frequency bands, a newly presented method for finer frequency distribution of the data was applied. This approach has been presented at various conferences within our scientific community [1, 2]. The fixed five frequency bands have been the subject of frequent criticism in the past [4, p. 917].

In this work, we evaluate the performance of machine learning algorithms on data from clinical routines. In this way we try to do justice to the complexity of the work of physicians, who are usually under time pressure, work in imperfect laboratory conditions and have less clear clinical pictures.

2 Methodology

The entire work was developed in line with NeuroIS guidelines [5]. The Design Science Framework was chosen as the scientific framework [6].

F. D. Davis et al. (Eds.): NeuroIS 2020, LNISO 43, pp. 276–283, 2020.
https://doi.org/10.1007/978-3-030-60073-0_32

In order to conduct our retrospective pilot study, an application for anonymous data collection was submitted to the Ethics Committee of the State Medical Association and approved. The study was conducted in December 2019. The data collected were all measured by the same EEG device and at 256 Hz with 21 channels. The length of the measurement varies per recording and is between 8 and 15 min. During the first few minutes a resting EEG is recorded with eyes open and closed, followed by hyperventilation in the middle of the recordings. To replicate previous work, only the resting-state EEG should be used [2]. Therefore, only 31.25 s of each recording were selected. To avoid artifacts that may appear in the beginning of the measurement, the first 8 s were excluded.

As medical data are involved, the data protection of patients was given the highest priority. The data was made completely anonymous. Only age, gender, detailed findings and raw EEG data were collected.

2.1 Step 1: Selection of Data for Consideration of the Study

To ensure the scientific quality of the study, strict criteria were established for the data set to be collected. These were defined in discussion with the medical management of the neurological department of the hospital.

Since the activities in the brain change over the course of life, an age restriction was chosen for the data set. Thus, subjects were not included in the dataset until adolescence or the age of 16. The age limit was 60 years. Not included in the data set were patients who, according to the findings, were already receiving medication as, according to current scientific discourse, these can influence the alpha frequency range [7]. Also excluded were findings with diseases such as multiple sclerosis or meningitis and EEG data with too many interference signals, where the quality of the data collected was insufficient. Only clear epilepsy diagnoses were included in the data set. The control group consisted of test persons suffering from headaches or sensory disturbances or migraines. Here, the EEG was found by medical personnel to be free of the disease.

Approximately 8,000 findings were sighted. Due to the age restriction and the exclusion criteria outlined above, 698 data sets were collected. After a further review of the data to ensure the unambiguousness of the findings and data quality, a data set with 185 subjects remained (Table 1).

Table 1. Demographics

	Epilepsy	Control
Age min.	16	17
Age max.	59	56
Age av.	38.19	37.35
Male	60	28
Female	36	61

2.2 Step 2: Data Pre-processing: Filter, ICA and Fine-Graded EEG-Spectrum

In a second step, the collected data were pre-processed. Before we transformed the data into a fine frequency spectrum, the data was filtered (bandpass filter between 0.5 Hz and 50.5 Hz). In addition, the linear decomposition approach by Bell and Sejnowski was used to clean the data [8]. Independent components within a data recording can be extracted by using their ICA. There are three requirements for performing an ICA: (i) the mixed medium is linear and propagation delays are negligible, (ii) the time course of the sources are independent from each other, and (iii) the number of sources is equal to the number of sensors. All these conditions apply to EEG data. The ICA is based on the central limit theorem: the sum of n independent variables gives a normal distribution. If a signal has no normal distribution, it must be an independent component, while a normally distributed signal is likely to be a mixture of several components. In this way, the EEG data can be cleaned of interference artifacts such as blinking. We used an ICA with 1-s windows and the hanning window approach.

Afterwards the fine frequency bands were calculated. For this purpose, the frequency range of 0.5 Hz to 50.5 Hz was divided into a 50-point spectrum of 1 Hz spectra each. This resulted in a matrix of 185 test subjects and 50 features each. The two classes "epilepsy" and "control" were added.

2.3 Step 3: Training and Testing of the Random Forest Classifier

The Random Forest algorithm was proposed by Breiman [9], and is a machine learning classifier which is based on an ensemble (a bag) of unpruned decision trees. A Random Forest contains a collection of tree predictors, each tree being based on independently selected vectors. The classification outputs of the individual trees are used to determine the overall classification. Ensemble methods are related to the concept that an aggregated decision from various experts is often superior to a decision from a single system. The final classification decision is built on the majority vote.

In order to obtain a robust model and avoid overfitting, the k-fold cross validation was applied. This method randomly divides the existing training data set into k equal parts. The training is executed on k-1 data, the validation is performed on the excluded data, which is used as a test data set. This process is repeated k times so that each part is used once as a test dataset [10]. For this work, a k of 10 was selected. For evaluation purposes, 20% of the data were held out. The final model was evaluated based on these 20% of the data, which were not used for training the algorithm.

For training, 148 subjects were used. The trained algorithm was subsequently tested on 37 subjects. Training was performed with ntree = 100. The depth of the tree structure was mtry = 4. The best features are determined by the Gini coefficient, which calculates the reduction of impurity of the relevant information that would result from splitting the data set according to a certain feature. Thus, the Gini coefficient is suitable for minimizing misclassifications.

3 Results

The classification results on the test data set of 37 subjects unknown to the algorithm can be viewed in the following confusion matrix (Table 2).

Table 2. Confusion matrix

		Reference	
		Epileptic	Control
Predicted	Epileptic	**12**	4
	Control	5	**16**

The algorithm was correct in 28 cases. Incorrect classification was made in 9 cases. Overall, the algorithm achieves a balanced accuracy of 0.756 on the test data.

4 Discussion

Considering that this is not study data, the accuracy of 0.756 seems excellent - especially since only the patients' resting-state EEG was used. Recent studies suggest that between 23% and 30% of epilepsy diagnoses in clinical practice are erroneous. With our algorithm, this error rate can also be roughly mapped.

The goal of our retrospective study was to test the performance of machine learning algorithms in a real environment. While in scientific studies excellent results in diagnosis can already be achieved by machine learning, we wanted to use a representative dataset to train our algorithms.

The approach common in science of using two very discriminating groups of test persons to collect data for studies is mainly used to gain knowledge about disease characteristics. Patterns that distinguish sick people from healthy people can be identified. These findings are relevant to decode diseases.

The frequently drawn conclusion that the trained algorithms can easily be transferred from the laboratory environment into practice seems to be a fallacy. In everyday clinical practice, there are hardly any examples of clearly healthy and clearly ill patients. The patient comes with an individual ailment and the examination is often performed by differential diagnosis.

So, we continue to see information technology and machine learning algorithms as powerful new tools to decode diseases. In addition, we see the technical possibilities offered by machine learning as a great opportunity to support physicians and to make everyday clinical work more efficient. However, the wealth of experience of doctors and patient conversations still seem to be a central element in the correct diagnosis of epilepsy.

We were able to learn the following important experiences regarding the implementation of machine learning algorithms in everyday practice, based in part on discussions with physicians:

– important information can be found especially in patient conversations
– the description of the symptoms and the patient's history helps in finding a diagnosis - this information is missing in the algorithm
– the medication of the patients can play a decisive role in the examination by EEG

– in the context of diagnosis, differential diagnosis is often used, which is difficult to map using algorithms
– the cooperation with physicians should be intensified in order to close the mutual information delta and create the best possible algorithms.

5 Limitation and Future Work

The main limitation is related to the fact that our ambitious work is a research-in-progress. Furthermore, it must be noted that the diagnoses used have been made by only one physician in each case and therefore no gold standard exists. Unfortunately, this can hardly be avoided in retrospective studies. In order to consolidate this result, further retrospective studies on dementia, strokes, Parkinson's and autism are to follow.

In future work we evaluate the robustness of the approach by assessing the influence of individual states and mental concepts such as cognitive workload [11–13], concentration [14] and mindfulness [15, 16] in multi-agent-settings [17–20]. Furthermore, we will triangulate psychophysiological [21–23] and physiological data (i.e., electroencephalographic data [24–27] and spectra [28], electrocardiographic data [29, 30], electrodermal activity [31], eye fixation [32, 33], eye pupil diameter [34–36], facial data [37]) to increase reliability. In addition, we will evaluate technology acceptance [38–41] and trust [42, 43] by physicians and patients and if the automated approach improves the coordination [44–50] between physicians more efficiently. Furthermore, we want to apply our identified resting-state EEG sub-bands to convolutional neural networks [51–54].

References

1. Rieg, T., Frick, J., Hitzler, M., Buettner, R.: High-performance detection of alcoholism by unfolding the amalgamated EEG spectra using the Random Forests method. In: Proceedings of HICSS-52, pp. 3769–3777 (2019)
2. Buettner, R., Rieg, T., Frick, J.: High-performance detection of epilepsy in seizure-free EEG recording: a novel machine learning approach using very specific EEG sub-bands. In: ICIS 2019 Proceedings of 40th International Conference on Information Systems, Munich, Germany, 15–18 December 2019
3. Riedl, R., Fischer, T., Léger, P.M.: A decade of NeuroIS research: status quo, challenges, and future directions. In: ICIS 2017 Proceedings of 38th International Conference on Information Systems, Seoul, South Korea, 10–13 December 2017 (2017)
4. Müller-Putz, G.R., Riedl, R., Wriessnegger, S.C.: Electroencephalography (EEG) as a research tool in the information systems discipline: foundations. Meas. Appl. CAIS **37**, 911–948 (2015)
5. vom Brocke, J., Liang, T.-P.: Guidelines for neuroscience studies in information systems research. JMIS **30**(4), 211–234 (2014)
6. Hevner, A.R., March, S.T., Park, J., Ram, S.: Design science in information systems research. MISQ **28**(1), 75–105 (2004)
7. Larsson, P.G., Kostov, H.: Lower frequency variability in the alpha activity in EEG among patients with epilepsy. Clin. Neurophysiol. **116**(11), 2701–2706 (2005)
8. Bell, A., Sejnowski, T.J.: An information-maximization approach to blind separation and blind deconvolution. Neural Comput. **7**(6), 1129–1159 (1996)

9. Breiman, L.: Random forests. Mach. Learn. **45**(1), 5–32 (2001)
10. Kohavi, R.: A study of cross-validation and bootstrap for accuracy estimation and model selection. In: International Joint Conference on Artificial Intelligence, pp. 1137–1145 (1995)
11. Buettner, R.: The relationship between visual website complexity and a user's mental workload: a NeuroIS perspective. In: LNISO, vol. 16, pp. 107–113 (2016)
12. Buettner, R.: A user's cognitive workload perspective in negotiation support systems: an eye-tracking experiment. In: Proceedings of PACIS 2016, no. 115 (2016)
13. Buettner, R., Timm, I.J., Scheuermann, I.F., Koot, C., Roessle, M.: Stationarity of a user's pupil size signal as a precondition of pupillary based mental workload evaluation. In: LNISO, vol. 25 (2017)
14. Buettner, R., Baumgartl, H., Sauter, D.: Microsaccades as a predictor of a user's level of concentration. In: LNISO, vol. 29, pp. 173–177 (2018)
15. Sauer, S., Lemke, J., Zinn, W., Buettner, R., Kohls, N.: Mindful in a random forest: assessing the validity of mindfulness items using random forests methods. Pers. Individ. Differ. **81**, 117–123 (2015)
16. Sauer, S., Buettner, R., Heidenreich, T., Lemke, J., Berg, C., Kurz, C.: Mindful machine learning: using machine learning algorithms to predict the practice of mindfulness. Eur. J. Psychol. Assess. **34**(1), 6–13 (2018)
17. Buettner, R.: A classification structure for automated negotiations. In: Proceedings of IEEE/WIC/ACM WI-IAT 2006, pp. 523–530 (2006)
18. Buettner, R., Kirn, S.: Bargaining power in electronic negotiations: a bilateral negotiation mechanism. In: EC-Web 2008 Proceedings. LNCS, vol. 5183, pp. 92–101 (2008)
19. Buettner, R.: Cooperation in hunting and food-sharing: a two-player bio-inspired trust model. In: Proceedings of BIONETICS 2009, 9–11 December, pp. 1–10. Avignon, France, (2009)
20. Landes, J., Buettner, R.: Argumentation-based negotiation? Negotiation-based argumentation! In: Proceedings of EC-Web 2012, pp. 149–162 (2012)
21. Buettner, R.: Asking both the user's brain and its owner using subjective and objective psychophysiological NeuroIS instruments. In: Proceedings of the ICIS 2017, 38th International Conference on Information Systems, Seoul, South Korea 10–13 December 2017 (2017)
22. Buettner, R.: Getting a job via career-oriented social networking markets: the weakness of too many ties. Electron. Markets **27**(4), 371–385 (2017)
23. Buettner, R.: Predicting user behavior in electronic markets based on personality-mining in large online social networks: a personality based product recommender framework. Electron. Markets **27**(3), 247–265 (2017)
24. Buettner, R., Hirschmiller, M., Schlosser, K., Roessle, M., Fernandes, M., Timm, I.J.: High-performance exclusion of schizophrenia using a novel machine learning method on EEG data. In: Proceedings of IEEE Healthcom 2019 (2019)
25. Buettner, R., Fuhrmann, J., Kolb, L.: Towards high-performance differentiation between narcolepsy and idiopathic hypersomnia in 10 minute EEG recordings using a novel machine learning approach. In: Proceedings IEEE Healthcom 2019 (2019)
26. Buettner, R., Beil, D., Scholtz, S., Djemai, A.: Development of a machine learning based algorithm to accurately detect schizophrenia based on one-minute EEG recordings. In: Proceedings of HICSS-53, pp. 3216–3225 (2020)
27. Buettner, R., Grimmeisen, A., Gotschlich, A.: High-performance diagnosis of sleep disorders: a novel, accurate and fast machine learning approach using electroencephalographic data. In: Proceedings of HICSS-53, pp. 3246–3255 (2020)
28. Buettner, R., Rieg, T., Frick, J.: Machine learning based diagnosis of diseases using the unfolded EEG spectra: towards an intelligent software sensor. In: LNISO, vol. 32, pp. 165–172 (2019)
29. Buettner, R., Bachus, L., Konzmann, L., Prohaska, S.: Asking both the user's heart and its owner: empirical evidence for substance dualism. In: LNISO, vol. 29, pp. 251–257 (2018)

30. Buettner, R., Schunter, M.: Efficient machine learning based detection of heart disease. In: Proceedings of IEEE Healthcom 2019 (2019)
31. Eckhardt, A., Maier, C., Buettner, R.: The Influence of pressure to perform and experience on changing perceptions and user performance: a multi-method experimental analysis. In: ICIS 2012 Proceedings of 33rd International Conference on Information Systems, Orlando, USA, 16–19 December 2012
32. Buettner, R.: Cognitive workload of humans using artificial intelligence systems: towards objective measurement applying eye-tracking technology. In: Proceedings of KI 2013. LNAI, vol. 8077, pp. 37–48 (2013)
33. Eckhardt, A., Maier, C., Hsieh, J.P.-A., Chuk, T., Chan, A.B., Hsiao, A.B., Buettner, R.: Objective measures of IS usage behavior under conditions of experience and pressure using eye fixation data. In: ICIS 2013 Proceedings of 34th International Conference on Information Systems, Milano, Italy, 15–18 December 2013
34. Buettner, R.: Social inclusion in eParticipation and eGovernment solutions: a systematic laboratory-experimental approach using objective psychophysiological measures. In: Proceedings of EGOV/ePart 2013. LNI, vol. P-221, pp. 260–261 (2013)
35. Buettner, R., Daxenberger, B., Eckhardt, A., Maier, C.: Cognitive workload induced by information systems: introducing an objective way of measuring based on pupillary diameter responses. In: Proceedings of Pre-ICIS HCI/MIS 2013, paper 20 (2013)
36. Buettner, R., Sauer, S., Maier, C., Eckhardt, A.: Towards ex ante prediction of user performance: a novel NeuroIS methodology based on real-time measurement of mental effort. In: Proceedings of HICSS-48, pp. 533–542 (2015)
37. Buettner, R.: Robust user identification based on facial action units unaffected by users' emotions. In: Proceedings of HICSS-51, pp. 265– 273 (2018)
38. Buettner, R., Daxenberger, B., Woesle, C.: User acceptance in different electronic negotiation systems - a comparative approach. In: Proceedings of ICEBE 2013, pp. 1–8 (2013)
39. Buettner, R.: Towards a new personal information technology acceptance model: conceptualization and empirical evidence from a bring your own device dataset. In: Proceedings of AMCIS 2015 (2015)
40. Buettner, R.: Analyzing the problem of employee internal social network site avoidance: are users resistant due to their privacy concerns?, In: Proceedings of HICSS-48, pp. 1819–1828 (2015)
41. Buettner, R.: Getting a job via career-oriented social networking sites: the weakness of ties. In: Proceedings of HICSS-49 , pp. 2156– 2165 (2016)
42. Meixner, F., Buettner, R.: Trust as an integral part for success of cloud computing. In: Proceedings of ICIW 2012, pp. 207–214 (2012)
43. Buettner, R.: The impact of trust in consumer protection on internet shopping behavior: an empirical study using a large official dataset from the European Union. In: BDS-2020 Proceedings (2020, in press)
44. Buettner, R.: The state of the art in automated negotiation models of the behavior and information perspective. ITSSA **1**(4), 351–356 (2006)
45. Buettner, R.: Electronic negotiations of the transactional costs perspective. In: Proceedings of IADIS 2007, vol. 2, pp. 99–105 (2007)
46. Buettner, R.: Imperfect information in electronic negotiations: an empirical study. In: Proceedings of IADIS 2007, vol. 2, pp. 116–121 (2007)
47. Landes, J., Buettner, R.: Job allocation in a temporary employment agency via multi-dimensional price VCG auctions using a multi-agent system. In: Proeedings of MICAI 2011, pp. 182– 187 (2011)
48. Buettner, R., Landes, J.: Web service-based applications for electronic labor markets: a multi-dimensional price VCG auction with individual utilities. In: Proceedings of ICIW 2012, pp. 168–177 (2012)

49. Buettner, R.: A systematic literature review of crowdsourcing research from a human resource management perspective. In: Proceedings of HICSS-48, pp. 4609–4618 (2015)
50. Rodermund, S., Buettner, R., Timm, I.J.: Towards simulation-based preplanning for experimental analysis of nudging. In: WI-2020 Proceedings (2020, in press)
51. Baumgartl, H., Buettner, R.: Development of a highly precise place recognition module for effective human-robot interactions in changing lightning and viewpoint conditions. In: Proceedings of HICSS-53, pp. 563–572 (2020)
52. Baumgartl, H., Tomas, J., Buettner, R., Merkel, M.: A deep learning-based model for defect detection in laser-powder bed fusion using in-situ thermographic monitoring. Progress Addit. Manufact. **5**(3), 277–285 (2020)
53. Buettner, R., Baumgartl, H.: A highly effective deep learning based escape route recognition module for autonomous robots in crisis and emergency situations. In: Proceedings of HICSS-52, pp. 659–666 (2019)
54. LeCun, Y., Bengio, Y., Hinton, G.: Deep learning. Nature **521**(7553), 436–444 (2015)

Ambient Facial Emotion Recognition: A Pilot Study

François Courtemanche, Elise Labonté-LeMoyne(✉), David Brieugne, Emma Rucco, Sylvain Sénécal, Marc Fredette, and Pierre-Majorique Léger

Tech3Lab, HEC Montreal, Montreal, QC, Canada
{francois.courtemanche,elise.labonte-lemoyne,david.brieugne, emma.rucco,sylvain.senecal,marc.fredette,pml}@hec.ca

Abstract. As technology evolves, studies of user emotion in naturalistic settings in an utetherd manner becomes more and more necessary. To achieve this goal, we present a proposed architecture for synchronized automatic facial emotion recognition and physiological recording in a mobile environment in an IS context. We describe a pilot study using this infrastructure and lessons learned for researchers who wish to employ this setup in the future.

Keyword: Automatic facial emotion recognition · Naturalistic setting

1 Introduction

Facial emotion recognition is the detection of a person's emotion as reflected from the muscular contractions of their face [1, 2]. The relationship between facial expressions and emotional state was established in the 1970s [3] and used to be studied by affixing EMG electrodes to a person's face [4]. Since then, the technology has evolved, and it is now possible to detect those muscle contractions and the emotional state they denote by recording a person's face and analyzing it with specialized software [5]. Facial emotion recognition has been used in NeuroIS for some time now [6–9], as it is a very useful method since it is not invasive [1]. Participants can interact freely with technological artefacts without any equipment on their bodies. It is even possible to simply record a participant's face with a computer's webcam and post process the videos to detect their emotional state. The method, however, has limits, mainly, the need for the participant to be immobile in front of the camera. In addition, a recent review identified the three main challenges of the method as: "illumination variation, subject-dependence, and head pose-changing" [10]. The ability to have participants interact with their environment as they wish is important for the future of neurophysiological research as technology becomes less and less tethered to computer screens. As such, calls for more naturalistic and in the wild NeuroIS research have been made in recent years [2, 11].

Due to these limitations, facial emotion recognition has not been used in contexts where people are mobile. For instance, facial emotion recognition cannot be used in a physical store to capture a consumer's facial emotion as she walks around the store

F. D. Davis et al. (Eds.): NeuroIS 2020, LNISO 43, pp. 284–290, 2020.
https://doi.org/10.1007/978-3-030-60073-0_33

shopping for an IT product. To our knowledge, only physiological data such as EDA has been used in stores to capture consumers' emotional reactions [12]. While useful, EDA data can only inform about emotional intensity (i.e., arousal), but not emotional valence, which facial emotion can provide. In order to address this methodological issue, we present a new method that: 1) captures facial emotions of a person while she roams freely in an environment of interest, 2) captures and precisely synchronizes complementary physiological signals such as EDA and EKG. This will allow studies of emotional responses in more naturalistic settings and more authentic responses from participants.

To test the proposed method, we conducted a pilot study in a store of a telecommunication service provider. While it can be used in various contexts other than retailing (e.g., education and training, health, public services), this context was deemed relevant for the purpose of the pilot study. There is a call for more research using psychophysiological data in store contexts to better understand consumer behaviors in stores [13] and in store pre-purchase information search has long been recognized in consumer research as an important element of consumer purchase decision [14]. Moreover, contemporary phenomenon such as showrooming reiterates consumers' need to interact with technological products and salespeople in physical stores [15].

2 Proposed Architecture for Synchronized Ambient Facial Emotion Recognition and Physiological Recording in Mobile Environment

Using an IoT architecture, we propose a synchronisation framework that relies on the exchange of Bluetooth Low Energy Beacons (BLE) amongst all the devices [16]. Each device receives and broadcasts different types of beacons. Two processes are executed simultaneously using beacons: devices synchronisation and participant localisation. An indirect approach is used to synchronise the data from the different devices (physiological, face images, and event markers) that broadcast synchronisation beacons (bs) [17]. The data files of all devices are synchronized by aligning the timestamps of these synchronisation beacons. Building upon [18], a linear regression method is then used to correct clock drifting effects in the different data streams and ensures that synchronization remains throughout the entire time course of a recording session. Event markers describing the different steps of a session are synchronised using the videos and expert live time stamping. The videos are synchronised to each other by using LED displays (LED1 to 3) placed in the environment that visually relay the synchronisation beacons.

In order to ensure that only the participant's emotions are recorded, the cameras need to avoid recording bypassers' faces. This is achieved using an indoor positioning system (IPS) based on BLE beacons. Using the received signal strength indication (RSSI) of the Bluetooth Low Energy beacons sent by the user device each camera was able to estimate its current distance with the participant. Each camera then broadcasted its estimated distance to the other cameras. Therefore, at any point in time, each camera knows its relative proximity with the participant and can decide to start or stop capturing images [19]. The localisation process is composed of the the following steps:

1. The physiological recording device is broadcasting positioning beacons (b_p).
2. Upon reception of b_p, each camera ($C_{1\ \text{to}\ 8}$) estimates its relative distance to the participant using RSSI (Received Signal Strength Indication) information of b_p.
3. Each camera broadcasts its estimated distance with the participants (b_d beacons).
4. Every 2 s, each camera computes the average distance between the participant and all the camera groups (e.g. group 1 distance = distance($C_1 + C_2 + C_3$)/3)
5. When a camera estimates that its group is the current closest one to the participant, it starts recording. Otherwise it stops recording (Fig. 1).

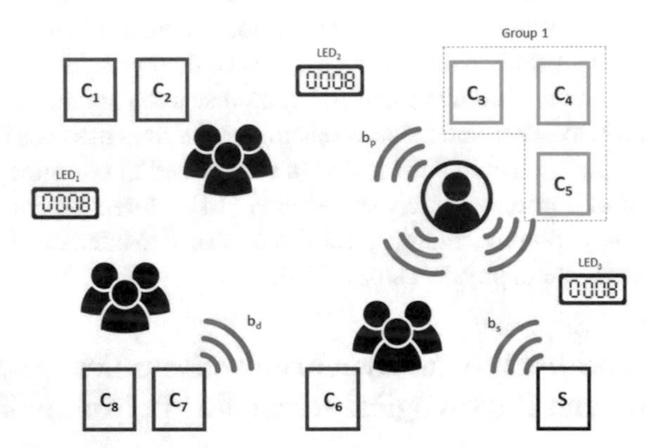

Fig. 1. Flowchart of the bluebox system. C_X: cameras to capture the participant's signal; b_X: broadcast synchronisation beacons to locate the participant and activate nearby cameras as well as input synchronisation marker in the signal feeds. The participant of interest is circled; LED_X: LED lights used as visual event markers in the video feeds.

3 Pilot Study

The objective of the study was to test the feasibility of capturing facial emotions and physiological data in an actual retail store context where participants can freely move and shop for a product or service. The study was approved by our institution's research ethics committee.

To test the proposed architecture, multiple cameras were installed in the various areas of a store. Cameras were bundled in each store area. Images were recorded under normal store lighting conditions. A semi-opaque curtain was added on one large window to avoid direct sun exposure. Video recordings were post-processed to eliminate empty images and to select the optimal camera feed when multiple cameras from the same store area recorded the same participant at the same time. In these cases, the weighted average of emotion prediction was used. Image quality was used as the weighting factor. Each participant's video recording was synchronized with their physiological data following Courtemanche and et al.'s [20] procedure. Finally, the recordings were processed in the FaceReader (Noldus, Netherlands) software in order to assess each participant's facial emotion.

The pilot study was performed in a physical store designed with different specific areas (e.g., product demonstration, waiting area). Twenty-eight adults (50% female, mean age: 39) were recruited to participate in the study. Before entering the store, they were equipped with specifically designed physiological sensors (EDA and EKG) and a Bluetooth emitter [16] (see Fig. 2). Then, their task was to shop for a specific new product offered by the service provider. Following the completion of their shopping task, participants were debriefed and received monetary compensation.

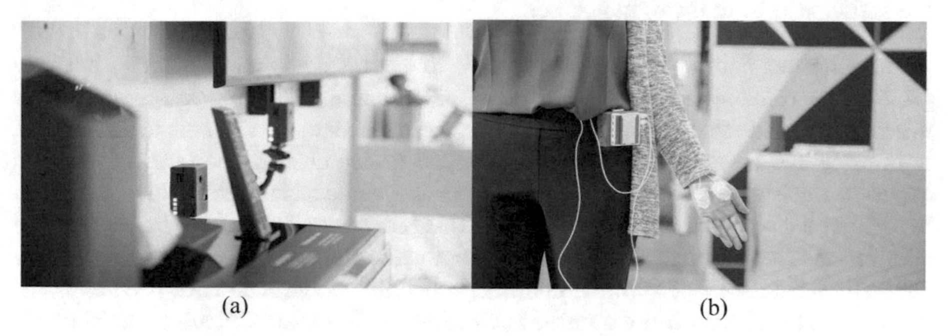

(a) (b)

Fig. 2. a) Camera setup; b) Physiological sensors with Bluetooth emitter.

4 Results and Discussion

The pilot study was successful in collecting facial data of participants freely interacting with products and salespeople in a physical store. The proposed method, in addition to providing ambient facial emotion recognition, provided researchers with complementary synchronised physiological data such as EDA and EKG. Combined, the data presents a clearer picture of the participants' emotional and cognitive reactions to in-store stimuli and interactions (e.g., arousal, cognitive load). Although the pilot study confirmed the feasibility of the proposed method, it also highlighted many limitations and areas for improvement, which are discussed next.

Environment. First off, the research environment should be mapped out to identify the most likely places for the participant to be located. Cameras will need to be positioned to capture the participant's face in these locations. A challenge here rests in the zoom adjustment of the camera, as it needs to be close enough to cover a large surface area but small enough to detect facial microexpressions. Evidently, more cameras are preferable. Our initial setup included 9 cameras, in the future we will at least double that number. Secondly, the setup we used did not allow live viewing. Research assistants positioned the cameras at a predetermined angle, but were unable to adjust them according to participant characteristics. Thus, taller or shorter than average participants' faces were often not captured and the data was useless. Live viewing of the camera feed will be a necessity in future setups. Thirdly, a significant challenge of facial expression analysis is illumination variation [10]. Variations in lighting can lead to misidentification of emotions. In a naturalistic environment with windows, time of day and weather can have a significant impact on the data. We recommend the use of curtains to standardise illumination and/or scheduling participants at similar moments of the day to lessen the variations.

User. As for the management of participants, it is a given that the participant's face will not be captured 100% of the time, thus the duration of the experiment and the number of participants recruited need to be adjusted. It will also help to have other people in the situation, a salesperson for example, or even a confederate, position themselves to help the participants be in the right position.

Task. Allowing participants and salespeople to interact freely and with few directives proves to be challenging for the segmenting of the interaction. In our case, we set up a microphone on the salesperson and had an expert member of the research team listening in in real-time. This person had preemptively identified different portions of the sales experience that were of interest for comparisons and would manually input event markers to identify the beginning of a new portion. This proved challenging for the expert as it required their constant focused attention, and while the use of a second expert may have alleviated the responsibility, interrater reliability would then have been an issue. A natural sales context is not a linear interaction and it proved difficult to differentiate when, for example, the consumer was providing their personal information to conclude the sale (transaction), but then asked another question about the product (demonstration). This also led to data samples that were of uneven lengths and sometimes a succession of too short samples which proved challenging for posthoc processing.

5 Conclusion

The results from this pilot test suggest that the proposed architecture is likely to offer significant benefits to study IS research questions in naturalistic contexts. While the current state of this methodology and associated apparatus needs further testing and adjustment to improve its reliability and accuracy, it is already possible to envision its usage beyond the test case described in this paper. As interactions with IT evolve to vocal and gesture-based interactions [21], this new method will enable the use of synchronized neurophysiological recording in contexts that have never been used before. For example, it could be possible to monitor synchronized physiological and facial data over long usage duration (for example, a full day) a user at his/her desk without having to put an artificial constraint on their positioning. It could also be possible to monitor the usage of IT by shop floor workers that are using a manufacturing execution system while doing their work on an assembly line. The same method could be used in a classroom to monitor learners doing teamwork with a tablet. Any context of IT usage, which needs free movement from the users could be within the scope of these new methods.

References

1. Riedl, R., Léger, P.M.: Fundamentals of NeuroIS. Springer, Heidelberg (2016) https://doi.org/10.1007/978-3-662-45091-8

2. vom Brocke, J., Hevner, A., Léger, P.M., Walla, P., Riedl, R.: Advancing a NeuroIS research agenda with four areas of societal contributions. Eur. J. Inf. Syst. **29**(1), 9–24 (2020). https://doi.org/10.1080/0960085X.2019.1708218
3. Ekman, P., Freisen, W., Ancoli, S.: Facial signs of emotional experience. J. Pers. Soc. Psychol. **39**(6), 1125 (1980). https://psycnet.apa.org/record/1981-25797-001
4. Jerritta, S., Murugappan, M., Nagarajan, R., Wan, K.: Physiological signals based human emotion recognition: a review. In: Proceedings - 2011 IEEE 7th International Colloquium on Signal Processing and Its Applications, CSPA 2011, pp. 410–415 (2011). https://doi.org/10.1109/CSPA.2011.5759912
5. Stöckli, S., Schulte-Mecklenbeck, M., Borer, S., Samson, A.C.: Facial expression analysis with AFFDEX and FACET: a validation study. Behav. Res. Methods **50**(4), 1446–1460 (2018). https://doi.org/10.3758/s13428-017-0996-1
6. Resseguier, B., Léger, P.M., Sénécal, S., Bastarache-Roberge, M.C., Courtemanche, F.: The influence of personality on users' emotional reactions. In: Lecture Notes in Computer Science (including subseries Lecture Notes in Artificial Intelligence and Lecture Notes in Bioinformatics), vol. 9752, pp. 91–98. Springer (2016). https://doi.org/10.1007/978-3-319-39399-5_9
7. Lamontagne, C., Sénécal, S., Fredette, M., Chen, S.L., Pourchon, R., Gaumont, Y., De Grandpré, D., Léger, P.M.: User test: how many users are needed to find the psychophysiological pain points in a journey map? In: Advances in Intelligent Systems and Computing, vol. 1018, pp. 136–142. Springer (2020). https://doi.org/10.1007/978-3-030-25629-6_22
8. Beauchesne, A., Sénécal, S., Fredette, M., Chen, S.L., Demolin, B., Di Fabio, M.L., Léger, P.M.: User-centered gestures for mobile phones: exploring a method to evaluate user gestures for UX designers. In: Lecture Notes in Computer Science (including subseries Lecture Notes in Artificial Intelligence and Lecture Notes in Bioinformatics), LNCS, vol. 11584, pp. 121–133. Springer (2019). https://doi.org/10.1007/978-3-030-23541-3_10
9. Giroux-Huppé, C., Sénécal, S., Fredette, M., Chen, S.L., Demolin, B., Léger, P.M.: Identifying psychophysiological pain points in the online user journey: the case of online grocery. In: Lecture Notes in Computer Science (including subseries Lecture Notes in Artificial Intelligence and Lecture Notes in Bioinformatics), LNCS, vol. 11586, pp. 459–473. Springer (2019). https://doi.org/10.1007/978-3-030-23535-2_34
10. Samadiani, N., Huang, G., Cai, B., Luo, W., Chi, C.H., Xiang ,Y., He, J.A.: A review on automatic facial expression recognition systems assisted by multimodal sensor data. Sensors **19**(8) 1863 (2019). https://doi.org/10.3390/s19081863
11. Labonte-Lemoyne, E., Courtemanche, F., Fredette, M., Léger, P.M.: How wild is too wild: lessons learned and recommendations for ecological validity in physiological computing research. In: International Conference on Physiological Computing Systems, pp. 123–130 (2018)
12. Groeppel-Klein, A.: Arousal and consumer in-store behavior. Brain Res. Bull. **67**(5), 428–437 (2005). https://doi.org/10.1016/j.brainresbull.2005.06.012
13. Lajante, M., Ladhari, R.: The promise and perils of the peripheral psychophysiology of emotion in retailing and consumer services (2018). https://doi.org/10.1016/j.jretconser.2018.07.005
14. Newman, J., Lockeman, B.D.: Measuring prepurchase information seeking. J. Consum. Res. (1975). https://www.academic.oup.com. https://academic.oup.com/jcr/article-abstract/2/3/216/1785238
15. Mehra, A., Kumar, S., Raju, J.S.: Management science competitive strategies for brick-and-mortar stores to counter "showrooming". **64**(7), 3076–3090 (2018). https://doi.org/10.1287/mnsc.2017.2764. https://pubsonline.informs.org
16. Courtemanche, F., Léger, P.-M., Sénécal, S., Georges, V.: Système d'acquisition physiologique sans-fil pour données utilisateur multimodales. Déclaration d'invention (2020)

17. Léger, P.M., Davis, F.D., Cronan, T.P., Perret, J.: Neurophysiological correlates of cognitive absorption in an enactive training context. Comput. Hum. Behav. **34**, 273–283 (2014). https://doi.org/10.1016/j.chb.2014.02.011
18. Courtemanche, F., Leger, P.-M., Senecal, S., Georges, V., Normandin, F., Fredette, M.: Système de synchronisation sans-fil pour acquisition de données utilisateur multimodales. Déclaration d'invention (2019)
19. Courtemanche, F., Léger, P.-M., Sénécal, S., Georges, V., Brieugne, D.: Système de caméras interconnectées pour reconnaissance ambiante des expressions faciales. Déclaration d'invention (2020)
20. Courtemanche, F., Léger, P.M., Dufresne, A., Fredette, M., Labonté-LeMoyne, É., Sénécal, S.: Physiological heatmaps: a tool for visualizing users' emotional reactions. Multimedia Tools Appl. **77**(9), 11547–11574 (2018). https://doi.org/10.1007/s11042-017-5091-1
21. Plummer, D.: Gartner's Top Strategic Predictions for 2020 and Beyond: Technology Changes the Human Condition (2019)

A Tri-Hybrid Brain-Computer Interface for Neuro-Information Systems

Daniel Godfrey(✉), Chantel Findlay, Dinesh Mulchandani, Ravishankar Subramanilyer, Colin Conrad, and Aaron Newman

Dalhousie University, Halifax, Canada
daniel.godfrey@dal.ca

Abstract. Brain-computer interfaces (BCIs) are computerized systems that convert brain activity into control commands to operate software or external devices. Though promising, BCIs currently have limited practicality and usership due to poor signal classification and large training data requirements. The present study aims to overcome both challenges by combining three brain signals. This paradigm could improve existing BCI technical efficacy, and extrapolate to applications where hands-free visual interfaces could equip users with communication and information resources that improve work processes.

Keywords: Brain-computer interface (BCI) · Event-related potential (ERP) · Machine learning · Hybrid-BCI · Visual interfaces

1 Introduction

1.1 Background

A brain-computer interface (BCI) is an electronic system that converts a user's brain signals into control commands that operate software or external devices. Early BCI technology allowed users to type words by attending to desired letters on an on-screen keyboard (i.e., a "P3 speller") [1]. Modern iterations grant users access to basic Facebook [2] and Windows Explorer [3] functions. Identity authentication security systems scan users' covert, idiosyncratic brain signals instead of fingerprints [4]. "The MOMENT" is a choose-your-own-adventure style film with multiple versions of each scene and a BCI that uses signals elicited by one scene to select which version of the next scene to play [5]. In the field of NeuroIS, BCI could be useful for monitoring user activity and responses during IT use. For example, Conrad and Newman [6] have explored the use of EEG in identifying mind wandering during online learning. These authors noted that if a reliable mind wandering signal is identified, it could be used in a BCI to adjust the delivery of learning content in real time. However, BCIs do not work equally well for all people, or at all for many. In fact, an estimated 15–30% of all people are "BCI illiterate," referring to a state of incompatibility between user and system [7]. Overcoming the signal variance problem would increase BCI efficacy for current users, and grant it to new ones.

F. D. Davis et al. (Eds.): NeuroIS 2020, LNISO 43, pp. 291–297, 2020.
https://doi.org/10.1007/978-3-030-60073-0_34

To this end, we propose a novel system that could leverage weaker, less frequent signals to perform as well as (or better than) current BCIs that require strong, frequent signals. When successful, this would decrease training requirements, increase BCI usership, and advance information system applications such as biometric security, entertainment, and more.

1.2 Hypothesis Development

Many BCI use event-related potentials (ERPs)—electroencephalographic (EEG) activity time-locked to particular events of interest—to detect brain signals related to user attention or intention (or both). Typical BCI application involves first a training, and then a usage phase. In the training phase, a paradigm is used to elicit examples of the ERP signal of interest, as well as samples of data where the signal is absent. The training data is then submitted to a machine learning algorithm (*classifier*) to identify features that discriminate between signal-present and signal-absent trials. This trained classifier is then applied in the usage phase to automatically classify brain signals and use them to control the interface.

For example, the seminal P3 speller BCI is so-named as it uses the P3 ERP—a positive deflection in EEG activity that occurs roughly 300 ms after the onset of a task-relevant, infrequent stimulus—as its dependent measure [1]. The P3 speller displays a grid of letters and numbers, and the user is instructed to attend to the character they wish to spell (the *target*) while the characters are highlighted in a random sequence. Since the target is task-relevant (attended by the user) and infrequent (due to the random sequence of highlighting all possible characters), a P3 is elicited when the target is highlighted, relative to when any other character is.

BCI performance is typically measured on the basis of the classifier's accuracy—the proportion of trials correctly classified. In the literature, 70% accuracy [8] is often considered "acceptable", although in real-world applications obviously higher accuracy is always desirable. Although the speller paradigm is an effective method to elicit P3 signals, no method works equally well (if at all) for everyone (the *universality problem* [9]). Signal strength and signal-to-noise ratio (SNR) vary both between users and within a user over time, both due to individual differences, and the presence of artifacts in the data such as those arising from muscle contractions, eye movements, and poor electrode contact (the *non-stationarity problem* [10]). The universality problem is in large part a product of the non-stationarity problem, and the latter is often a matter of weak signals and/or low SNR throughout the training data.

Previous studies have approached the SNR issue by increasing signal strength. For example, the P3 speller paradigm has undergone several successful modifications in recent years, including stimuli variations designed to elicit stronger ERP signals. It has been found that flashing a familiar face over target characters in the P3 speller generated a larger P3 amplitude than did classic character flashing [11], and larger still when the face is inverted [12] or green [13]. This was possibly caused by the overlapping presence of another ERP, the N170. As its name implies, N170 peaks approximately 170 ms after the appearance of a face, and is negative over temporal-parietal scalp regions [14]. The N170 is accompanied by a "vertex positive potential" (VPP) maximal over the top of the head, similar to the location of the P3 [11]. This increased accuracy of a P3 speller that uses faces may be due to the salience of the face, and/or to the summation of the P3 and VPP components.

Notwithstanding signal summation, hybrid BCIs with dual-input classifiers [see 15 for a recent review] can be created to accurately determine a user's target object with less training data with convergent validity from a combination of two different signal types, each from a different scalp region (e.g., the vertex P3 and temporal-parietal N170). Many of these hybrid systems use steady-state visually evoked potentials (SSVEPs)—oscillations in occipital neural activity entrained by flickering visual stimuli [16]—as one of their input signals. In one study [17], a P3-SSVEP hybrid BCI achieved 93% mean accuracy with minimal training data, though these results were similar to the SSVEP-only (89%) and P3-only (90%) comparison iterations investigated in the study.

We aim to extend these approaches by using three distinct signals as input for a novel tri-hybrid BCI paradigm. Using a visual oddball paradigm, we will present participants with custom face stimuli to elicit P3 and N170 signals, and flicker each object at a different frequency to elicit SSVEPs. We will vary face colour (Caucasian, Green) and orientation (Upright, Inverted) characteristics to test which stimulus condition elicits the strongest signals, and measure each signal at a different region of interest (ROI). We hypothesize that (1) the custom stimuli will elicit strong P3, N170, and SSVEP signals, particularly in the Green Inverted condition; (2) the three-class algorithm will achieve very high classification accuracy; and (3) very little training data will be required to optimize performance.

Below, we describe our custom stimuli, methodology, and planned analyses, and outline potential information system applications.

2 Method

2.1 Participants

We will recruit 35 participants from our university to participate in a BCI pilot study. Participants will be excluded if they report having any condition which affects brain activity (e.g., neurological disorders; psychoactive medications) or have a hairstyle that impedes EEG recording (e.g., dreadlocks). All participants will provide written informed consent consistent with the declaration of Helsinki, and all procedures will be approved by the Dalhousie University Research Ethics Board.

2.2 Stimuli

Stimuli will consist of six vector images of cartoon pirate faces purchased from Shutterstock.com [18] and six grey silhouettes (one of each face) created using Adobe Illustrator CC 2019 software. Cartoon faces have been shown to elicit N170s as effectively as photographs [19], and were chosen for the present study in effort to decrease user fatigue by increasing user engagement. Stimuli will be presented using custom-built software written in Python (v3.7.3) with the PyGame library (v1.9.6 [20]) open source software.

2.3 Procedure

After informed consent, EEG setup (see below), participants will be seated 70 cm in front of a computer screen angled perpendicular to their gaze. When participants are ready to begin, the experimenter will dim the laboratory lights and start the stimulus. Participants will be instructed to try to sit still, fix their gaze on target objects, and avoid blinking during trials.

Stimuli will be presented in a series of experimental blocks, one block for each condition (caucasian upright, green upright, caucasian inverted, and green inverted faces). Each block will consist of six, 42 s trials. Each trial will begin with the display of six solid grey silhouettes of cartoon pirate heads arranged in a circular array, equidistant from a central red fixation cross, centered on a computer screen (see Fig. 1). After 3 s, the fixation cross will disappear, and one of the six silhouettes will turn solid red to highlight that trial's target object. After another 3 s, the red silhouette will return to grey. One at a time, and in random order, a cartoon pirate face will appear in place of its corresponding silhouette. Each face will appear for 1 s and flicker at a different frequency (clockwise, from top: 7 Hz, 11 Hz, 8 Hz, 15 Hz, 9 Hz, and 13 Hz). Each pirate face will appear six times per trial, and never twice consecutively. After the end of the trial, there will be a 5 s pause, followed by the start of the next trial. There are 6 trials per block, such that each pirate face serves as the target (in random order) once per block. A three-minute rest period will be placed between each block to reduce fatigue and discomfort, during which participants will be encouraged to relax and blink freely. In total, the experiment will last roughly 30 min.

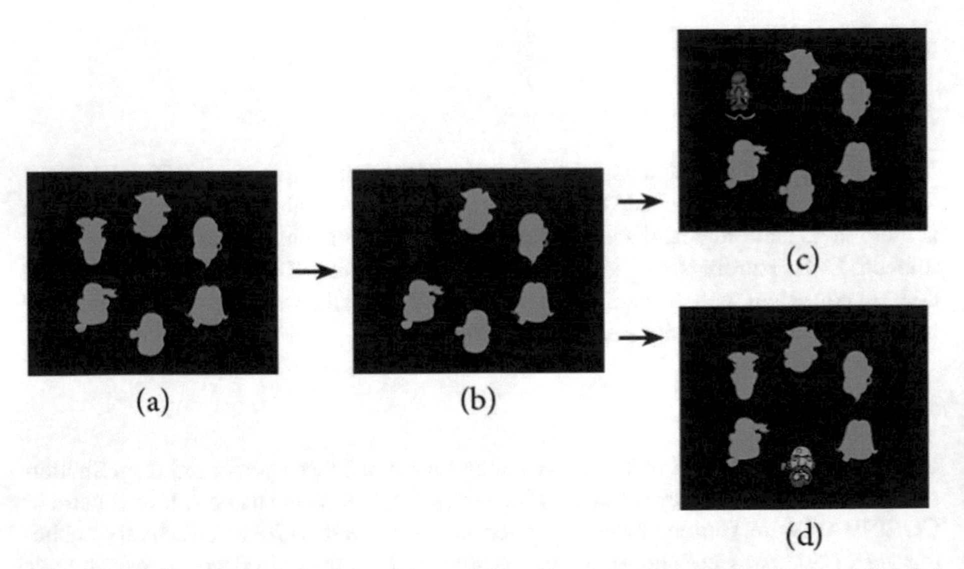

Fig. 1. Example stimulus presentation per trial. In order: (a) Solid fixation cross (3 s); (b) Solid target object identification (3 s); (c) Target pirate face is randomly flashed or (d) Non-target pirate face is randomly flashed (1 s per flash $\times$ 6 flashes per pirate $\times$ 6 pirates = 36 s).

2.4 Data Collection

Data will be acquired using a BrainVision V-Amp EEG system [21], BrainVision Recorder software, and actiCAP Control software. N170 ERPs will be elicited by the face stimuli, P300s by unpredictable (i.e., random) and rare (1:5 target to non-target flash ratio) appearances of target faces, and SSVEPs by flickering the stimuli. Sixteen electrodes will be placed at locations relative to the international 10–20 EEG montage system, selected to coincide with the maximum amplitude of the targeted EEG effects: P3 signal from Pz, Cz, C3, and C4; N170 from TP9 and TP10; and SSVEP from Oz, O1, and O2. We will then test a variety of classifiers (detailed below) to ascertain which achieves the highest accuracy. Results will indicate which stimulus condition and classifier we will take forward into our next phase--real time BCI usage.

3 Planned Analyses

3.1 Data Processing

Data will be processed in the MNE-python (v0.20.0 [22]) analysis and visualization library, using standard procedures described elsewhere [23]. The dependent measures we will extract from the data for hypothesis testing and classification will be focused on the expected ERP components, the P3 and N170, as well as the SSVEP. Because this is an exploratory study aimed at validating our approach, we will determine the precise details of these dependent measures based on visual inspection of the data once it is collected (as experimental parameters can influence the timing and scalp location of ERP effects). However, based on past literature we anticipate that the P3 will be quantified in the 200–400 ms time window as the adaptive mean positive amplitude (average over 50 ms centered on the peak positive value) at electrode Cz; the N170 will be quantified as the adaptive mean negative amplitude at electrodes TP9 and TP10 (temporal-parietal) and the associated VPP as adaptive mean positive amplitude at Cz, from 125–205 ms; and the SSVEP will be quantified as power centered on the flicker frequency of the target stimulus over the 1 s epoch during which each stimulus flickers.

3.2 Hypothesis Testing

To determine the detectability of each EEG signal of interest at the individual subject level, we will use permutation *t*-tests between pairs of conditions of interest (specifically, target vs. Non-target, upright vs. Inverted targets, caucasian vs. Green targets; caucasian upright targets vs. Green inverted targets) for the electrodes and time windows of interest as well as additional electrodes and time windows that may be identified as relevant through visual inspection of the data. In addition, to determine the efficacy of our manipulations at the group level we will perform linear mixed effects modelling with fixed effects of target status (target/non-target), orientation (upright/inverted), color (caucasian/green), and scalp electrode location, along with random by-subject intercepts, and random electrode-by-subject slopes.

3.3 Classification

For N170 and P3 signals, we will test three time domain classification algorithms—dynamic logistic regression [24], support vector machines [25], and linear discriminant analysis [26]—with six-fold cross validation to determine which gives optimal performance.

4 Applications for Information Systems

We posit that our tri-hybrid paradigm could improve existing BCI technology by increasing efficacy while decreasing training requirements. Once validated, this system could augment even the most sophisticated heads-up displays (HUDs) such as Google Glass [27] with image-based menus and thought-controlled navigation and selection features. This may be particularly helpful in high-stakes work processes or other settings where rapid, hands-free access to information or communication tools would be exceedingly helpful. Furthermore, this paradigm could be extrapolated to rapid target identification in virtual reality or in interaction with computer assistive avatars.

Future work could investigate applications of this BCI in the aforementioned domains, particularly in situations where eye tracking is not feasible or where EEG is already in use for other purposes. This paradigm might be specifically applied to the design of interactive avatars which make use of face designs in a goal-oriented task, such as visual search, immersive online learning contexts, or during interactive collaboration. With improvements to BCI signals and classification, it may also be possible to implement BCI beyond a lab setting.

References

1. Farwell, L.A., Donchin, E.: Talking off the top of your head: toward a mental prosthesis utilizing event-related brain potentials. Electroencephalogr. Clin. Neurophysiol. **70**(6), 510–523 (1988)
2. Randolph, A. B., Warren, B.: Facebrain: a P300 BCI to Facebook. In: Davis, F.D., Riedl, R., vom Brocke, J., Léger, P.M., Randolph A.B. (eds.) Information Systems and Neuroscience. Lecture notes in Information Systems and Organization, pp. 101–109. Springer, Cham (2019)
3. Bai, L., Yu, T., Li, Y.: A brain computer interface-based explorer. J. Neurosci. Methods **244**, 2–7 (2015)
4. Sun, Y., Lo, F.P.W., Lo, B.: EEG-based user identification system using 1D-convolutional long short-term memory neural networks. Expert Syst. Appl. **125**, 259–267 (2019)
5. Ramchurn, R., Wilson, M. L., Martindale, S., Benford, S.: #Scanners 2 - the MOMENT: a new brain-controlled movie. In: Extended Abstracts of the 2018 CHI Conference on Human Factors in Computing Systems (CHI EA 2018), Paper D210, pp. 1–4. Association for Computing Machinery, New York (2018)
6. Conrad, C., Newman, A.J.: Measuring the impact of mind wandering in real time using the P1-N1-P2 auditory evoked potential. In Davis, F., Riedl, R., vom Brocke, J., Léger, P.-M., Randolph, A. (eds.) Information Systems and Neuroscience: NeuroIS Retreat 2018, vol. 29, no. 2, pp. 37-45. Springer, Cham (2018)
7. Allison, B., Wolpaw, E., Wolpaw, J.: Brain–computer interface systems: progress and prospects. Expert Rev. Med. Devices **4**(4), 463–474 (2007)

8. Kübler, A., Neumann, N., Kaiser, J., Kotchoubey, B., Hinterberger, T., Birbaumer, N.P.: Brain-computer communication: self-regulation of slow cortical potentials for verbal communication. Arch. Phys. Med. Rehabil. **82**(11), 1533–1539 (2001)
9. Allison, B., Neuper, C.: Could anyone use a BCI? In: Karat, J., Vanderdonckt, J. (eds.) Brain-Computer Interfaces: Applying Our Minds to Human-Computer Interaction, pp. 35–54. Springer (2010)
10. Lotte, F., Congedo, M., Lécuyer, A., Lamarche, F., Arnaldi, B.: A review of classification algorithms for EEG-based brain–computer interfaces. J. Neural Eng. **4**, 1–24 (2007)
11. Kaufmann, T., Schulz, S., Grünzinger, C., Kübler, A.: Flashing characters with famous faces improves ERP-based brain-computer interface performance. J. Neural Eng. **8**(5), 056016 (2011)
12. Zhang, Y., Zhao, Q., Wang, X., Cichocki, A.: A novel BCI based on ERP components sensitive to configural processing of human faces. J. Neural Eng. **9**(2), 026018 (2012)
13. Li, Q., Liu, S., Li, J., Bai, O.: Use of a green familiar faces paradigm improves P300-speller brain-computer interface performance. PLoS ONE **10**(6), 1–15 (2015)
14. Eimer, M.: Event-related brain potentials distinguish processing stages involved in face perception and recognition. Clin. Neurophysiol. **111**(4), 694–705 (2000)
15. Choi, I., Rhiu, I., Lee, Y., Yun, M.H., Nam, C.S.: A systematic review of hybrid brain-computer interfaces: taxonomy and usability perspectives. PLoS ONE **12**(4), 1–35 (2017)
16. Herrmann, C.: Human EEG responses to 1–100 Hz flicker: resonance phenomena in visual cortex and their potential correlation to cognitive phenomena. Exp. Brain Res. **137**, 346–353 (2001)
17. Chang, M.H., Lee, J.S., Heo, J., Park, K.S.: Eliciting dual-frequency SSVEP using a hybrid SSVEP-P300 BCI. J. Neurosci. Methods **258**, 104–113 (2016)
18. One line man.: Royalty-free stock vector ID: 156955199 (2020). https://www.shutterstock.com/image-vector/vector-illustration-pirate-faces-easyedit-layered-156955199
19. Chen, L., Jin, J., Zhang, Y., Wang, X., Cichocki, A.: A survey of the dummy face and human face stimuli used in BCI paradigm. J. Neurosci. Methods **239**, 18–27 (2015)
20. Shinners, P.: PyGame - Python Game Development (Version 1.9.6) (2011). https://www.pygame.org
21. Brain Products (2020). https://www.brainproducts.com/productdetails.php?id=15
22. Gramfort, A., Luessi, M., Larson, E., Engemann, D., Strohmeier, D., Brodbeck, C., Goj, R., Jas, M., Brooks, T., Parkkonen, L., Hämäläinen, M.: MEG and EEG data analysis with MNE-Python. Front. Neurosci. **7**(267), 1–13 (2013). ISSN 1662-453X
23. Conrad, C., Agarwal, O., Calix Woc, C., Chiles, T., Godfrey, D., Krueger, K., Marini, V., Sproul, A., Newman, A.J.: On using Python to run, analyze, and decode EEG experiments. In: Information Systems and Neuroscience: NeuroIS Retreat 2019, pp. 287–293 (2020)
24. Penny, W. D., Roberts, S. J.: Dynamic logistic regression. In: International Joint Conference on Neural Networks. Proceedings (Cat. No. 99CH36339), vol. 3, pp. 1562–1567 (1999)
25. Cortes, C., Vapnik, V.: Support-vector networks. Mach. Learn. **20**, 273–297 (1995)
26. Pedregosa, F., Varoquaux, G., Gramfort, A., Michel, V., Thirion, B., Grisel, O., Blondel, M., Prettenhofer, P., Weiss, R., Dubourg, V., Vanderplas, J., Passos, A., Cournapeau, D., Brucher, M., Perrot, M., Duchesnay, É., Braun, M.: Scikit-learn: machine learning in Python. J. Mach. Learn. Res. **12**, 2825–2830 (2011)
27. Google Glass (2020). https://www.google.ca/glass/start

The Association Between Information Security and Reward Processing

Robert West(✉) and Kaitlyn Malley

Department of Psychology and Neuroscience, DePauw University, Greencastle, USA
robertwest@depauw.edu, kaitlynm424@gmail.com

Abstract. Insider threat represents a significant source of violations of information security within corporations and government entities. Therefore, gaining a clearer understanding of the factors that moderate insider threat is important for the information systems community. The current study builds upon previous behavioral research demonstrating that an imbalance between one's sensitivity to gains and losses may contribute to violations of information security by using event-related brain potentials (ERPs) to examine the association between information security decision-making and the neural correlates of the processing of gains and losses. The ERP data revealed that the amplitude of differences in neural activity between gains and losses was greater in individuals who were more likely to endorse violating information security, and that this association was observed regardless of whether gains or losses resulted from an active choice or were outside of the control of the individual.

Keywords: Information security · Reward processing · Feedback processing

1 Introduction

Insider threat represents a significant source of violations of information security, accounting for roughly 50% of security violations [1], and many organizations realize the extent of the problem. For instance, a 2015 survey [2] revealed that 89% of respondents indicated that their organizations were at risk from insider threat. The potential for insider threat has led to the widespread implementation of deterrence training programs. However, some research has demonstrated that programs grounded in the threat of punishment may not decrease intentions to commit, or instances of, violations of information security [3]. In contrast, meta-analytic work has revealed that the strongest predictors of compliance with security policies are related to the values and characteristics of employees [4]. This and similar findings highlight the importance of identifying individual differences that moderate compliance with, and violations of, information security policy. Motivated by previous behavioral findings, the current study was designed to examine the relationship between violations of information security in a hypothetical decision-making task and the neural correlates of reward processing using event-related brain potentials (ERPs).

F. D. Davis et al. (Eds.): NeuroIS 2020, LNISO 43, pp. 298–306, 2020.
https://doi.org/10.1007/978-3-030-60073-0_35

Some behavioral evidence indicates that individual differences in the perceptions of rewards and punishers may be a predictor of individuals' propensity to violate information security. Meta-analytic work reveals that perceptions of rewards and punishment embodied in security policies may uniquely account for 2%–6% of compliance with information security policy [4]. Furthermore, Hu et al. [5] demonstrated that low self-control may serve to enhance the perception of benefits and attenuate the perception of sanctions in the context of information security. This finding is consistent with the broader literature related to criminality demonstrating that poor self-control can be a general predictor of deviant behavior resulting from an imbalance in the weighting of immediate gains and longer term losses [6]. The swiftness, certainty, and severity of sanctions (i.e., punishment) affect the perceived risk of violations of information security [6]. However, it is also known that the threat of punishment may not be effective in reducing violations of information security [5].

The Information Security Paradigm (ISP) has been used in previous research to study the behavioral and neural correlates of decision-making related to information security [7]. In the task, a set of scenarios are posed to individuals and they are asked to decide whether a fictitious IS specialist should take an action described in the scenarios. Some of the scenarios reflect benign actions (e.g., meet a friend for lunch), while others embody a violation of information security policy or practice (e.g., unauthorized access of a secure server). Performance on the task is sensitive to factors that are known to moderate cyber-deviance (e.g., severity of the violation) [8] or that is associated with violations of information security (e.g., individual differences in self-control) [3]. Given these findings, it seems that the ISP provides a reasonable starting point to explore the association between individual differences in decision-making related to information security and reward processing.

Event-related potentials (ERPs) have been used in a variety of contexts to study the neural correlates of reward processing in simple gambling and reinforcement learning tasks [9]. The 2-doors task is one example of a gambling task used in this domain. In this task two doors are presented and individuals are instructed to select the door they believe hides a small reward (e.g., 50 cents) and avoid the door that hides a small loss (i.e., 25 cents). The difference in the value of gains and losses is designed to account for the general fact that losses are more salient than gains in decision-making [10]. After a door is selected, individuals receive feedback indicating the outcome of the choice (i.e., a win or loss) represented by an upward or downward pointing arrow, respectively. In the current study, we used the modified 2-doors task wherein each trial begins with a cue indicating whether the participant or computer will make the choice of a door for the current trial. Participants win or lose money on both person and computer selected trials. This modification of the task allows us to disentangle the effects of choice or agency and outcome on the ERPs related to feedback processing, and has proven useful in understanding the relationship between various individual differences and reward processing (e.g., self-control [11], socioeconomic status [12], depression [13]).

In the 2-doors task the ERPs elicited by feedback begin to differ between wins and losses over the midline frontal-central region of the scalp around 200 ms after feedback is presented. Between 200–300 ms the ERPs are more positive going for gains relative to losses representing the Reward Positivity (RewP) or Feedback Negativity (FN) [10,

14]. The RewP reveals a clear effect of agency, being greater in amplitude for person select trials than for computer select trials. Following the RewP, the ERPs become more positive for losses than gains representing the frontal P3. The frontal P3 typically peaks around 400 ms after feedback is presented. This component is much greater in amplitude for person select trials than for computer select trials, and in some studies is limited to losses that result from the action of the individual [14]. These midline frontal components are accompanied by transient and sustained ERP components over the lateral frontal-temporal and occipital regions [14].

The imbalance in the perception of reward and punishment related to individual differences in self-control in the context of violations of information security policy [7], is interesting within the broader literature related to Neural Imbalance Theory (NIT) [15] and the relationship between individual differences in self-control and the neural correlates of reward processing. NIT represents a general theory of risky decision-making wherein variation in the contribution of subcortical and cortical systems underpinning reward processing and cognitive control is thought to moderate the frequency of risky decisions. For instance, the violation of an information security policy might arise from strong input from the ventral striatum reward system making the potential gain associated with a violation particularly salient, coupled with weak input from control systems supported by the medial and lateral frontal cortex resulting a failure to weigh the longer term costs (i.e., penalties) associated with violating an information security policy.

In the context of the current study, our group has observed that individual differences in self-control are associated with two effects in the modified 2-doors task [11]. First, low self-control is associated with an attenuation of the RewP for both person and computer select trials, demonstrating a general decrease in the sensitivity to gains with low self-control. Consistent with the finding, other evidence indicates that individuals that engage in risky decision-making (i.e., driving behavior) also demonstrate an attenuation in the amplitude of the RewP relative to non-risky drivers [16]. Second, individuals with high self-control distinguish between gains and losses for both person and computer select trials, while individuals with low self-control distinguish between gains and losses for person select trials, but not computer select trials. Given these findings and the association between individual differences in self-control and violations of information security [5], we predicted that individuals more likely to violate information security would demonstrate decreased in the amplitude of the RewP for gains and a reduction in the differences between the ERPs elicited by gains and losses in the computer select trials.

2 Method

2.1 Participants

Forty-eight individuals completed the study; the EEG data for one participant had a high level of artifact and two participants did not complete the ISP. For the 45 participants with complete data, the average age was 19.5 yr (range 18–33), there were 31 females, 13 males, and one unidentified, 33 of the participants were White-European, 6 were Asian, 5 were Laninx, and 1 was Black-African American.

2.2 Materials and Procedure

Individuals completed demographic and individual difference measures after providing informed consent, and then EEG data were recorded while individuals completed the ISP and the modified 2-doors task. The individual difference measures included the Self-Control Scale [7], a 24 item scale measuring self-reported impulsivity, risk tasking, self-centeredness, preference for simple tasks, preference for physical activity, and tempe (emotion regulation); the CES-D measuring depressive symptoms, and a measure of sensitivity to punishment and reward. The ISP included 45 scenarios that were equally divided between control, minor violation, and major violation scenarios. In the ISP individuals read a scenario and then responded to a prompt with a 1–4 scale. The complete materials for the ISP can be obtained at (osf.io/f9dbv). In the modified 2-doors task individuals either won 50 cents or lost 25 cents on each of the 80 trials. For half of the trials a cue indicated that the subject should select one of the two doors; for the remaining trials a cue indicated that the computer would select the door for that trial. Individuals were instructed that they won or lost money on all trials, regardless of whether they or the computer made the choice. Follow the presentation of the doors, visual feedback was provided (win - upward pointing green arrow, loss -downward pointing red arrow).

2.3 EEG Recording and Analysis

The EEG was recorded from a 32-channel active electrode system from Brain Vision (Fpz ground, Cz reference, 500 Hz). The recording montage included 30 scalp electrodes and 2 electrodes placed below the eyes to assist in compensating for blinks and eye movements. EEGLAB [17] and ERPLAB [18] were used for data processing. Blinks and saccades were corrected using ICA and a $\pm$ 100 μV threshold was used to reject trials with remaining artifacts. The ERPs were averaged for −200 to 2000 ms around onset of the feedback cues in the 2-doors task for subject wins, subject losses, computer wins, and computer losses.

Task Partial Least Squares (PLS) Analysis [19] was used to examine differences in the mean amplitude of the ERPs elicited by wins and losses in the person and computer select trials for groups of individuals that varied in performance on the ISP. This approach allowed us to consider the data for all conditions, time points, and electrodes within a single set of analyses. The inclusion of the full data matrix, avoids the need to make what can be somewhat arbitrary decision regarding specific time points to include in an analysis as can be the case when using measures of mean amplitude. The TaskPLS Analyses were run with the ERP module of the PLSGUI (https://www.rotman-baycrest.on.ca) under Matlab R2013b. The analyses utilized 1000 permutation or bootstrap samples to evaluate the significance and stability of the latent variables that represent patterns of mean differences in ERP amplitude across conditions and groups, and electrode saliences that represent where in time and space (i.e., at which electrode(s)) mean differences are observed.

3 Results

3.1 Behavioral Data

The choice data for the ISP revealed an effect of scenario, $F(2,88) = 190.25$, $p < .001$, with individuals being more likely to respond 'Yes' for control scenarios, M = 2.91, SD = .32, than for minor violation scenarios, M = 1.76, SD = .42, t = 14.68, $p < .001$, or major violation scenarios, M = 1.68, SD = .45, t = 14.12, $p < .001$; individuals were also less likely to say 'Yes' to major violation scenarios than to minor violation scenarios, t = 2.46, p = .05. The response time data also revealed an effect of scenario, $F(2,88) = 9.67$, $p < .001$, with individuals taking longer to respond to minor violation scenarios, M = 2.78 s, SD = 1.07, that to control scenarios, M = 2.45 s, SD = .97, t = 4.20, $p < .001$, or major violation scenarios, M = 2.43, SD = 1.33, t = 4.73, p = .002, the difference between control and major violation scenarios was not significant, t = .12, p = 1.00.

Response choice for minor and major violation scenarios was highly correlated, r = .86, $p < .001$, while choice for the violation scenarios was not significantly related to choice on the control scenarios (minor r = .001, p = .99; major r = −.15, p = .33). Given the strong correlation between choice for the two violation scenarios, a composite representing the average of the minor and major scenarios was used for further analyses. The composite measure of violations was significantly correlated with impulsivity and self-centeredness from the self-control scale (Table 1), consistent with the association between violations of information security and self-control observed in previous research [3].

Table 1. Pearson correlations between composite of choice for Minor and Major Violation scenarios and subscales of the Self-Control scale.

	Impulsivity	Risk taking	Self centered	Simple task	Physical activity	Tempe
Violation	.31*	.18	.34*	.15	.03	−.05

Note: *p < .05.

3.2 ERP Data

To examine the relationship between information security violations and the ERP correlates of reward processing we split the subjects into three groups each including 15 individuals based upon tertiles of the violation composite. Scores for the low group ranged from 1.20–1.27, scores for the middle group ranged from 1.67–1.80, and scores for the high group ranged from 2.10–2.43. Two TaskPLS analyses were considered, one that contrasted wins and losses for the person select trials and one that contrasted wins and losses for the computer select trials. These analyses included data for 30 electrodes and 0–1000 ms after onset of the feedback stimulus, as beyond 1000 ms differences in the ERP largely seemed to have been resolved.

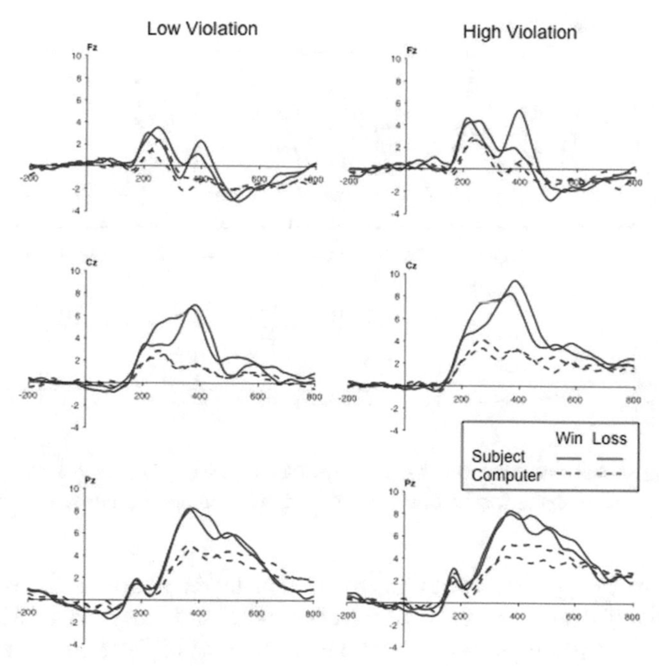

Fig. 1. Grand-averaged ERPs at electrodes Fz, Cz, and Pz for the low and high information security violation groups demonstrating the greater amplitude for the RewP and frontal P3 in high than low violation individuals over the frontal and central regions, but not the parietal region (parietal P3).

The analysis of the person select trials revealed one significant latent variable (66.79%, $p = .001$) that represented an interaction between outcome and violation group, where the strength of the contrast was similar for the low and middle violation groups and then increased for the high violation group (Fig. 2a). The latent variable reflected the RewP/FN and frontal P3 over the midline frontal-central region and transient and sustained ERP components over the posterior regions. The analysis of the computer select trials also revealed one significant latent variable (63.64%, $p = .001$) that represented an interaction between outcome and violation group. For this analysis the strength of the contrast increased from the low violation group to the middle and high violation groups (Fig. 2b). The latent variable reflected the RewP/FN, but not the frontal P3, in addition to sustained ERP activity over the occipital-parietal region (Fig. 2b).

4 Discussion

The behavioral data for the study were generally consistent with evidence from previous research. Choice behavior in the ISP was sensitive to the severity of the violation of information security converging with the broader literature on cyber-deviance [8], and was correlated with impulsivity and self-centeredness [20]. This later finding is consistent with the idea that aspects of poor self-control (i.e., impulsivity and self-centeredness) are one factor that may contribute to violations of information security [3, 7, 20].

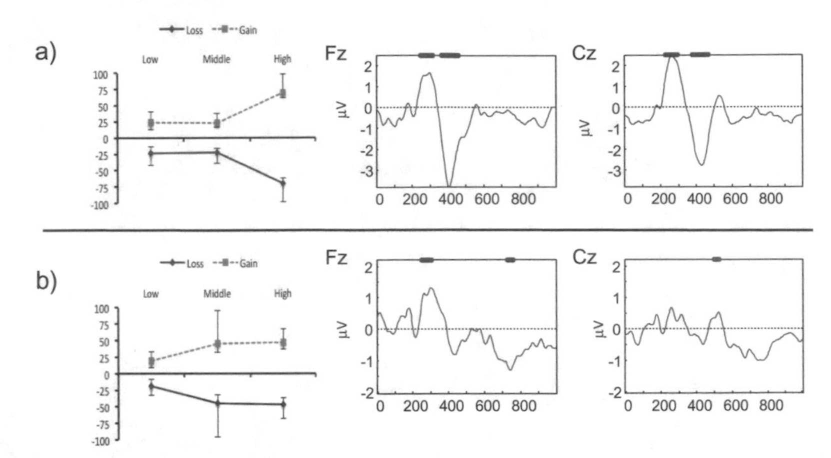

Fig. 2. Design scores and electrode saliences at electrodes Fz and Cz for the person (a) and computer (b) select analyses for the low, middle, and high violation groups.

Counter to our hypothesis that was grounded in the literature examining the relationship between self-control and the ERP correlates of reward processing, the PLS analyses revealed that the strength of the effect of outcome on the ERPs increased as individuals became more likely to endorse a violation of information security. This effect could be seen for both the ERPs related to gains (RewP) and losses (frontal P3) over the midline frontal-central region. Based upon these data, it seems that violations of information security may be associated with a general increase in ones sensitivity to both positive and negative outcomes, rather than an imbalance between the processing of gains and losses. A supplemental analysis of the parietal P3 revealed that the amplitude of this component was insensitive to individual differences in violations of information security. These data are important, as they demonstrate that the effects observed over the midline frontal-central region are selective to feedback processing, and do not simply reflect a nonspecific increase in neural activity in the high violation group.

The failure of the ERP data to support our hypotheses that were grounded in previous research exploring the relationship between self-control and reward processing [11, 16] is interesting when considered within the context of our behavioral data. These data did reveal reliable correlations between two aspects of self-control (i.e., impulsivity and self-centeredness) and choice behavior in the ISP, this finding indicates that decision-making in the current task and sample was related to individual differences in self-control as has been reported in previous research [5]. Together, the behavioral and ERP data may indicate that the relationship between the neural correlates of feedback processing and information security decision-making in the current study arise from attributes of the participants that are not driven by individual differences in self-control, since the effects are the opposite of what was anticipated based upon the extant literature. Following the findings Crum et al. [4], these data may indicate that considering individual differences in the processing of gains and losses could provide unique information when attempting to understanding the factors that contribute to compliance with, and violation of, information security policies.

There are some limitations of the study that should be considered. First, the sample was relatively small (i.e., 45) and the primary findings were inconsistent with our hypotheses that were grounded in the extant literature. This might lead one to have concern regarding the robustness or replicability of the results. To this point, we have collected data for a second study with a larger sample using the modified 2-doors task and a different version of the ISP. The pattern of data for this study is similar to what is reported here, so it seems that the association between information security decision-making and reward processing reported here is likely robust. Second, in the best case, the ISP measures the intention to violate information security policy rather than actual violations of these policies. This could lead onc to wonder whether the associations reported herein would in fact translate to real-world decision-making. Cram et al. [4] reported that similar factors tended to predict intended and actual compliance with information security policy, possibly providing some reassurance that the current findings could provide insight that is useful to information security professional. Third, the current participants were primarily female and undergraduate students. Given this, one line of future research could be to examine the relationships observed in the current study in a more representative sample with a balanced gender distribution and a broader range of professionals with greater work experience wherein issues relevant to information security are more salient.

In conclusion, the current findings demonstrate the utility of incorporating methods and data from neuroscience when considering factors that contribute to violations of information security policy [7]. Additionally, these data provide novel evidence revealing that the relationship between reward processing and information security may not simply reflect a shared relationship with individual differences in self-control. Future studies may seek to identify self-report or behavioral measures of reward sensitivity that could be used in large-scale individual difference studies wherein the collection of EEG/ERP data may not be practical given sample size or approaches to data collection (e.g., online surveys).

References

1. Richardson, R.: CSI Computer Crime and Security Survey (2011). https://www.GoSCI.com
2. Vormetric Data Security: Vormetric Insider Threat Report (2015). https://enterprise-encryption.vormetric.com/rs/vormetric/images/
3. Hu, Q., Zhang, C., Xu, Z.: Moral beliefs, self-control, and sports: effective antidotes to the youth computer hacking epidemic. Paper Presented at 45th Hawaii International Conference on Systems Science (2012)
4. Cram, W.A., D'Arcy, J., Proudfoot, J.G.: Seeing the forest and the trees: a meta-analysis of the antecedents to information security policy compliance. MIS Q. **43**, 525–554 (2019)
5. Hu, Q., Xu, Z., Dinev, T., Ling, H.: Does deterrence work in reducing information security policy abuse by employees? Commun. ACM. **54**, 55–60 (2011)
6. Vance, A., Siponen, M.: Is security policy violations: a rational choice perspective. J. Organ. End User Comput. **24**, 21–41 (2012)
7. Hu, Q., West, R., Smarandescu, L.: The role of self-control in information security violations: insights from a cognitive neuroscience perspective. J. Manage. Inform. Syst. **31**, 6–48 (2015)

8. Venkatraman, S., Cheung, C.M.K., Lee, Z.W.Y., Davis, F.D.: The "Darth" side of technology use: an inductively derived typology of cyberdeviance. J. Manage. Inform. Syst. **35**, 1060–1091 (2018)
9. Walsh, M.W., Anderson, J.R.: Learning from experience: event-related potential correlates of reward processing, neural adaptation, and behavioral choice. Neurosci. Biobehav. Rev **36**, 1870–1884 (2012)
10. Proudfit, G.H.: The reward positivity: from basic research on reward to a biomarker for depression. Psychophysiology **52**, 449–459 (2015)
11. Munoz, A., Freeman, E., Budde, E., West, R.: The influence of socioeconomic status on the neural correlates of feedback processing. J. Cogn. Neurosci. Suppl. 46 (2019)
12. West, R., Freeman, E., Munoz, A., Budde, E.: The influence of agency and self-control on the processing of gains and losses. J. Cogn. Neurosci. Suppl. 46 (2019)
13. Zhu, S., Budde, E., Malley, K., Ash, C., Dapore, A., West, R.: Agency matters: the interaction between agency and depressive symptoms on the neural correlates of reward processing. Unpublished data
14. West, R., Bailey, K., Anderson, S., Kieffaber, P.D.: Beyond the FN: a spatio-temporal analysis of the neural correlates of feedback processing in a virtual Blackjack game. Brain Cogn. **86**, 104–115 (2015)
15. Reyna, V.F.: Neurobiological models of risky decision-making and adolescent substance abuse. Curr. Addict. Rep. **5**, 128–133 (2018)
16. Ba, Y., Zhang, W., Pend, Q., Salvendy, G., Crundall, D.: Risk-taking on the road and in the mind: behavioral and neural patterns of decision making between risky and safe drivers. Ergonomics **59**, 27–38 (2016)
17. Delorme, A., Makeig, S.: EEGLAB: an open source toolbox for analysis of single-trial EEG dynamics. J. Neurosci. Meth. **143**, 9–21 (2004)
18. Lopez-Calderon, J., Luck, S.J.: ERPLAB: an open-source toolbox for the analysis of event-related potentials. Front. Hum. Neurosci. **8**, 213 (2014)
19. Lobaugh, N.J., West, R., McIntosh, A.R.: Spatiotemporal analysis of experimental differences in event-related potential data with partial least squares. Psychophysiology **38**, 517–530 (2001)
20. Zhang, L., Smith, W., McDowell, W.: Examining digital piracy: self-control, punishment, and self-efficacy. Inf. Resourc. Manag. J. **22**, 24–44 (2009)

Explaining Inconsistent Research Findings on the Relationship Between Age and Technostress Perceptions: Insights from the Neuroscience Literature

René Riedl[1,2](✉) and Karin VanMeter[3,4]

[1] University of Applied Sciences Upper Austria, Steyr, Austria
rene.riedl@fh-steyr.at
[2] Johannes Kepler University, Linz, Austria
[3] Austrian Biotech University of Applied Sciences, Tulln, Austria
karin.vanmeter@fh-hagenberg.at
[4] University of Applied Sciences Upper Austria, Hagenberg, Austria

Abstract. Technostress (TS) research has been conducted since the early 1980s. With regard to the relationship between user age and TS perceptions, research findings are inconsistent. While some scholars argued, and empirically showed, that older users are more prone to experience TS, other studies report opposite results. In this paper, we briefly review the literature and summarize major empirical findings on the relationship between age and TS, thereby documenting the inconsistency of results. Based on this review, we outline neurophysiological insights which might serve as an explanation for the mixed evidence. Specifically, we outline insights from physiology and brain research which describes neurobiological changes in normal aging (i.e., changes that are unrelated to pathologies). We focus on age-related changes related to the human stress system as we expected that these alterations might predominantly contribute to a better understanding of age-related differences in TS perceptions. We close this paper with a concluding comment.

Keywords: Age · Aging · Autonomic nervous system · Brain · Digital stress · Stress · Technostress

1 Introduction

Technostress (hereafter TS) refers to stress that results from both the use of information and communication technologies (ICTs) and the pervasiveness and expectations of ICT use in business and society [1, 2]. TS has been studied since the early 1980s [3], and hence a large body of research exists. Research indicates that TS negatively influences technology-supported performance [4, 5] and user satisfaction [6, 7]. Moreover, a review of the negative neurophysiological consequences of ICT use indicates that several detrimental processes may occur in a user's body, including elevations of stress hormones

F. D. Davis et al. (Eds.): NeuroIS 2020, LNISO 43, pp. 307–320, 2020.
https://doi.org/10.1007/978-3-030-60073-0_36

(adrenaline, noradrenaline, cortisol), increased activity of the sympathetic branch of the autonomic nervous system (e.g., increase in heart rate, reduction in heart rate variability, skin conductance elevation), and even changes in neural processes in the brain are reported [2].

Because the current wave of digitalization and digital transformation implies increasing future rates of ICT penetration in business and society, it is very likely that TS will remain a highly relevant phenomenon. This foreseeable trend is accompanied by another trend, namely the rapid aging of the workforce in industrialized countries, referred to as "graying of the workforce" [8]. If both trends are considered in conjunction, a fundamental question arises: *Is there a relationship between age and TS?*

Most people would expect that such a relationship exists. One widely held notion points towards a positive relationship: The older a person, the higher the TS level. A major argument underlying this notion is that the older generations (born in the 1960s or earlier) neither grew up with ICT nor used it frequently (if at all), and this may affect beliefs about, and attitudes towards, ICT [9]. Most people born in the 1990s at least had a chance to get in touch with ICT during their childhood and adolescence, both in business and at home, as the Personal Computer (PC) became pervasive in the 1980s. Yet, general usage of ICT, if compared with today, was much lower. Individuals born in the 2000s or later were exposed to, and used, ICT more than ever before, predominantly as a consequence of the availability of the Internet and corresponding applications, including the smartphone which began its "success story" with the introduction of the iPhone in 2007. A seminal research agenda paper by Tams et al. [8] starts with a vignette that supports the "have-not-grown-up-with-IT" argument, they write: "*Benita Oldfellow, who just turned 68, has to use information and communication technologies (ICTs) in support of her work as an accountant. She has a positive attitude toward the technology and believes it may be useful to her job, but she faces great trouble in using it effectively – in contrast to her grandson Frank who grew up with ICTs […]* (p. 284, italics in original).

However, recent evidence indicates that smartphone use as predominantly practiced by the younger generations is related to increased stress (for a review, see [10]). Thus, recently a phenomenon referred to as smartphone use disorder began to develop, and it has been shown that use of applications such as WhatsApp or Facebook, among many others, may lead to notable stress reactions in users and to a reduction in life satisfaction [11]. What follows is that it is also possible that a negative relationship between age and TS could exist today: The younger a person, the higher the TS level. All in all, therefore, *the relationship between age and TS is not clear today.* Some arguments may support a positive relationship, while other arguments may substantiate a negative relationship (Note that some papers indicate that eventually no relationship exists at all, see [12]).

Against this background, we reviewed papers to develop a picture regarding the empirical evidence on the relationship between age and TS *perceptions*. What follows is that our focus is on studies in which users' (techno)stress, or a related construct (e.g., communication load), was measured through a self-report. Thus, both survey and experimental research is covered by our review.

As the findings of our review show, research findings are inconsistent (for details, see Sect. 2). We used this inconsistency as a starting point to search for neurophysiological insights which might explain the mixed evidence. To this end, we studied literature describing the neurophysiological changes in normal aging (i.e., changes that are unrelated to pathologies). In particular, we studied age-related changes related to the human stress system as we expected that these alterations might specifically contribute to a better understanding of age-related differences in TS perceptions (Sect. 3). We close this paper with a concluding comment (Sect. 4).

2 Evidence on the Relationship Between Age and TS

On February 1, 2020, we conducted several search queries in Web of Science, one of the worldwide largest databases with scientific articles, based on the following keywords: "technostress"/"computerstress"/"digital stress" (specification "Title") AND "age" (specification "Topic"). In total, we identified eight papers of which five were considered as relevant, based on studying the full text of the papers [1, 13–16]. In addition to these 5 papers, we identified 7 further relevant papers [7, 17–22], based on forward and backward searches and personal communication in the scientific community. In Table 1, in ascending order based on publication date, we summarize the main findings of the 12 papers. We cite relevant text passages literally and indicate our conclusion in bold.

In essence, the findings of our review indicate that 4 studies report a direct positive influence of age on TS (i.e., the older, the more TS), while 3 studies report a direct negative influence (i.e., the older, the less TS). The remaining studies either report an indirect effect of age on stress-related outcomes (e.g., Maier et al. [7]; mediator: social networking site usage characteristics) or moderation effects (e.g., Tams et al. [16]; age moderates the effect of interruption characteristics on stress perceptions, whereat age increases the effect). Moreover, our analyses reveal that even one and the same research group found inconsistent findings (see [18] and [22]). Also, our analyses do *not* suggest that the relationship between age and TS has changed systematically over the years. Rather, we observe that what earlier studies already found in the 1980s and 1990s, namely that age may positively affect TS, was replicated in more recent studies in the 2010s. Finally, the more recent studies have a focus on more complex moderation effects rather than on direct effects of age on TS.

Altogether, based on the evidence presented, we observe that research findings on the relationship between age and TS perceptions are inconsistent. While inconsistent research findings can be attributed to a number of factors (e.g., those related to methodology, research context, or sample characteristics), in the following we discuss neurophysiological insights that support either a positive, or a negative, relationship between age and TS.

Table 1. Review of papers on the relationship between age and TS perceptions

Source and sample	Main finding
Elder et al. [19] N = 403; white-collar workers in government offices; USA	"The percentage of technostressed respondents older than 50 years of age is more than double the percentage of respondents under 30." The percentages of technostressed people in the four investigated age groups are: <30 years = 14.8%, 30–39 y. = 13.0%, 40–50 y. = 20.4%, >50 y. = 31.9%. (p. 19) **Conclusion: Age → TS (+)**
Birdi and Zapf [17 N = 134; office workers in 12 companies; Germany	"The present study examined the reactions of older and younger workers to the situation of encountering an error during computer-based work. It was expected that older workers would have a stronger negative emotional reaction to such an error […] this was found to be the case." (p. 309) **Conclusion: Age → TS (+)**
Ragu-Nathan et al. [1] N = 608; end users of ICT in five organizations; USA	"With respect to age, we expected that it would not affect technostress, whereas our findings show that older people experience less technostress. This result can be explained by possible greater organizational tenure of older employees, leading to more organization-specific experience and better understanding of how to assimilate the stress creating effects of ICTs in their work context. This finding may be sample specific." (pp. 429/430) **Conclusion: Age → TS (−)**
Sahin and Coklar [22] N = 765; Internet users; Turkey	"This finding [… indicates] that users of ages 26 and above have lower technostress levels due to use of technology than users under 20 years of age. The effect of age on technology use generally shows that younger individuals have a more positive attitude regarding the use of technology […] However, it appears that regarding technostress, age has an inverse effect. The lower technostress levels of 26 year or older participants may be attributed to their lifelong experiences. These experiences may have had an effect on the results." (p. 1441) **Conclusion: Age → TS (−)**
Coklar and Sahin [18] N = 287; Internet users; Turkey	"An interesting finding is that technostress levels appear to rise with age […] technostress levels of social networking users may change based on their age […] This finding shows that users aged 31 and above have greater technostress levels than those aged 20 and below." (p. 179) **Conclusion: Age → TS (+)**

(*continued*)

Table 1. (*continued*)

Source and sample	Main finding
Tarafdar et al. [21] N = 233; information systems users in 2 government organizations; USA	"Older professionals experience less technostress. Intuitive reasoning suggests that younger people, being more familiar with technology, would experience less technostress. However, it is also true that older people are better able to handle stress in general and computer-related changes because of maturity. Also, the older employees in our sample had greater organizational tenure, which means they had more organization-specific experience and better understanding of how to assimilate the stress-creating effects of IS in their work context. Further, having possibly greater power within the workplace, they are likely to have had more choice and latitude in their use of IT. Hence, they were possibly able to appropriately pace out IT-related change and learning activities, thus experiencing lower technostress." (p. 119) **Conclusion: Age → TS (–)**
Jena and Mahanti [20] N = 116; Internet users; India	"The results also show that the older [users] experience more technostress than younger [users …] the obvious reasoning suggests that the younger people are more familiar with latest technology and hence would experience less technostress." **Conclusion: Age → TS (+)**
Maier et al. [7] N = 571; Internet users; Germany	"Although age has no significant effect on social overload, bivariate correlations indicate a relationship between an SNS [Social Networking Site] user's age and the two SNS usage characteristics [UC] extent of usage and number of friends in SNS […] On the basis of this, it might be justified to say that the effect is mediated on social overload [SO] through these two variables." (p. 475) **Conclusion: Age → UC → SO (+)**
Reinecke et al. [15] N = 1,557; Internet users; Germany	"As predicted […] communication load [CL] was positively related to perceived stress [PS] in the age range of 50 to 85 years […] but not in the subsamples of participants aged 14 to 34 years […] and 35 to 49 years […] this pattern of results suggests a moderation effect of age on the effect of communication load on perceived stress […] As predicted […] Internet multitasking [IM] was significantly related to perceived stress in the age range of 14 to 34 years […] and 35 to 49 years […] but not in the subsample of participants ages 50 years and above." (pp. 103–104) "While the present study clearly underlines the universal relevance of ICT-related stress over the life span, our findings also demonstrate marked discrepancies between different age groups both with regard to the effects of communication load and Internet multitasking on psychological health." (p. 106) **Conclusion: CL → PS (Age ↑), IM → PS (Age ↓)**

(*continued*)

Table 1. *(continued)*

Source and sample	Main finding
Tams et al. [16] N = 128, subjects in experiment, half younger (on average 21 years) and half older people (on average 71 years), North America	"[A]ge acts as a moderator of the interruption-stress relationship due to age-related differences in inhibitory effectiveness, computer experience, computer self-efficacy, and attentional capture. We refer to these age-related differences as concentration, competence, confidence, and capture [...] We tested our model through a laboratory experiment with a 2 × 2 × 2 mixed-model design, manipulating the frequency with which interruptions [IR] appear on the screen and their salience (e.g., reddish colors). We found that age acts as a moderator of the interruption-stress link [measured as perceived stress, PS] due to differences in concentration, competence, and confidence, but not capture." (p. 857) **Conclusion: IR → PS (Age ↑)**
Hauk et al. [13] N = 1,216; employees; longitudinal data collection (T1, T2, T3); Germany, Austria, Switzerland	"Contrary to our hypothesis, age was negatively related to techno-stressors at T2 [...] and not related to techno-stressors at T1 and T3 [...] there was no significant relationship between age and the level of techno-stressors after controlling for job-related ICT dependency. Therefore, in occupational settings higher age is not per se connected to an increased prevalence of situations in which ICT-related demands exceed workers' abilities to meet them." (pp. 9/14) **Conclusion: Age → TS (−), Age ↔ TS (n.s.)**
Marchiori et al. [14] N = 927; users in 14 public institutions; Brazil	"Regarding the effect of the age of the workers on technostress [...] we tested hypotheses that when compared to younger workers, older workers are more affected by (a) techno-overload, (b) techno-invasion, (c) techno-complexity, and (d) techno-uncertainty. In this sense, the evidence supported hypothesis H1c and did not support hypotheses H1a, H1b, and H1d. More precisely, we detected a difference related to the age of users: the results suggest that older users tend to perceive the organizational technology environment (techno-complexity) to be more complex than younger users." (pp. 8/9) **Conclusion: Age → TS (+) (but only for techno-complexity)**

3 Neurophysiological Changes in Normal Aging

Normal aging causes various forms of neurophysiological degeneration as well as some cognitive decline [23], obviously with individual differences. The overall reduction in brain volume and weight during the normal aging process has been well documented with a particular *decline in the prefrontal cortex (PFC)* [24]. Essig [25] indicates that with a maximum weight in the third decade of age, a gradual decline of brain volume takes place thereafter. Other studies report similar findings. For example, it has been found that the volume of the brain and/or its weight declines with age at a rate of approximately 5% per decade after age 40, with the actual rate of decline eventually increasing with age particularly over 70 [26, 27]. Baron and Godeau [28], reviewing original empirical studies, indicate that decline in postmortem brain volume with age is smaller than reported in other studies (around 1.5% per decade or even smaller). Independent from whether changes in brain volume that result from aging occur at a rate of 1% or 5%, it is an established fact that volume changes can be attributed, among others, to loss of neurons, reduced dendritic arborization, decreased spine density and loss of synapses, as well as *reductions in the amount of white matter* [28].

As summarized in Riedl and Léger ([29], p. 37), the PFC constitutes the neural basis for the implementation of several mental processes. First, PFC is critical for the neural implementation of inhibitory control, a cognitive process that makes possible that an individual is able to inhibit impulses in order to select more appropriate behaviors. A recent TS paper found that inhibitory effectiveness is better in younger than in older users, and relates this finding to alterations in the aging brain ("As people age, the frontal lobe's connections with other brain areas change and the frontal lobe shrinks, resulting in an inhibitory deficit" [16], p. 866). Moreover, Riedl and Léger outline that the dorsolateral PFC is critical for general thought processes, working memory, and fairness perceptions, while the ventromedial PFC is important for executive functions and abstract thinking. Several of these mental processes play an important role in the perception and evaluation of potential stressors [30, 31], including those related to use of ICT [2]. Perception and evaluation of stressors influence stress consequences such as health issues [32].

White matter forms a large portion of the deep brain areas and the outer parts of the spinal cord. It is composed of myelinated axons allowing rapid communication between the regions of gray matter. The myelin sheaths in the central nervous system are produced by oligodendrocytes (glial cells) and function in the protection of the axons as well as allowing saltatory conduction of axon potentials. Degenerative changes of myelin during the normal aging process could lead to cognitive decline. However, there are some indications that the number of oligodendrocytes increases during aging, suggesting that some myelination is still possible [33].

Thus, white matter alterations directly affect information transmission in the brain, and possibly also information processing and information integration, because white matter subserves the functioning of gray matter (the latter consists primarily of neural cell bodies and is located in different brain areas). Such alterations, in turn, affect cognitive functions. Therefore, it is possible that TS research findings like "age acts as a moderator of the interruption-stress relationship" ([16], p. 857) can be traced back to anatomical and/or functional alterations in the PFC and white matter.

Even without any disease processes, the anatomical changes in the aging brain do not occur at the same extent in all brain regions, just as it is the case throughout the aging body in general. As indicated, while the PFC does seem to be affected the most and earliest, the *hippocampus, insula, thalamus,* and *basal ganglia,* among others, are affected as well [24, 25]. As summarized in Riedl and Léger ([29], Chapter 2), the mentioned brain areas contribute to the neural implementation of the following mental processes, among others: learning and memory (hippocampus), homeostasis and emotion (insula), regulation of alertness, attentiveness, and sleep (thalamus), and reward processing, learning, and motivation (basal ganglia). Because all these processes are related to stress perceptions [2, 30, 31], including stress in the context of ICT use, it is evident that declines in these areas could explain why older people experience more, or less, TS than younger people.

In stressful situations, two major biological systems become activated: sympathetic-adrenomedullary (SA) and hypothalamic-pituitary-adrenocortical (HPA). The former system controls the stress response and describes the "[o]utflow of sympathetic autonomic nervous system [ANS] that triggers rapid physiological and behavioral reactions to imminent danger or stressors," while the latter system "describes the complex chain of physiological events that characterizes [...] stress response systems" ([34], pp. 147–148).

Regarding the SA system, several studies indicate an *increase in basal plasma noradrenaline levels* with age suggesting that healthy aging is associated with elevated basal sympathetic activity; the reactivity of the sympathetic and parasympathetic branches of the ANS, however, seem to be reduced [35]. What follows is that some effects of aging, at least theoretically, should result in increased physiological stress (e.g., higher basal plasma noradrenaline levels, reduced reactivity of the parasympathetic ANS), while other effects could lead to attenuated stress (e.g., reduced reactivity of the sympathetic ANS). In this context, it is critical to emphasize that noradrenaline, much more than adrenaline, acts as a neurotransmitter in the human nervous system. What follows is that changes in basal plasma noradrenaline levels with age necessarily have strong effects on the brain and ANS, thereby likely affecting stress perceptions.

Regarding the HPA system, Sampedro-Piquero et al. [36] write: "During aging, excessive activation of the HPA axis and hypersecretion of glucocorticoids can lead to dendritic atrophy in neurons in the hippocampus, resulting in impairment in learning and memory functions. Damage or loss of hippocampal neurons would result in impaired feedback inhibition of the HPA axis and glucocorticoid secretion, leading to further damage caused by elevated glucocorticoid concentrations. This feed-forward effect on hippocampal neuronal loss is known as the glucocorticoid cascade hypothesis" (p. 285). Research also indicates that *diurnal cortisol increases* with age and *glucocorticoid sensitivity decreases* with age, resulting in a higher peak of cortisol levels and a longer stress response [37].

In their review, Sampedro-Piquero et al. [36] also refer to neurotransmitters and neuropeptides. Serotonin levels, a substance related to emotion, mood, anxiety, impulsiveness, sleep, and stress ([38]; [29], p. 33), change with age [39, 40]. Based on this evidence, Sampedro-Piquero et al. [36] argue that "it is highly possible that age-dependent *changes in the serotonergic system* contribute to the vulnerability to stress in old age" (p. 286, italics added). Further substances discussed in the review by Sampedro-Piquero et al. are, among others, brain-derived neurotrophic factor (BDNF, "[a]ging itself appears to be associated with *decreased BDNF signaling* in the brain and depressed patients are also characterised by low blood BDNF levels", italics added) and GABA ("the primary inhibitory neurotransmitter in the brain, progressively decreases during aging [...] Dysfunction of the GABAergic system is heavily implicated in anxiety and depression [...] Therefore, *decreased GABA in the brain* may contribute to problematic mood and anxiety symptoms during aging", italics added) ([36], p. 286).

Also, it is reported that *dopamine levels decline* by around 10% per decade from early adulthood. Regarding the specific underlying mechanisms, Peters [24] indicates, based on a review of original studies, that it is possible that "dopaminergic pathways between the frontal cortex and the striatum decline with increasing age, or that levels of dopamine itself decline, synapses/receptors are reduced or binding to receptors is reduced" (p. 85). There is evidence for a relation between HPA axis responses and the dopaminergic system [41]. Moreover, it was found that a positive interaction exists between dopamine and oxytocin [42], a neuropeptide that has been shown to positively affect human social behavior and trust among humans (for a review, see [43]). Because a negative relationship between oxytocin and the stress hormone cortisol is well established in the literature [44], and considering the positive interaction between oxytocin and dopamine, what follows is that decline in dopamine levels with aging presumably increases stress through HPA axis alterations.

Finally, another factor to consider in the normal aging process and stress are telomeres, the caps that protect the end of chromosomes. With each cell division these telomeres become a little shorter, providing less protection to the chromosome. Blackburn and her team described the role of the telomeres in the aging process and how *stress can shorten the telomeres, thus accelerating aging and also influencing the response to stress* [45, 46]. What follows is the following causal chain: Stress → Telomeres → Aging → Stress (then the loop begins again). Thus, it is possible that changes in telomeres, which are both a consequence of stress and an (indirect) stress determinant, could help explain why older users experience more TS than younger ones.

In Table 2, we summarize major findings regarding neurophysiological changes in normal aging with effects on stress. This list is not exhaustive. Yet, the indicated changes constitute a starting point to better understand the review findings on the relationship between age and TS (Sect. 2).

So far, our discussion in Sect. 3 has dealt with neurophysiological changes in normal aging that predominantly explain why older users experience *more* TS than younger ones [Age → TS (+)]. As briefly discussed in the following paragraph, the literature on changes in normal aging has identified several *factors that may significantly slow down aging processes*, thereby attenuating negative consequences such as those related to stress. Thus, the factors that we outline in the following, referred to as "protective factors" [24], can explain why some older users experience little TS, or even no TS, and that is consistent with the studies in Sect. 2 which indicate that age negatively relates to TS [Age → TS (−)], that is, older users experience less TS than younger ones. Importantly, genetic predisposition, along with interaction of that predisposition with an individual's environmental influences, may not only affect stress reactions and perceptions (as discussed in a review by Riedl [2] on TS), but also neurophysiological changes in normal aging. However, the "protective factors" below can be directly influenced by an individual. Thus, users—independent from their age—should be aware of the fact that consideration of these factors is likely to reduce stress in general, and hence also TS.

First, *diet* significantly affects aging. Evidence indicates that a diet higher in energy and/or lower in antioxidants accelerates aging; moreover, it is reported that energy restriction may prolong life [47–50]. Second, moderate, low, or no *alcohol* intake may help

Table 2. Selected neurophysiological changes in normal aging with effects on stress

Physiological system	Main finding
Brain	Decline mainly in the prefrontal cortex (PFC), but also in other brain regions, including hippocampus, insula, thalamus, and basal ganglia. Reductions in the amount of white matter
Hormones, Neurotransmitters, Neuropeptides	Increase in basal plasma noradrenaline levels and in diurnal cortisol levels. Glucocorticoid sensitivity decreases. Changes in the serotonergic system. Decreased BDNF signaling and decreased GABA in the brain. Reduced dopamine levels
Genes	Stress can shorten the telomeres, thereby accelerating aging, which, in turn, affects the response to stress. A telomere is a region of repetitive nucleotide sequences at each end of a chromosome, which protects the end of the chromosome from deterioration or from fusion with close-by chromosomes

to avoid accelerated aging. Specifically, Peters [24] reviewed papers and reports that "moderate drinkers show reduced white matter lesions, infarcts, and even dementia" (p. 86). Third, *exercise* is also beneficial and it was found that it improves neurocognitive function with aging; moreover, aerobic fitness reduces brain tissue loss in aging humans [51, 52]. Fourth, it is reported that "increased *cognitive effort* in the form of education or occupational attainment" ([24], p. 84, italics added) may also be a good defense against the alterations in an aging brain.

4 Concluding Comment

TS research has been conducted since the early 1980s. A fundamental research question in this field is whether a relationship between user age and TS perceptions exists, and if so, what the exact nature of this relationship is. To date, research findings are inconsistent. While some studies showed that older users are more prone to experience TS, other studies report opposite results. Some papers found no relationship at all. As discussed in Riedl et al. [53], NeuroIS research should also consider application of existing neuroscience findings to interpret, or reinterpret, existing behavioral findings or findings that are based on self-reports. Here, we discussed insights on neurophysiological changes in normal aging in order to provide *possible explanations* for the inconsistent research findings on the relationship between user age and TS perceptions. As deductive argumentation (based on neurophysiological insight in our case, as reviewed in Sect. 3) cannot substitute for original empirical research, it is hoped that the present paper instigates future studies which empirically examine, based on neuroscience methods, what we derived theoretically in this publication.

Complementary to this "neuroscience direction" of research, purely behavioral research could also contribute to a better understanding of the inconsistent research findings on the relationship between age and TS perceptions. To resolve the conflicting findings in prior work effectively, it would be necessary to develop a theoretical model examining the moderating factors that determine when the relationship between age and perceived TS is negative and when it is positive. Moreover, a mediation model could be developed that explains which mediating factors create a positive or negative indirect relationship, when the direct effect has the opposite directionality. As a starting point, we refer the reader to research by Tams and colleagues, which may serve as an example for what we call "high-caliber" theorizing, see [8] and [16].

It is hoped that the present paper instigates future studies at the nexus of TS and user age, both behavioral and neurophysiological in nature. It will be rewarding to see what insight future research will reveal.

Acknowledgement. This research was funded by the Austrian Science Fund (FWF) as part of the project "Technostress in organizations" (project number: P 30865) at the University of Applied Sciences Upper Austria.

References

1. Ragu-Nathan, T.S., Tarafdar, M., Ragu-Nathan, B.S., Tu, Q.: The consequences of technostress for end users in organizations: conceptual development and empirical validation. Inf. Syst. Res. **19**, 417–433 (2008). https://doi.org/10.1287/isre.1070.0165
2. Riedl, R.: On the biology of technostress: literature review. DATA BASE Adv. Inf. Syst. **44**, 18–55 (2013)
3. Brod, C.: Managing technostress: optimizing the use of computer technology. Pers. J. **61**, 753–757 (1982)
4. Tarafdar, M., Pullins, E.B., Ragu-Nathan, T.S.: Technostress: negative effect on performance and possible mitigations. Inf. Syst. J. **25**, 103–132 (2015). https://doi.org/10.1111/isj.12042
5. Tams, S., Hill, K., Guinea, A., Thatcher, J., Grover, V.: NeuroIS—alternative or complement to existing methods? Illustrating the holistic effects of neuroscience and self-reported data in the context of technostress research. J. Assoc. Inf. Syst. **15**, 723–753 (2014). https://doi.org/10.17705/1jais.00374
6. Fuglseth, A.M., Sørebø, Ø.: The effects of technostress within the context of employee use of ICT. Comput. Hum. Behav. **40**, 161–170 (2014). https://doi.org/10.1016/j.chb.2014.07.040
7. Maier, C., Laumer, S., Eckhardt, A., Weitzel, T.: Giving too much social support: social overload on social networking sites. Eur. J. Inf. Syst. **24**, 447–464 (2015). https://doi.org/10.1057/ejis.2014.3
8. Tams, S., Grover, V., Thatcher, J.: Modern information technology in an old workforce: toward a strategic research agenda. J. Strateg. Inf. Syst. **23**, 284–304 (2014). https://doi.org/10.1016/j.jsis.2014.10.001
9. Morris, M.G., Venkatesh, V.: Age differences in technology adoption decisions: implications for a changing work force. Pers. Psychol. **53**, 375–403 (2000). https://doi.org/10.1111/j.1744-6570.2000.tb00206.x

10. Vahedi, Z., Saiphoo, A.: The association between smartphone use, stress, and anxiety: a meta-analytic review. Stress Heal. **34**, 347–358 (2018). https://doi.org/10.1002/smi.2805
11. Sha, P., Sariyska, R., Riedl, R., Lachmann, B., Montag, C.: Linking internet communication and smartphone use disorder by taking a closer look at the Facebook and WhatsApp applications. Addict. Behav. Rep. **9**, 100148 (2019). https://doi.org/10.1016/j.abrep.2018.100148
12. Krishnan, S.: Personality and espoused cultural differences in technostress creators. Comput. Hum. Behav. **66**, 154–167 (2017). https://doi.org/10.1016/j.chb.2016.09.039
13. Hauk, N., Göritz, A.S., Krumm, S.: The mediating role of coping behavior on the age-technostress relationship: a longitudinal multilevel mediation model. PLoS ONE **14**, e0213349 (2019). https://doi.org/10.1371/journal.pone.0213349
14. Marchiori, D.M., Mainardes, E.W., Rodrigues, R.G.: Do individual characteristics influence the types of technostress reported by workers? Int. J. Hum.-Comput. Interact. **35**, 218–230 (2019). https://doi.org/10.1080/10447318.2018.1449713
15. Reinecke, L., Aufenanger, S., Beutel, M.E., Dreier, M., Quiring, O., Stark, B., Wölfling, K., Müller, K.W.: Digital stress over the life span: the effects of communication load and internet multitasking on perceived stress and psychological health impairments in a german probability sample. Media Psychol. **20**, 90–115 (2017). https://doi.org/10.1080/15213269.2015.1121832
16. Tams, S., Thatcher, J.B., Grover, V.: Concentration, competence, confidence, and capture: an experimental study of age, interruption-based technostress, and task performance. J. Assoc. Inf. Syst. **19**, Article 2 (2018)
17. Birdi, K.S., Zapf, D.: Age differences in reactions to errors in computer-based work. Behav. Inf. Technol. **16**, 309–319 (1997). https://doi.org/10.1080/014492997119716
18. Çoklar, A.N., Şahin, Y.L.: Technostress levels of social network users based on ICTs in Turkey. Eur. J. Soc. Sci. **23**, 171–182 (2011)
19. Elder, V.B., Gardner, E.P., Ruth, S.R.: Gender and age in technostress: effects on white collar productivity. Gov. Financ. Rev. **3**, 17–21 (1987)
20. Jena, R.K., Mahanti, P.K.: An empirical study of technostress among Indian academicians. Int. J. Educ. Learn. **3**, 1–10 (2014). https://doi.org/10.14257/ijel.2014.3.2.01
21. Tarafdar, M., Tu, Q., Ragu-Nathan, T.S., Ragu-Nathan, B.S.: Crossing to the dark side. Commun. ACM **54**, 113 (2011). https://doi.org/10.1145/1995376.1995403
22. Şahin, Y.L., Çoklar, A.N.: Social networking users' views on technology and the determination of technostress levels. Procedia Soc. Behav. Sci. **1**, 1437–1442 (2009). https://doi.org/10.1016/j.sbspro.2009.01.253
23. Park, D.C.: The basic mechanisms accounting for age-related decline in cognitive function. In: Park, D.C., Schwarz, N. (eds.) Cognitive Aging: A Primer, pp. 3–21. Psychology Press, New York (2000)
24. Peters, R.: Ageing and the brain. Postgrad. Med. J. **82**, 84–88 (2006). https://doi.org/10.1136/pgmj.2005.036665
25. Essig, M.: Normal aging of the brain. In: Proceedings of the International Society for Magnetic Resonance in Medicine, vol. 19 (2011)
26. Svennerholm, L., Boström, K., Jungbjer, B.: Changes in weight and compositions of major membrane components of human brain during the span of adult human life of Swedes. Acta Neuropathol. **94**, 345–352 (1997). https://doi.org/10.1007/s004010050717
27. Scahill, R.I., Frost, C., Jenkins, R., Whitwell, J.L., Rossor, M.N., Fox, N.C.: A longitudinal study of brain volume changes in normal aging using serial registered magnetic resonance imaging. Arch. Neurol. **60**, 989 (2003). https://doi.org/10.1001/archneur.60.7.989
28. Baron, J.C., Godeau, C.: Human aging. In: Toga, A.W., Mazziotta, J.C. (eds.) Brain Mapping – The Systems. Academic Press (2000)

29. Riedl, R., Léger, P.-M.: Fundamentals of NeuroIS – Information Systems and the Brain. Springer, Heidelberg (2016)
30. Lazarus, R.S.: Psychological Stress and the Coping Process. McGraw-Hill, New York (1966)
31. Lazarus, R.S., Folkman, S.: Stress, Appraisal, and Coping. Springer, New York (1984)
32. McEwen, B.S.: Protective and damaging effects of stress mediators: central role of the brain. Dialogues Clin. Neurosci. **8**, 367–381 (2006)
33. Peters, A.: The effects of normal aging on myelin and nerve fibers: a review. J. Neurocytol. **31**, 581–593 (2002). https://doi.org/10.1023/A:1025731309829
34. Gunnar, M., Quevedo, K.: The neurobiology of stress and development. Annu. Rev. Psychol. **58**, 145–173 (2007). https://doi.org/10.1146/annurev.psych.58.110405.085605
35. Seals, D.R., Esler, M.D.: Human ageing and the sympathoadrenal system. J. Physiol. **528**, 407–417 (2000). https://doi.org/10.1111/j.1469-7793.2000.00407.x
36. Sampedro-Piquero, P., Alvarez-Suarez, P., Begega, A.: Coping with stress during aging: the importance of a resilient brain. Curr. Neuropharmacol. **16**, 284–296 (2018). https://doi.org/10.2174/1570159X15666170915141610
37. Hibberd, C., Yau, J.L.W., Seckl, J.R.: Glucocorticoids and the ageing hippocampus. J. Anat. **197**, 553–562 (2000). https://doi.org/10.1046/j.1469-7580.2000.19740553.x
38. Fuller, R.: The involvement of serotonin in regulation of pituitary-adrenocortical function. Front. Neuroendocr. **13**, 250–270 (1992)
39. Miura, H., Qiao, H., Ohta, T.: Influence of aging and social isolation on changes in brain monoamine turnover and biosynthesis of rats elicited by novelty stress. Synapse **46**, 116–124 (2002). https://doi.org/10.1002/syn.10133
40. Rehman, H.U.: Neuroendocrinology of ageing. Age Ageing **30**, 279–287 (2001). https://doi.org/10.1093/ageing/30.4.279
41. Rostène, W., Sarrieau, A., Nicot, A., Scarceriaux, V., Betancur, C., Gully, D., Meaney, M., Rowe, W., De Kloet, R., Pelaprat, D.: Steroid effects on brain functions: an example of the action of glucocorticoids on central dopaminergic and neurotensinergic systems. J. Psychiatry Neurosci. **20**, 349–356 (1995)
42. Baskerville, T.A., Douglas, A.J.: Dopamine and oxytocin interactions underlying behaviors: potential contributions to behavioral disorders. CNS Neurosci. Ther. **16**, e92–e123 (2010). https://doi.org/10.1111/j.1755-5949.2010.00154.x
43. Riedl, R., Javor, A.: The biology of trust: integrating evidence from genetics, endocrinology, and functional brain imaging. J. Neurosci. Psychol. Econ. **5**, 63–91 (2012). https://doi.org/10.1037/a0026318
44. Heinrichs, M., Baumgartner, T., Kirschbaum, C., Ehlert, U.: Social support and oxytocin interact to suppress cortisol and subjective responses to psychosocial stress. Biol. Psychiatry **54**, 1389–1398 (2003). https://doi.org/10.1016/S0006-3223(03)00465-7
45. Blackburn, E., Epel, E.: The Telomere Effect. Hachette Group, Inc. (2017)
46. Epel, E.S., Blackburn, E.H., Lin, J., Dhabhar, F.S., Adler, N.E., Morrow, J.D., Cawthon, R.M.: Accelerated telomere shortening in response to life stress. Proc. Natl. Acad. Sci. **101**, 17312–17315 (2004). https://doi.org/10.1073/pnas.0407162101
47. Ames, B.N.: Optimal micronutrients delay mitochondrial decay and age-associated diseases. Mech. Ageing Dev. **131**, 473–479 (2010). https://doi.org/10.1016/j.mad.2010.04.005
48. Joseph, J.A., Shukitt-Hale, B., Casadesus, G., Fisher, D.: Oxidative stress and inflammation in brain aging: nutritional considerations. Neurochem. Res. **30**, 927–935 (2005). https://doi.org/10.1007/s11064-005-6967-4
49. Mattson, M.P.: Will caloric restriction and folate protect against AD and PD? Neurology. **60**, 690–695 (2003). https://doi.org/10.1212/01.WNL.0000042785.02850.11
50. Mattson, M.P., Chan, S.L., Duan, W.: Modification of brain aging and neurodegenerative disorders by genes, diet, and behavior. Physiol. Rev. **82**, 637–672 (2002). https://doi.org/10.1152/physrev.00004.2002

51. Colcombe, S.J., Erickson, K.I., Raz, N., Webb, A.G., Cohen, N.J., McAuley, E., Kramer, A.F.: Aerobic fitness reduces brain tissue loss in aging humans. J. Gerontol. Ser. A Biol. Sci. Med. Sci. **58**, M176–M180 (2003). https://doi.org/10.1093/gerona/58.2.M176
52. Kramer, A.F., Hahn, S., Cohen, N.J., Banich, M.T., McAuley, E., Harrison, C.R., Chason, J., Vakil, E., Bardell, L., Boileau, R.A., Colcombe, A.: Ageing, fitness and neurocognitive function. Nature **400**, 418–419 (1999). https://doi.org/10.1038/22682
53. Riedl, R., Davis, F.D., Banker, R.D., Kenning, P.H.: Neuroscience in Information Systems Research: Applying Knowledge of Brain Functionality Without Neuroscience Tools. Springer, Heidelberg (2017)

Smartphone Pathology, Agency and Reward Processing

Bridget Kirby(✉), Ashley Dapore, Carl Ash, Kaitlyn Malley, and Robert West

Department of Psychology and Neuroscience, DePauw University, Greencastle, USA
bridgekirby5@gmail.com, {ashleydapore_2021,carlash_2021,
robertwest}@depauw.edu, kaitlynm424@gmail.com

Abstract. Smartphones have become ubiquitous in society; for instance, 81% of Americans report they own at least one device. Along with an increase in smartphone use, there is growing concern surrounding the pathological use of these devices. Pathological smartphone use is associated with elevated anxiety, sleep disturbance, and increased impulsivity. Given these concerns, the current study examined the relationship between pathological smartphone use and the neural correlates of reward processing in a college-aged sample. The amplitude of neural activity elicited by gains and losses was negatively correlated with pathological smartphone use when individuals were the choice agent, but not when a computer was the choice agent. These data reveal that overlapping neural systems may contribute to pathological technology use and other forms of addictive behavior and substance abuse.

Keywords: Smartphone pathology · Feedback processing

1 Introduction

Smartphone use is pervasive in today's society and has become integral to how we socialize and operate in our everyday lives. 81% of Americans owned smartphones as of 2018, which increased from only 35% in 2011 [1]. We have become so closely attached to our smartphones that 46% of people say they cannot live without them [2]. Pathological smartphone use has been linked to increased neuroticism, impulsivity, sleep disturbance, anxiety, and depression [3]. Due to these significant issues and the fact that 20–40% of individuals may be pathological smartphone users, emerging research seeks to identify the correlates and causes of pathological smartphone use [3]. To this end, the current study builds upon research examining the psychological and neural correlates of pathological technology use (i.e., social media and video games) [4] by examining the relationship between pathological smartphone use and the neural correlates of reward processing using event-related brain potentials (ERPs).

Substantial research has been devoted to identifying the neural correlates of substance abuse and chemical dependence, while less is known about the neural underpinning of the pathological use of technology. For instance, in an influential review paper Goldstein

F. D. Davis et al. (Eds.): NeuroIS 2020, LNISO 43, pp. 321–329, 2020.
https://doi.org/10.1007/978-3-030-60073-0_37

and Volkow [5] demonstrated that there was significant overlap in the neural structures that revealed differences between individuals experiencing addiction and controls related to cognition, emotion, and motivation on the one hand, and feelings of craving, intoxication, and withdrawal on the other. These differences were widely distributed across the medial, lateral, and orbital surface of the prefrontal cortex [5]. While other research has consistently demonstrated that altered activity within the reward system involving the ventral striatum and medial frontal cortex are associated with the presence of substance abuse [6].

In the context of technology use, many of the symptoms of pathological smartphone use (e.g., impulsivity, depression) are associated with altered neural activity related to the medial and lateral prefrontal cortex. As an example, depression is associated with attenuated ERPs related to the processing of rewards and positive outcomes over the medial frontal cortex in late childhood to adulthood [7, 8]. Also, there is some evidence indicating that the influence of depression and impulsivity on reward processing may interact [9, 10]. Together, the general effects of addiction on reward processing [5, 6] and the association between pathological smartphone use and psychiatric correlates of addiction [3] led us to the current exploration of the relationship between pathological smartphone us and reward processing.

The findings of studies examining the pathological use to social networking sites (SNS) [4] and video games [11] are consistent with the idea that pathological technology use is associated with alterations of the processing of gains and losses in addition to differences in decision-making that involve the reward system. For instance, Meshi et al. [4] reported a negative correlation between the pathological use of SNS and performance on the Iowa Gambling Task (IGT) that could reflect either a failure to learn from negative outcomes over time or a greater sensitivity to gains than losses within the task. At the neural level, Turel et al. [12] demonstrated pathological SNS use was associated with a reduction in volume of the posterior insula that contributes interoceptive information to decision-making, and that this reduction was related to steeper delay discounting rates in individuals with greater SNS pathology. Together these findings may reveal a shift in subjective value related to pathological SNS use wherein individuals place greater value on immediate rewards. Complimenting the evidence from studies of pathological SNS use, pathological video gaming is associated with poor decision-making in the IGT and diminished learning from positive and negative feedback in the Probabilistic Selection Task [11]. Related to this finding, high levels of action gaming – that is correlated with pathological gaming – are associated with a reduction in the amplitude of ERPs related to negative feedback resulting from the action of the player (i.e., busts) in a virtual Blackjack game [13]. While action gaming was not related to the processing of wins or losses, that did not result from the direction action of the individual (i.e., dealer wins).

To explore the association between pathological smartphone use and the neural correlates of reward processing we used ERPs in combination with a modified 2-doors gambling task. The 2-doors task is a simple gambling paradigm that has been widely used in basic research to explore the neural correlates of reward processing [14], and in applied research to examine the effects of psychopathology (e.g., depression and anxiety) on reward processing [7]. In the modified 2-doors task, a cue is presented at the beginning of each trial indicating whether the participant or computer will make the selection for the current trial. Individuals win or lose some money on each trial regardless of whether they or the computer chooses for the given trial. This manipulation allows one to consider the effect of choice or agency on the ERPs related to processing gains and losses, and has been useful in exploring the relationship between reward processing and various individual differences including socioeconomic status [15], self-control [16], and depression [17].

Differences in the ERPs related to outcome (i.e., wins vs. Losses) in the modified 2-doors task emerge around 200 ms after feedback is presented over the frontal-central region of the scalp. The Reward Positivity (RewP) represents greater positivity for gains than for losses, while the Feedback Negativity (FN) represents greater negativity for losses than for gains, both occurring between 200–300 ms [14]. Following the RewP/FN, the ERPs become more positive for losses than for gains over the frontal-central region reflecting the frontal P3 [16, 17]. The amplitude of the RewP and FN are modulated by agency (i.e., greater for subject than other select trials), while the frontal P3 appears to be limited to person select trials [16, 18] only being observed when the participant is the agent of choice.

Given the associations between pathological smartphone use and impulsivity and depression [3], the association between these individual differences and reward processing [7, 9], the relationship between pathological technology use (i.e., SNS and video games) and reward processing as well as decision-making [4, 13], we predicted that elevated pathological smartphone use would be associated with a decrease in the amplitude of the RewP for wins and the frontal P3 for losses in the person select condition. Based upon evidence from the video game literature [13], we can also predict that the association between pathological smartphone use and reward processing will be limited to the person select trials. Furthermore, based upon work related to depression, we predicted that pathological smartphone use would not be related to the amplitude of the parietal P3; revealing a selective effect of pathological smartphone use on the neural correlates of reward processing rather than a general effect of information processing and decision-making.

2 Method

2.1 Participants

Sixty-nine individuals who were undergraduate students at DePauw University participated in the study (age M = 19.5 years, 15 male, 53 female, 1 unidentified) for course credit. Data for 4 participants were incomplete (i.e., either the survey data were missing or there was a high degree of artifact in the EEG data) and not included in the analyses.

2.2 Materials and Procedure

Participants provided informed consent and then competed a survey including the demographic and individual difference measures. These measures included the Smartphone Addiction Scale-Short Version (SAS-SV) [19], the Self-Control Scale [20], the CES-D to measure depressive symptoms [21], and short measures of Internet [22] and video game pathology. The Smartphone Addiction Scale-Short Version [19], is a 10-item scale that measures the physical, psychological, and social implications of smartphone use. Some examples of items are "Missing planned work due to smartphone use, using my smartphone longer than intended." and "Having my smartphone in mind even when I am not using it".

For the modified 2-doors task, subjects were presented with a cue that indicated whether they or the computer would choose one of two doors for the trial. Half of the 80 trials were person select trials and half were computer select trials. Individuals either won 25 cents or lost 12.5 cents for each trial, regardless of whether they or the computer chose the door. The difference in the value of gains and losses was implemented to control for loss aversion within the paradigm with the goal of balancing the subjective value of wins and losses on average across participants [14]. An upward pointing green arrow indicated a win, and a red downward pointing arrow indicated a loss. Unknown to the subjects, outcomes in the task were random so that all individuals won $5 and were paid in cash at the end of the task.

2.3 EEG Recording and Analysis

The EEG was recorded from a 32 channel actiCHamp system using the Brain Vision Recorder software and a standard 32 channel actiCAP scalp montage where CP5-CP6 were replaced with active electrodes located below the eyes to monitor blinks and vertical saccades. During recording, the electrodes were grounded to electrode Fpz and referenced to electrode Cz, for data analysis the data were re-referenced to the average reference. The EEG was digitized at 500 Hz and then bandpass filtered between .1–30 Hz using an IIR filter implemented in ERPLAB (5.1.1.0) [23] for the analyses. Ocular artifacts associated with blinks and saccades were corrected with ICA implemented in EEGLAB (13.6.5b) [24]. Trials contaminated by other artifacts were rejected before averaging using a $\pm 100\ \mu V$ threshold. The ERPs were averaged for -200 to 1500 ms around onset of the feedback cues for subject wins, subject losses, computer wins, and computer losses.

3 Results

For the SAS-SV the median score was 22 and the range of scores was 9 to 44, revealing a fairly broad distribution in the level of pathological smartphone use across the sample. Scores on the SAS-SV were highly correlated with pathological use of the Internet (Table 1), and moderately correlated with video game pathology. Consistent with the extant literature [3], elevated pathological smartphone use was associated with both greater impulsivity and higher levels of depressive symptoms.

To determine whether the data for the modified 2-doors task studies were consistent with previous research [17], we first examined the effects of outcome and agency on the RewP, frontal P3, and parietal P3 in a set of ANOVAs. Consistent with previous research, at electrode FC1, the effect of outcome was greater on the RewP for the person select trials than for computer select trials (outcome $\times$ agent $F(1,64) = 15.08$, $p < .001$). At electrode Fz the frontal P3 was also greater for person select trials than computer select

Table 1. Correlations between pathological smartphone use and individual difference measures.

Internet	Video game	Impulsivity	Depression	Simple tasks
.60**	.38**	.29*	.29*	.43**

Note: p < .05 *, p < .01 **

trials (outcome × agent, $F(1,64) = 24.97$, $p < .001$). In contrast, at electrode Pz the parietal P3 revealed an effect of agent ($F(1,64) = 144.79$, $p < .001$), and no effect of outcome ($F < 1.00$).

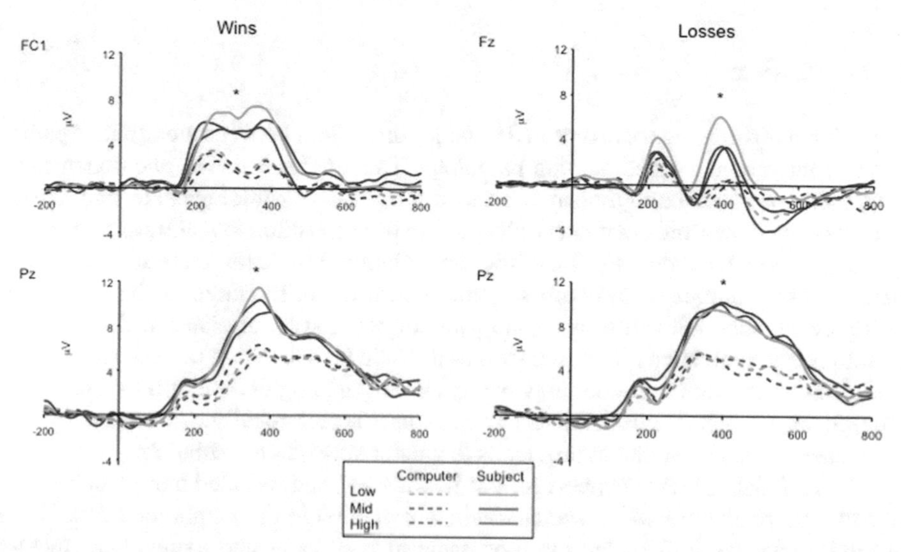

Fig. 1. Grand-averaged ERPs for wins and losses on the person and computer select trials demonstrating the association between smartphone pathology and the RewP and frontal P3 for person select trials; and the absence of an association for the parietal P3. The Low, Mid, and High groups represent tertials of the distribution of SAS-SV scores.

In exploring the association between smartphone pathology and the ERP data (Fig. 1), we observed that the ERPs tended to be similar for individuals with moderate and high levels of pathology relative to individuals with low levels of pathology. Therefore, for the analyses of the brain behavior relationships we considered both linear and quadratic (i.e., ERP amplitude squared) associations.

The RewP and parietal P3 for wins are presented in the left panel of Fig. 1. For the RewP, person select trials revealed a linear association between smartphone pathology and ERP amplitude that was marginally significant ($r = -.19$, $p = .06$); and a significant quadratic effect ($r = -.26$, $p = .02$). For the computer select trials, the correlation between the RewP and smartphone pathology was not significant ($r = -.13$, $p = .147$). The parietal P3 and smartphone pathology were not significantly correlated for either trial type (person $r = -.11, p = .19$, computer $r = -.05, p = .35$). These findings support the hypothesis that increasing levels of pathological smartphone use would be associated with a reduction in the amplitude of the RewP for person select trials.

The frontal and parietal P3 for losses are presented in the right panel of Fig. 1. For the frontal P3, smartphone pathology was significantly correlated with ERP amplitude for the person select trials (linear $r = -.30, p = .009$; quadratic $r = -.29, p = .01$); while these correlations were not significant for computer select trials ($r = .14, p = .13$). For loss trials the parietal P3 was not significantly correlated with smartphone pathology in either person ($r = .08, p = .73$) or computer ($r = .03, p = .60$) select trials. These findings are consistent with the hypothesis that increasing levels of pathological smartphone use would be associated with a reduction in the amplitude of the frontal P3 for person select trials.

4 Discussion

The individual difference data revealed reliable correlations between smartphone pathology and Internet and video gaming pathology. This finding leads to two conclusions. First, there appears to be significant overlap in symptoms of pathological technology use across the three measures that may reflect a core predisposition to pathological technology use; second, the moderate size of these correlations also leads to the suggestion that there are likely unique contributors to pathological use of the three media. Consistent with previous research [3], smartphone pathology was also correlated with depressive symptoms and impulsivity. In future studies it would be interesting to explore the direction of this association (i.e., does impulsivity lead to pathological smartphone use or does smartphone use increase impulsivity) or whether it is reciprocal (i.e., does smartphone use increase depression that in turn leads to greater smartphone pathology).

The ERP data provide support for our hypotheses, and revealed that greater pathological smartphone use was associated with a reduction in the amplitude of the RewP and the frontal P3, influencing the processing of both gains and losses, that emerged even at moderate levels of pathology. These findings may indicate that alterations of reward processing are observed in individuals expressing levels of smartphone pathology that are below the threshold of what would be considered disruptive or clinically relevant in previous research [19]. Converging with evidence from a study examining the relationship between action video gaming and reward processing [13], the influence of pathological smartphone use was limited to person select trials where the participant was an active agent of choice for the RewP and frontal P3, and did not extend to the parietal P3. Together these findings reveal that pathological smartphone use is not associated with a general disruption of information processing, but is specifically related to the attenuation of neural activity associated with the processing of gains and losses that are tied to the actions of the individual. Future studies could seek to further explore the boundary conditions of the association between pathological smartphone use and the neural correlates of cognitive, affective, and social information processing and decision-making.

The current findings demonstrate that pathological smartphone use was associated with a decrease in the amplitude of neural activity related to the processing of both gains and losses when the individual was the agent of choice in the modified 2-doors task. These findings compliment those of other recent studies demonstrating that various forms of digital technology use are associated with alterations in the processing of positive and negative outcomes that may serve to guide decision-making over time. Specifically, our data are consistent with evidence indicating that video game pathology is associated with decreases in the efficient processing of both positive and negative feedback related to gambling and reinforcement learning [11]. At the neural level, there also seems to be some overlap in the relationship between pathological smartphone use and high levels of action video game play, as both are associated with a decrease in the amplitude of ERPs elicited by losses resulting from one's own actions [13]. Our findings are also consistent with the finding that the pathological use of social networking sites may be associated with disruptions of feedback processing that guide decision-making in the Iowa Gambling Task [4]. Together, these data may reveal a behavioral and physiological phenotype of individuals that are susceptible to the development of pathological technology use that transcends the specific device, platform, or medium embodied within a given piece of technology.

Some commentators have questioned the appropriateness or utility of terms like smartphone addiction or addiction to other forms of digital technology (e.g., social network sites, video games) [25]. In the context of pathological smartphone use, these authors argue that the current literature is insufficient to demonstrate that pathological smartphone use rises to the level of an addiction in the way that substance abuse or gambling might; and that it is unclear whether smartphones merely provide a vehicle to support the pathological use of other forms of technology. For instance, using one's phone to play games on the Internet or surf social media sites. Our data do reveal moderate to strong correlations between smartphone, Internet, and video game pathology; so it seems possible that in the current study the smartphone pathology scale is serving as a surrogate for the pathological use of technologies that can be accessed via the device. This may to some degree provide an explanation of the higher rates of pathological smartphone use that are described in the literature relative to pathological Internet and gaming, as measuring smartphone pathology may be capturing different samples of individuals. One avenue of future research could be to attempt to tease apart the various elements of smartphone use that contribute to smartphone pathology in order to gain a deeper and more nuanced understanding of the phenomenon.

There are some limitations of the study that are worth considering. First, the participants included in the study were generally recruited through convenience sampling and were drawn from a pool of students taking introductory psychology and more advanced content courses. Relevant to this concern, a screening sample used to recruit participants with elevated scores on the SAS-SV revealed that approximately 20% of the sample would have met criteria used in previous studies for smartphone addiction, a value that is similar to some published research [3]. This may indicate that our sample is similar to those reported in prior research in this area. Second, the modified 2-doors task does not provide meaningful behavioral data, so it is impossible to determine how pathological smartphone use may have contributed to decision-making guided by feedback processing. This could be addressed in future research using tasks with well characterized behavioral outcomes such as the IGT, probabilistic selection task, or the balloon analogue task. Third, ERPs do not provide direct measures of subcortical structures that contribute to reward processing, therefore, future work in this area may benefit from the use of fMRI to provide a more complete view of activity within the reward network related to pathological smartphone use. Finally, the sample was primarily female, and this likely limited the correlation between smartphone pathology and video game pathology that is more common in males [11]. Therefore, it may be worthwhile to obtain larger and more gender balanced samples in future research to estimate the true size of this relationship.

In conclusion, the current study revealed reliable correlations between pathological smartphone use and ERP activity related to processing both gains and losses in a simple gambling task when the individual was an active agent in determining the outcome of the trial. In contrast, pathological smartphone use was not associated with the parietal P3. Together, these findings demonstrate a specific association between pathological smartphone use and reward processing rather than a more general association with information processing and decision-making. Finally, consistencies between our findings related to pathological smartphone use and those from other digital technologies including social networking sites and video games may reveal a general behavioral and physiological phenotype that is marked by the disruption of reward processing and underpins the pathological use of technology.

References

1. Pew Research Center Infographic (2019). https://www.pewresearch.org/internet/fact-sheet/mobile/
2. Ning, W., Davis, F., Taraban, R.: Smartphone addiction and cognitive performance of college students. In: Twenty-Forth Americas Conference on Information Systems (2018)
3. De-Sola Gutiérrez, J., Rodríguez de Fonseca, F., Rubio, G.: Cell phone addiction: a review. Front. Psychiatry **7**, 175 (2016)
4. Meshi, D., Elizarova, A., Bender, A., Verdejo-Garcia, A.: Excessive social media users demonstrate impaired decision making in the Iowa Gambling Task. J. Behav. Addict. **8**, 169–173 (2019)
5. Goldstein, R.Z., Volkow, N.D.: Dysfunction of the prefrontal cortex in addiction: neuroimaging findings and clinical implications. Nat. Rev. Neurosci. **12**, 652–669 (2011)
6. Bechara, A.: Decision making, impulse control and loss of willpower to resist drugs: a neurocognitive perspective. Nat. Neurosci. **8**, 1458–1463 (2005)
7. Foti, D., Hajcak, G.: Depression and reduced sensitivity to non-rewards versus rewards: evidence from event-related potentials. Biol. Psychol. **81**, 1–8 (2009)
8. Bress, N.B., Smith, E., Foti, F., Klein, N.D., Hajcak, G.: Neural response to reward and depressive symptoms in late childhood to early adolescence. Biol. Psychol. **89**, 152–162 (2012)
9. Novak, B.K., Novak, K.D., Lynam, D.R., Foti, D.: Individual differences in the time course of reward processing: stage-specific links with depression and impulsivity. Biol. Psychol. **119**, 79–90 (2016)
10. Oumeziane, B.A., Foti, D.: Reward-related neural dysfunction across depression and impulsivity: a dimensional approach. Psychophysiology **53**, 1174–1184 (2016)
11. Bailey, K., West, R., Kuffel, J.: What would my avatar do? Gaming, pathology, and risky decision making. Front. Psychol. **4**, 609 (2013)

12. Turel, O., He, Q., Brevers, D., Bechara, A.: Delay discounting mediates the association between posterior insular cortex volume and social media addiction symptoms. Cogn. Affect. Behav. Neurosci. **18**, 694–704 (2018)
13. Bailey, K., West, R.: Did I do that? The association between action video gaming experience and feedback processing in a gambling task. Comput. Hum. Behav. **69**, 226–234 (2017)
14. Proudfit, H.G.: The reward positivity: from basic research on reward to a biomarker for depression. Psychophysiology **52**, 449–459 (2015)
15. Munoz, A., Freeman, E., Budde, E., West, R.: The influence of socioeconomic status on the neural correlates of feedback processing. J. Cogn. Neurosci. (Suppl. 46) (2019)
16. West, R., Freeman, E., Munoz, A., Budde, E.: The influence of agency and self-control on the processing of gains and losses. J. Cogn. Neurosci. (Suppl. 46) (2019)
17. Zhu, S., Budde, E., Malley, K., Ash, C., Dapore, A., West, R.: Agency matters: the interaction between agency and depressive symptoms on the neural correlates of reward processing. Unpublished data
18. West, R., Bailey, K., Anderson, S., Kieffaber, P.D.: Beyond the FN: a spatio-temporal analysis of the neural correlates of feedback processing in a virtual Blackjack game. Brain Cogn. **86**, 104–115 (2014)
19. Kwon, M., Kim, D., Cho, H., Yang, S.: The smartphone addiction scale: development and validation of a short version for adolescents. PLoS ONE **8**(12) e83558 (2013)
20. Hu, Q., West, R., Smarandescu, L.: The role of self-control in information security violations: insights from a cognitive neuroscience perspective. J. Manag. Inf. Syst. **31**, 6–48 (2015)
21. Easton, W.W., Smith, C., Ybarra, M., Muntaner, C., Tien, A.: Center for epidemiologic studies depression scale: review and revision (CESD and CESD-R). In: Maruish, M.E (ed.) The Use of Psychological Testing for Treatment Planning and Outcomes Assessment: Instruments for Adults, pp. 363–377. Lawrence Erlbaum Associates Publishers, New Jersey (2004)
22. Pawlikowski, M., Alstötter-Gleich, C., Brand, M.: Validation and psychometric properties of a short version of Young's Internet Addiction Test. Comput. Hum. Behav. **29**, 1212–1223 (2013)
23. Lopez-Calderon, J., Luck, S.J.: ERPLAB: an open-source toolbox for the analysis of event-related potentials. Front. Hum. Neurosci. **8**, 213 (2014)
24. Delorme, A., Makeig, S.: EEGLAB: an open source toolbox for analysis of single-trial EEG dynamics. J. Neurosci. Meth. **143**, 9–21 (2004)
25. Panova, T., Carbonell, X.: Is smartphone addiction really an addiction? J. Behav. Addict. **7**, 252–259 (2018)

Hedonic Multitasking: The Effects of Instrumental Subtitles During Video Watching

Félix Giroux[1(✉)], Jared Boasen[1,2], Sylvain Sénécal[1], and Pierre-Majorique Léger[1]

[1] Tech3Lab, HEC Montréal, Montréal, QC, Canada
{felix.giroux,jared.boasen,sylvain.senecal,pierre-majorique.leger}@hec.ca

[2] Faculty of Health Sciences, Hokkaido University, Sapporo, Japan

Abstract. Hedonic videos are often accompanied by instrumental text/subtitles. However, the neurophysiological effects of task-switching between instrumental text processing and audiovisual processing during hedonic video watching remains unexplored. We investigated these effects using changes in pupil diameter and theta-band spectral power as indices of cognitive load, and self-reported and automated facial expression-based measures of emotion. We found that the presence of instrumental subtitles subjectively improved comprehension without negatively impacting cognitive and emotional response. These findings contribute to the literature by showing that in this specific context, multitasking does enhance the user experience. Further research is needed to explore the extent to which instrumental text is useful, and how its effects change in accordance with audio presence during passive hedonic experiences across various contexts.

Keywords: Multitasking · Hedonic · Subtitles · Facial expression · Emotion · Cognitive load · Eye tracking

1 Introduction

Switching between tasks involving text processing and those involving non-textual audiovisual processing is a form of multitasking which is commonplace in our modern technological world, and of great interest to information systems (IS) research [1]. Typically, tasks involving text processing will interfere with other tasks and increase cognitive processing load. This in turn has been associated with attenuated task performance [2], reduced emotional responsiveness to hedonic activities [3], increased pupil diameter [4], and even sustained increases in electroencephalographic (EEG) theta and alpha band spectral power [5].

However, there are occasions where text processing can instrumentally support another task. One example of this is reading intralingual subtitles during video watching. Although the presence of intralingual subtitles during hedonic video watching can reportedly erode comprehension of the imagery content, reading subtitles in this context

F. D. Davis et al. (Eds.): NeuroIS 2020, LNISO 43, pp. 330–336, 2020.
https://doi.org/10.1007/978-3-030-60073-0_38

has nevertheless been associated with increased comprehension of dialogue and local narrative coherence [5, 6]. Furthermore, reading intralingual subtitles during academic video watching has also reportedly been associated with increased comprehension and reduced cognitive load as indexed via pupil diameter [7]. Although these studies show that multitasking in such context has benefits for the viewer, they do not simultaneously take into account emotional, cognitive and behavioral responses of users. Thus, our understanding regarding the effect of subtitles on cognitive load and emotional response remains limited. Considering the ubiquitous instrumental use of subtitles in online advertising videos, educational training, news, games, and movies, further IS research targeting subtitles is highly desirable.

The present study takes a step in this direction by investigating the effect of intralingual subtitles during hedonic video watching. Using eye-tracking, EEG, automated facial expression-based emotion, and self-reported emotion, we compared neuropsychophysiological responses between scenes with and without subtitles. Given the aforementioned evidence on subtitle reading during hedonic video watching, we hypothesized that subtitle presence would improve dialogue comprehension, but would distract from image content and thereby reduce emotional responsiveness to it. Correspondingly, for subtitled compared to non-subtitled conditions, we predicted: smaller pupil diameter, a sign of lower cognitive load; lower EEG theta-band spectral power; and lower AFE-based emotion for image content. Our study contributes to IS research by advancing our understanding of how instrumentally supportive text information affects users during multisensory hedonic experiences.

2 Methods

2.1 Subjects and Stimuli

Subjects comprised 17 healthy young adults (M:8, F:9; mean age $\pm$ std: 23.7 $\pm$ 3.5 years). They sat in a cinema seat and watched 16 scenes from a 720p high definition cinematic recording of Georges Bizet's opera, Carmen[1]. Opera was chosen as it is intrinsically hedonic, and the display of subtitles is traditional [8], making it a relevant real-world case study and therefore congruent with the future direction of IS research [9]. The scenes were randomly presented with or without subtitles (four trials each per subject) to control for variance in emotionality. The video was presented centrally on the screen (hereafter: image zone) and flanked on top and bottom by black bars. Subtitles appeared in the bottom bar (hereafter: subtitle zone). Subjects viewed the stimuli on a 70 $\times$ 120 cm high definition TV at a distance of 262 cm, with the audio presented in 2.0 stereo. Subtitles were intralingual and native to the subjects.

2.2 Measurements and Processing

Gaze fixation location and pupil diameter were recorded at a 60 Hz sampling rate with Smart Eye Pro 6.2 (Gothenburg, Sweden). Image and subtitle zone fixations were delineated for each subject individually according to the Y-axis value between the two zones

[1] Produced by Opéra Royal de Wallonie-Liege, directed by S. Scappuci, with staging from H. Brockhaus (2018).

that had the lowest frequency of fixations. No fixation outliers were excluded on either the x or the y axis. Cognitive load was indexed via pupil diameter. Because, pupil diameter varies across individuals [10], mean pupil diameter in each condition was normalized as a percent change from the mean pupil diameter across both conditions. Synchronous with eye-tracking, the facial expressions of subjects were continuously sampled at 30 fps with a high definition webcam using MediaRecorder (Noldus, Wageningen, Netherlands). Based on these recordings, the AFE detection software FaceReader 7.0 (Noldus, Wageningen, Netherlands) was used to assess the frequency (%) of one positive (*happiness*) and four negative basic emotions *(anger, disgust, sadness, and scared)*. Subjective emotional valence toward the stimuli was assessed after viewing each scene using a nine-point version of the Self-Assessment Manikin (SAM) pictogram scale ranging from the feeling of *sadness* (1) to *happiness* (9) [11]. Finally, in post-experiment interview, subjects were asked their preference for presence or absence of subtitles and the reason for their response. 32-channel electroencephalographs (EEG) and electrode positions were additionally recorded for all subjects throughout the experiment at 500 Hz (BrainProducts). EEG signal was cleaned with independent component analysis and band-pass filtered from 1–40 Hz. Source-level Hilbert transforms were applied to extract mean theta (5–7 Hz) activity in 62 cortical areas (Mindboggle Atlas) for each condition based on 10 s post-scene windows normalized as a percent deviation from the mean activity over 10 s pre-scene windows for each condition.

2.3 Statistical Analyses

A three-way repeated measures analysis of variance (RM ANOVA) was performed to assess differences in the frequency of AFE-based emotions according to fixation zone and subtitle condition (2: conditions $\times$ 2: zones $\times$ 5: AFE-based emotion types). Paired t-tests were used to compare the difference in subjective emotional valence, and the difference in image zone PCPD, between subtitled and non-subtitled conditions. Note that we restricted our analyses of PCPD to the image zone as pupil diameter changes according to brightness of the point of focus [12], thereby making comparisons between the bright image zone and the dark subtitle zone inappropriate. Two-way RM ANOVA was performed to assess differences in cortical theta activity between conditions (2: conditions $\times$ 62: brain areas). All statistical analyses were performed using SPSS software version 22 (IBM, Armonk, NY, USA), with a threshold for statistical significance set at $p \leq 0.05$.

3 Results

Three-way RM ANOVA revealed a marginally significant main effect of subtitle presence ($F_{(1, 16)} = 4.469$, $p = .051$), with higher AFE-based emotional frequency in the subtitled versus non-subtitled condition. There was also a significant main effect of AFE-based emotion type ($F_{(4, 64)} = 26.316$, $p < .001$), indicating differences in the occurrence frequency of AFE-based emotions according to their type. There was no significant main effect of fixation zone ($F_{(1, 16)} = .220$, $p = .225$), nor was there a significant interaction between subtitle presence and fixation zone ($F_{(1, 16)} = 1.929$, $p = .184$). However, there was a significant interaction between subtitle presence and AFE-based emotion type ($F_{(4, 64)} = 4.878$, $p = .002$), with simple main effects testing revealing significantly greater frequency sadness in the subtitled versus non-subtitled condition ($p = .033$). Meanwhile, there was a significant three-way interaction between subtitle presence, fixation zone, and AFE-based emotion type ($F_{(4, 64)} = 4.366$, $p = .003$). Simple-main effects testing revealed no difference in the frequency of any AFE-based emotion type in the image zone between subtitled and non-subtitled conditions. However, frequency of sadness was significantly higher in the subtitle zone compared to the image zone in subtitled conditions ($p = .007$) (see Fig. 1). Paired t-test revealed no difference in subjective emotional valence between subtitled and non-subtitled conditions ($t = .046$, $p = .964$). In fact, post-experiment interviews revealed that 15 out of 17 subjects preferred subtitles as they improved narrative comprehension. That said, paired t-test revealed no significant difference in image zone PCPD between the subtitled and non-subtitled condition ($t = -.022$, $p = .983$). Meanwhile, two-way RM ANOVA on source-level theta activity revealed no main effect of condition ($F_{(1, 17)} = 1.991$, $p = .176$) or brain area ($F_{(61, 1037)} = .741$, $p = .931$), nor interaction between condition and brain area ($F_{(61, 1037)} = .826$, $p = .827$).

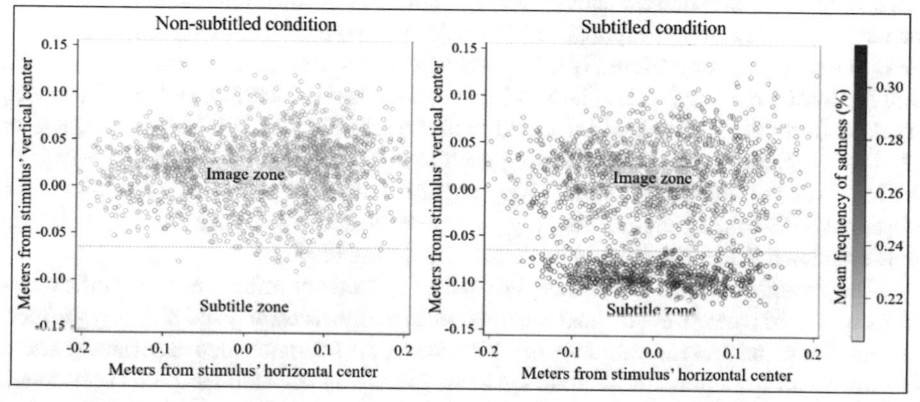

Fig. 1. One subject's aggregated fixations associated with automatic facial expression-based sadness. Fixations are displayed in 2D plans. Mean sadness frequency in each condition's zone of interest is indicated by the darkness of the colored circles.

4 Discussion

The present study explored task-switching between instrumental text processing and hedonic audiovisual processing by examining the effect of intralingual, native-language subtitle presence on neuropsychophysiological response to viewing cinematic opera. Gaze fixation measures clearly indicated that subjects looked at the subtitles (Fig. 1). In line with our hypothesis, subjects reported that the presence of subtitles improved their subjective comprehension of the opera. However, the absence of correspondingly significantly smaller pupil diameter and lower EEG theta-band spectral power for subtitled compared to non-subtitled conditions indicated that whatever improvements in comprehension, reading subtitles did not translate to any marked changes in cognitive processing burden. Meanwhile, a lack of significant differences in self-reported valence and AFE-based emotions associated with the image zone between conditions similarly indicated that the improvement in subjective comprehension due to subtitle presence did not come at the cost of emotional responsiveness to audiovisual content. In short, the subtitles appeared to serve their instrumental purpose without significantly disrupting the hedonic experience.

Interestingly, our analysis of AFE-based emotion indicated a significantly higher frequency of sadness in the subtitled compared to the non-subtitled condition, which simple-main effects testing revealed to be driven specifically by a high detection rate of AFE-based sadness for fixations in the subtitle zone. Although the opera, Carmen, is a tragedy, increased sadness due to text information is suspicious, particularly when there was no difference in self-reported emotion between conditions. Riedl and Léger (2016) stressed the importance of considering the impact of non-emotional cognitive processes on facial expressions, and perhaps herein lies the explanation [13]. AFE software are based on the Facial Action Coding System (FACS), where emotion is determined according to specific muscle activity configurations or action units (AUs) [14, 15]. The principal muscle activity responsible for the AU related to AFE-based measures of sadness is the corrugator muscle [16]. Reports have implicated greater corrugator muscle activity during reading in association with decreased text readability, and with increased mental effort during sustained information processing [17, 18, 19]. Meanwhile, Kaiser (2014) observed the mere act of reading itself to induce corrugator muscle activity [20]. Although our results did not indicate any cognitive differences between conditions, the observed AFE-based phenomenon, together with the above evidence strongly points to non-emotional drivers of facial muscle activity during reading.

The present study has limitations. For instance, it did not address more specific questions such as to what extent the text information is really necessary, and how do its effects change due to the presence/absence of audio during the hedonic video experience. These are important questions for hedonic systems, for as van der Heijden (2004) observed, subjective enjoyment and ease of use more strongly determine intent to use/consume than subjective usefulness [21]. Additionally, our neurophysiological measures were not sensitive enough to detect differences in correspondence with those regarding subjective comprehension. To better support these subjective results, future studies could benefit from an EEG approach based on eye-fixation related potentials, which could provide insight into cognitive processing load at the time of each eye fixation [22].

In conclusion, the present study provides first-time combined evidence that task-switching between instrumental subtitle processing and audiovisual processing during hedonic video watching subjectively improves comprehension without negatively impacting cognitive and emotional response. Our findings also demonstrate the importance of multitasking studies relying on AFE-based emotion to differentiate measures according to task due to the potential influence of non-emotional cognitive drivers of facial muscle activity. This study highlights the importance and need for further neurophysiological investigations into instrumental text processing during hedonic video watching, not only in cinema, but also in other contexts such as educational training, online advertising, gaming, and even live contexts.

References

1. Cameron, A.F., Webster, J.: Relational outcomes of multicommunicating: integrating incivility and social exchange perspectives. Organ. Sci. **22**(3), 754–771 (2011)
2. Courtemanche, F., Labonté-LeMoyne, E., Léger, P.M., Fredette, M., Senecal, S., Cameron, A.F., Faubert, J., Bellavance, F.: Texting while walking: an expensive switch cost. Accid. Anal. Prev. **127**, 1–8 (2019)
3. Tchanou, A.-Q., Giroux, F., Léger, P.-M., Senecal, S., Ménard, J.-F.: Impact of information technology multitasking on hedonic experience. In: Association for Information Systems. SIGHCI Proceedings (2018)
4. Katidioti, I., Borst, J.P., Taatgen, N.A.: What happens when we switch tasks: pupil dilation in multitasking. J. Exp. Psychol. Appl. **20**(4), 380–396 (2014)
5. Lavaur, J.M., Bairstow, D.: Languages on the screen: is film comprehension related to the viewers' fluency level and to the language in the subtitles? Int. J. Psychol. **46**(6), 455–462 (2011)
6. Lee, M., Roskos, B., Ewoldsen, D.R.: The impact of subtitles on comprehension of narrative film. Media Psychol. **16**(4), 412–440 (2013)
7. Kruger, J-L., Hefer, E., Gordon, M.: Measuring the impact of subtitles on cognitive load: eye tracking and dynamic audiovisual texts. In: Proceedings of Eye Tracking South Africa, Cape Town, pp. 29–31 (2013)
8. Orero, P., Matamala, A.: Accessible opera: overcoming linguistic and sensorial barriers. Perspect. Stud. Transl. **15**(4), 262–277 (2008)
9. vom Brocke, J., Hevner, A., Léger, P.M., Walla, P., Riedl, R.: Advancing a NeuroIS research agenda with four areas of societal contributions. Eur. J. Inf. Syst. **29**(1), 9–24 (2020)
10. Tsukahara, J.S., Harrison, T.L., Engle, R.W.: The relationship between baseline pupil size and intelligence. Cogn. Psychol. **91**, 109–123 (2016)
11. Bradley, M.M., Lang, P.J.: Measuring emotion: the self-assessment manikin and the semantic differential. J. Behav. Ther. Exp. Psychiatry **25**(1), 49–59 (1994)
12. Holmqvist, K., Nyström, M., Andersson, R., Dewhurst, R., Jarodzka, H., Van de Weijer, J.: Eye Tracking: A Comprehensive Guide to Methods and Measures. Oxford University Press, Oxford (2011)
13. Riedl, R., Léger, P.M.: Fundamentals of NeuroIS. Studies in Neuroscience, Psychology and Behavioral Economics. Springer, Heidelberg (2016)
14. Ekman, P.: Universals and cultural differences in facial expressions of emotion. In: Cole, J. (ed.) Nebraska Symposium on Motivation, pp. 207–283. University of Nebraska Press, Lincoln (1972)
15. Ekman, P., Friesen, W.V.: Facial Action Coding System: A Technique for the Measurement of Facial Movement. Consulting Psychologists Press, Palo Alto (1978)
16. Ekman, P., Friesen, W.V., Hager, J.C.: The Facial Action Coding System CD-ROM. Research Nexus, Salt Lake City, UT (2002)
17. Larson, K., Hazlett, R.L., Chaparro, B.S., Picard, R.W.: Measuring the aesthetics of reading. In: People and Computers XX — Engage, pp. 41–56 (2004)
18. van Boxtel, A., Jessurun, M.: Amplitude and bilateral coherency of facial and jaw-elevator EMG activity as an index of effort during a two-choice serial reaction task. Psychophysiology **30**(6), 589–604 (1993)

19. Waterink, W., van Boxtel, A.: Facial and jaw-elevator EMG activity in relation to changes in performance level during a sustained information processing task. Biol. Psychol. **37**(3), 183–198 (1994)
20. Kaiser, M.: The prediction of an emotional state through physiological measurements and its influence on performance. Thesis. Utrecht University (2014)
21. van Der Heijden, H.: User acceptance of hedonic information systems. MIS Q. **28**(4), 695–704 (2004)
22. Léger, P.M., Sénecal, S., Courtemanche, F., de Guinea, A.O., Titah, R., Fredette, M., Labonte-LeMoyne, É.: Precision is in the eye of the beholder: application of eye fixation-related potentials to information systems research. J. Assoc. Inf. Syst. **15**, 651–678 (2014)

Neurophysiological Reactions to Social Media Logos

Michael Matthews[1(✉)], Thomas Meservy[1], Kelly Fadel[2], and Brock Kirwan[1]

[1] Brigham Young University, Utah, USA
{michaeljmatthews,tmeservy,kirwan}@byu.edu
[2] Utah State University, Utah, USA
kelly.fadel@usu.edu

Abstract. Use of social media is pervasive, and research has identified many benefits that arise from the affordances of social media platforms. Yet, social media use can often interfere with other areas of life, a condition termed social media overuse. Prior literature suggest that social media use can satisfy basic hedonic needs, triggering activation of the brain's reward processing network and inducing higher levels of use. In this paper we propose a study to examine the relationship between social media cues and the neurological reactions that underlie social media overuse. The study employs the affect misattribution procedure (AMP) to elicit spontaneous reactions to cues from three task conditions (social media logos, office images, non-social media company logos). Participants complete the study while situated in a magnetic resonance imaging (MRI) machine. We postulate an interaction between task condition and reported social media cravings/usage in rewards network regions of the brain.

Keywords: Social media · Social media overuse · Affect misattribution procedure · Rewards processing network

1 Introduction

The rise and broad adoption of social media is one of the most expansive phenomena of the last two decades. As such, the topic of social media usage has become a major stream of study within information systems (IS) research. Previous IS research has explored mappings between psychological needs and social media affordances [1] how social media consumption and production lead to various gratifications [2], and the underlying motivations behind social media usage [3]. In addition to the many proposed advantages of social media, researchers have begun to more thoroughly examine its "dark side" [4]. Individuals who turn to social media to satisfy basic wants and needs can fall into a behavioral pattern of social media overuse, which can be destructive on a personal level [5] and also disruptive on a societal level [6].

Research has explored many of the elements of social media overuse including antecedents [7], consequences [8], and intervention options [9]. From a neurological

F. D. Davis et al. (Eds.): NeuroIS 2020, LNISO 43, pp. 337–343, 2020.
https://doi.org/10.1007/978-3-030-60073-0_39

perspective, scholars have also begun to explore the effects of social media overuse on morphology of the brain network [10] and alterations of the brain anatomy [11]. Nevertheless, while there is a rich body of behavioral and psychological literature surrounding social media overuse [12], our understanding of how the brain is implicated in social media overuse is less developed than that in other domains of compulsive behavior research (e.g., gaming, pornography, gambling, alcohol, etc.). In the IS field, the application of NeuroIS methods and tools [13] can help to enhance our understanding of the neurological drivers of social media overuse.

In this paper, we propose a study which will examine how exposure to visual social media cues can trigger spontaneous hedonic reactions (manifested both behaviorally and neurologically) that derive from and reinforce social media overuse behavior. We designed a custom experimental instrument that implements the Affective Misattribution Procedure (AMP) [14] to measure how affective reactions to prime stimuli are implicitly transferred to neutral target stimuli. Specifically, we examine how affective reactions to target stimuli differ when they are associated with social media cues (i.e., logos of social media platforms) vs. when they are associated with other types of control cues, particularly among individuals who report high levels of social media use and craving. By implementing this instrument within an fMRI experimental protocol, we aim to capture the neurological activation patterns that are associated with spontaneous hedonic reactions to social media cues. We anticipate that the findings of this research stream will enrich our theoretical understanding of the drivers of social media overuse.

1.1 Theoretical Development

People use social media for both utilitarian and hedonic purposes; however, research indicates that hedonic use of social media use, such as entertainment or sharing experiences with friends are among the most important drivers of social media use [15]. People engage in hedonic social media use because it helps to satisfy basic human needs such as the need to connect with other individuals or feel a sense of self-worth [1]. The gratification of these needs induces a pleasurable psychological and emotional response that can relieve stress or boredom and help to maintain emotional equilibrium, etc. However, this pleasurable response can also induce users to overextend their use behavior [12].

Classical conditioning models demonstrate that exposure to visual stimuli can evoke feelings of pleasure and a reinforced desire to engage in the hedonic experience related to the pleasurable experience [17]. In a recent study that explored this phenomenon in the domain of social media, van Koningsbruggen et al. [18] reported the results of an experiment that tested hedonic reactions to Facebook cues cue using the AMP, an approach for measuring implicit affective reactions to a stimulus [14]. In their experiment, participants who reported heavy Facebook use were briefly exposed to a prime stimulus consisting of either a Facebook logo or a control image (e.g., picture of an office product such as a stapler) followed by a neutral target image consisting of a Chinese pictograph. Participants then rated the pictograph as either pleasant or unpleasant. According to the AMP, affective reactions to the briefly displayed prime stimulus will transmit to the target stimulus as participants "misattribute the spontaneous affective reactions triggered by the prime pictures to their evaluations of the ambiguous pictographs shown milliseconds after the prime" [18, p. 335]. Results of their experiment confirmed that participants who reported higher levels of Facebook use and craving exhibited more positive affective responses to target stimuli associated with Facebook cues than those associated with control cues, leading the authors to conclude that "frequent social media users' spontaneous hedonic reactions in response to social media cues might contribute to their difficulties in resisting desires to use social media" [18, p. 334].

The present study seeks to both replicate and extend the findings of van Koningsbruggen et al. [18] using a NeuroIS methodology. Based on previous results, we anticipate that individuals who report high levels of social media use and craving will exhibit greater spontaneous hedonic reactions to social media cues than to other types of control cues. However, leveraging our fMRI experimental design, we extend our exploration beyond that of van Koningsbruggen [18] in three important ways. First, we employ an fMRI experimental protocol to complement behavioral results with neurological data. Because spontaneous hedonic reactions to social media cues observed during the AMP presumably arise from associations between these cues and the gratifications that social media provides [18], we examine whether exposure to these cues stimulates activation of the brain's reward processing network, including areas such as the bilateral nucleus accumbens, bilateral midbrain, bilateral ventromedial prefrontal cortex, and left orbitofrontal cortex [19]. Second, to rule out the possible alternative explanation that such reactions are associated with mere familiarity, we enhance the design of our experimental instrument to include logos of highly recognized, non-social media companies to see whether neural activation differences exist between these types of stimuli. Finally, to broaden the external validity of our findings, we extend our social media stimuli beyond Facebook to include logos associated with three additional top social media platforms.

2 Proposed Method

AMP is designed to measure implicit responses [14] and performs favorably in comparison to other traditional implicit methods [20]. As can be seen in Fig. 1, AMP displays a randomized prime image for 75 ms, followed by a blank screen (125 ms), an abstract or unfamiliar image (in this case a random Chinese character for 100 ms), and then displays a mask screen until soliciting a response from the user. Before and during the experiment, the participant is asked to focus on the Chinese character and to evaluate whether they found it pleasant or unpleasant. The amount of time participants take to evaluate each character may vary. After inputting their response, the next iteration begins. However, to better differentiate between events in the fMRI study, we also display a fixation cross for a random amount of time between 1.0 and 1.5 s.[1] Thus, the experiment implements an event-related design with the study's four separate stimulus types randomly intermixed with one another. By placing the focus on the Chinese character—and due to the rapid nature of the experiment—users implicitly report to the previously displayed prime images. In other words, AMP posits a "carry over" effect from the prime image which is manifested in the participants' reporting of the pleasantness of the Chinese character.

[1] The effective "jitter-range" will be calculated as the minimum and maximum delay between trials of the same type (e.g., while some trials within the same category (non-social media company, social media, control, and filler) are displayed in back-to-back sequences, others are separated by many trials).

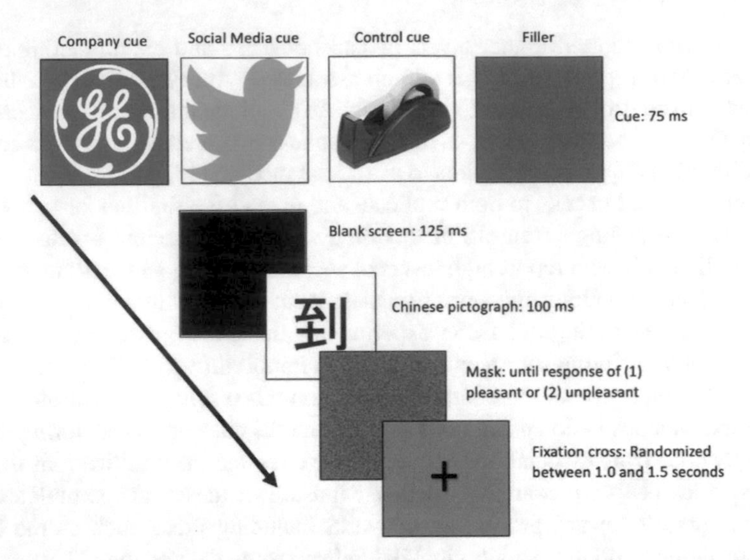

Fig. 1. Graphical representation of the AMP procedure

Our implementation of AMP includes four sets of prime images. Prime images are composed of 10 non-social media company icons, 10 images of each of four social media platforms (Instagram, Facebook, Twitter, and Snapchat) and 10 office product images. Each of these image sets contains 10 unique images related to that category. For example, the office product set had images such as a stapler, marker, etc. The social media platforms had 10 distinct—but related—images within each set. During the study individuals are presented with the logos from the social media platform that they use most (For example, a participant who uses Snapchat most will be exposed to the set non-social media company icons, the set of office product images, and 10 unique images related to Snapchat during the experiment. In this instance, the user would not be exposed to any Facebook, Twitter, or Instagram logos). After matching their social media usage to the correct prime image, the instrument administers several practice rounds to acclimate the participants to the experiment design. Exposure to company cues, control cues, and the filler image were held constant between all conditions. The only variation was the category of social media cues that the users was exposed to, but each participant was exposed to 10 distinct images related to the social media platform they used the most. Once they complete the practice sets, participants begin the main experimental procedure, which consists of 10 images of each of the prime categories in two back-to-back iterations for a total of 80 post-practice exposures. In this case, individuals would see each of the images (10 company, 10 social media image, 10 non-social media company) exactly twice.

After completing the AMP experiment, participants are asked various questions regarding their social media usage and social media cravings. These survey questions are adapted from the van Koningsbruggen et al. [18] study and serve as our independent variables.

2.1 fMRI Implementation

We follow guidelines established in NeuroIS literature [13, 21, 22] for conducting and analyzing the data from the fMRI portion of the study. All imaging is conducted in a Siemens TIM-Trio 3.0T MRI scanner at a private university in the western United States. At the NeuroIS 2020 Retreat, we intend to report data on 35 participants who will have undergone the experiment while situated in a MRI machine. Participants are excluded if they are MRI-incompatible, if they are left-handed, are familiar with Chinese characters, non-native English speakers, had a history of head injury, or had a history of physiological, psychological, or neurological disorders. Participants are compensated with $20 or a digital file with a 3D model of their brain. This study, including the compensation options, has been approved by the Institutional Review Board (IRB) where the data is collected.

Upon arriving at the MRI facility, participants are given a brief safety explanation and then asked to indicate which social media platforms they use most frequently. They are then placed in the MRI machine to complete the experimental protocol. After undergoing a structural scan, participants then perform the practice trials and main trials. Upon completion, participants exit the machine and are complete the post-exit survey regarding basic demographics, social media usage, and social media cravings.

3 Expected Outcomes

The primary measurement of this study is the hemodynamic response in relation to activation of the rewards processing center of the brain, primarily between the ventral tegmental area (VTA) in the midbrain and other regions of the brain that are the recipients of the neurotransmitter dopamine. Since our hypothesis focuses on the brain's rewards processing network, we will create anatomical regions of interest (ROIs) that corresponded to that network. Masks will be generated from Neurosynth [23] using a threshold of $z > 6$. As mentioned previously, four ROIs have been identified as potential areas of interest for our current study: the bilateral nucleus accumbens, bilateral midbrain, bilateral ventromedial prefrontal cortex, and left orbitofrontal cortex. For each of these four regions, we anticipate an interaction between task condition and social media use, as well as between task condition and reported social media cravings. Specifically, we anticipate that participants who report high levels of social media use and craving will exhibit higher activation levels in areas associated with the reward network when presented with social media cues vs. control cues (non-social media company logos, office product images). Statistically significant interactions would suggest that any or all of the selected ROI differentially process the different types of logos based on social media use. Relevant findings would thus imply that social media stimuli evoke a hedonic response among frequent users and/or individuals with high levels of cravings due to physiological reactions. These results would highlight the mediating factors behind previously observed spontaneous hedonic reactions [18] and would corroborate behavioral findings with a unique NeuroIS lens.

References

1. Karahanna, E., Xu, S.X., Xu, Y., Zhang, N.: The needs-affordances-features perspective for the use of social media. MIS Q. Manag. Inf. Syst. **42**, 737–756 (2018). https://doi.org/10.25300/MISQ/2018/11492
2. Meservy, T.O., Fadel, K., Nelson, B., Matthews, M.: Production vs. consumption on social media: a uses and gratifications perspective. In: AMCIS 2019 Proceedings (2019)
3. Whiting, A., Williams, D.: Why people use social media: a uses and gratifications approach. Qual. Mark. Res. Int. J. **16**, 362–369 (2013). https://doi.org/10.1108/QMR-06-2013-0041
4. Tarafdar, M., Gupta, A., Turel, O.: The dark side of information technology use. Inf. Syst. J. **23**, 269–275 (2013). https://doi.org/10.1111/isj.12015
5. Lee-Won, R.J., Herzog, L., Park, S.G.: Hooked on facebook: the role of social anxiety and need for social assurance in problematic use of facebook. Cyberpsychol. Behav. Soc. Netw. **18**, 567–574 (2015)
6. Zivnuska, S., Carlson, J.R., Carlson, D.S., et al.: Social media addiction and social media reactions: the implications for job performance. J. Soc. Psychol. **159**, 746–760 (2019). https://doi.org/10.1080/00224545.2019.1578725
7. Turel, O., Serenko, A.: Developing a (bad) habit: antecedents and adverse consequences of social networking website use habit. In: 17th Americas Conference on Information Systems 2011, AMCIS 2011, pp. 705–712 (2011)
8. Moqbel, M., Kock, N.: Unveiling the dark side of social networking sites: personal and work-related consequences of social networking site addiction. Inf. Manag. **55**, 109–119 (2018). https://doi.org/10.1016/j.im.2017.05.001
9. Hou, Y., Xiong, D., Jiang, T., et al.: Social media addiction: its impact, mediation, and intervention. Cyberpsychology **13** (2019). https://doi.org/10.5817/CP2019-1-4
10. He, Q., Turel, O., Bechara, A.: Association of excessive social media use with abnormal white matter integrity of the corpus callosum. Psychiatry Res. Neuroimaging **278**, 42–47 (2018). https://doi.org/10.1016/J.PSCYCHRESNS.2018.06.008
11. He, Q., Turel, O., Bechara, A.: Brain anatomy alterations associated with Social Networking Site (SNS) addiction. Sci. Rep. **7**, 45064 (2017). https://doi.org/10.1038/srep45064
12. Kuss, D.J., Griffiths, M.D.: Online social networking and addiction—a review of the psychological literature. Int. J. Environ. Res. Public Health **8**, 3528–3552 (2011). https://doi.org/10.3390/ijerph8093528
13. Dimoka, A., Pavlou, P.A., Davis, F.D.: NeuroIS: the potential of cognitive neuroscience for information systems research. Inf. Syst. Res. **22**, 687–702 (2011). https://doi.org/10.1287/isre.1100.0284
14. Payne, K., Lundberg, K.: The affect misattribution procedure: ten years of evidence on reliability, validity, and mechanisms. Soc. Pers. Psychol. Compass **8**, 672–686 (2014). https://doi.org/10.1111/spc3.12148
15. Olivia Valentine: Top 10 reasons for using social media - GlobalWebIndex. In: globalwebindex (2018). https://blog.globalwebindex.com/chart-of-the-day/social-media/. Accessed 25 Feb 2020
16. Hofmann, W., Friese, M., Strack, F.: Impulse and self-control from a dual-systems perspective. Perspect. Psychol. Sci. **4**, 162–176 (2009). https://doi.org/10.1111/j.1745-6924.2009.01116.x
17. Stein, L.: Habituation and stimulus novelty: a model based on classical conditioning. Psychol. Rev. **73**, 352–356 (1966). https://doi.org/10.1037/h0023449
18. Van Koningsbruggen, G.M., Hartmann, T., Eden, A., Veling, H.: Spontaneous hedonic reactions to social media cues. Cyberpsychol. Behav. Soc. Netw. **20**, 334–340 (2017). https://doi.org/10.1089/cyber.2016.0530

19. Liu, X., Hairston, J., Schrier, M., Fan, J.: Common and distinct networks underlying reward valence and processing stages: a meta-analysis of functional neuroimaging studies. Neurosci. Biobehav. Rev. **35**, 1219–1236 (2011)
20. Cameron, C.D., Brown-Iannuzzi, J.L., Payne, B.K.: Sequential priming measures of implicit social cognition: a meta-analysis of associations with behavior and explicit attitudes. Pers. Soc. Psychol. Rev. **16**, 330–350 (2012). https://doi.org/10.1177/1088868312440047
21. Vom Brocke, J., Liang, T.P.: Guidelines for neuroscience studies in information systems research. J. Manag. Inf. Syst. **30**, 211–234 (2014)
22. Riedl, R., Davis, F.D., Hevner, A.R.: Towards a NeuroIS research methodology: intensifying the discussion on methods, tools, and measurement. J. Assoc. Inf. Syst. **15**, 1–35 (2014)
23. Neurosynth: reward. https://neurosynth.org/analyses/terms/reward/. Accessed 4 Mar 2020

On the Modulation of Perturbation-Evoked Potentials After Motor Reaction in a Human-Machine Interaction Setup

Gernot R. Müller-Putz(✉), Melanie Stockreiter, Jonas C. Ditz, and Valeria Mondini

Graz University of Technology, Graz, Austria
gernot.mueller@tugraz.at

Abstract. Passive brain-computer interfaces (pBCIs) can be used to inform humans about their current mental state or in human-machine interaction (HMI) scenario. We introduced the perturbation-evoked potential (PEP) in the context of pBCI and are further investigating how neural correlates in a HMI could interact. In the current study we investigate the neural correlates after perturbation followed by motor reaction. We found that the PEP as well as a movement-related cortical potential appear, and that the latter has an influence on the shape of the PEP.

Keywords: Passive brain-computer interface · Perturbation-evoked potential · Human-machine interaction · Rehabilitation · Assistive device

1 Introduction

A brain-computer interface (BCI) is a system that translates mentally modulated brain activity into one or more control commands for computers, machines or devices. Since the beginning of BCI research in the early 1990s, BCIs have mainly been researched to compensate for lost neural functionality and restore or replace several key abilities in people with severe motor deficits [1–3]. BCI systems for the application to people with disabilities are usually described as active aBCI or reactive (rBCI) BCIs. These terms reflect the goal of the systems to provide users with autonomous control over a target computer or machine, or to react on external stimuli provided, e.g., by a spelling device.

In recent years, BCI systems to enrich human-machine interaction with implicit information about the user's state have been researched – the so-called passive BCIs (pBCIs) [4]. A pBCI system, mainly based on electroencephalography (EEG) [5], relies on activity pattern that are not voluntarily modulated by users, but automatically occur due to changes of their internal or external parameters, or changes in the situation. Proposed systems used e.g. mental workload [6], error-related potentials (ErrP) [7], or other brain states [8]. The implicit information about the user's mental state encoded in these activity patterns is used to evoke a reaction of the controlled machine or computer, e.g. Zander et al. used a pBCI to implicitly control movement of a cursor based on workload [6].

F. D. Davis et al. (Eds.): NeuroIS 2020, LNISO 43, pp. 344–349, 2020.
https://doi.org/10.1007/978-3-030-60073-0_40

A relatively new neural correlate in this area is the so-called perturbation-evoked potential (PEP). Such a PEP is elicited whenever a person loses balance. In order to use PEPs for a pBCI, a robust detection and classification of this activity pattern have to be possible. Our group performed a first study in this direction recently [9, 10]. The general goal is to use the PEP in a pBCI system, i.e. the real-time detection of this correlate in a human-machine interaction scenario. Specifically, we are aiming at applying this form of BCI in a clinical and aviation environment (cf. [9]). In the clinical setting, detecting the perturbation would permit to promptly correct the position when e.g. wearing an exoskeleton for walking. In an aviation scenario, it would permit to give the plane a faster reaction than it could be done with a motor reaction (e.g. during glider flying).

In this work, we further investigate the neural correlates of users when they react to a perturbation, and give an outlook on how we are going to use this in a further PEP-enriched HMI.

2 Perturbation Evoked Potential

A number of EEG studies have revealed that full-body perturbations are reflected by a specific cortical activity [11–15], called perturbation potential (PEP). In a recent work [10], we investigated the PEP from persons sitting in a car chair and getting either perturbed to their left or to their right side. We could also show that the PEPs are almost identical, independently from the side of perturbation. We were finally able to classify, in a group of 15 participants and in a single trial basis, the PEP against rest with an average accuracy of 85% [10]. However, the main aim of this work was not to study neurophysiology, but to detect this event-related potential (ERP) in an ongoing EEG. Since we are going to apply the PEP in an environment where motor reaction will be necessary, we investigate here how a motor reaction changes the PEP.

3 Experiment and Data Analysis

Participants: EEG recordings were performed with 10 healthy volunteers (5 males and 5 females). The age ranged from 22 to 26 (mean 23.90 $\pm$ 1.29) years. All participants provided written informed consent prior to the experiment. After data examination, three participants were excluded from further analysis due to technical problems during recordings.

Experiment: Participants were sitting (see Fig. 1) on a fully removed car-chair, which was assembled on a plate and connected to a bar to tilt the participant 5° to the left and to the right. The experiment was split in two different tasks, with a total of 160 trials. The first part consisted of 80 trials of one tilt movement each, i.e. the participant was rapidly tilted 40 times to the left and 40 to the right. In the second part of the experiment, the plate was again similarly tilted (40 times left and 40 times right), but the participants were asked to react with a joystick positioned on their laps. Whenever the participants were tilted to the left/right, they had to move the stick to the opposite direction.

During the whole experiment, participants had to focus on a fixation cross on the wall, to reduce eye-blinking artifacts. Rest-EEG was collected for 3 min before the experiment started, for 5 min following the first 80 trials, and finally 3 min in the end.

EEG Recording: The participants were equipped with a 32 channels EEG cap (F1, FFC1h, FC3, FC1, FCC1h, C5, C3, C1, CCP1h, CP3, CP1, P1, Pz, P2, CPz, CP2, CP4, CCP2h, Cz, C2, C4, C6, FCC2h, FCz, FC2, FC4, FFC2h, Fz, F2, Fpz (GND) reference left mastoid, LiveAmp BrainProducts GmbH). Sampling rate was 500 Hz. Additionally to EEG, the in-built accelerometer and the joystick data was recorded.

Signal Processing: The EEG was filtered with a causal bandpass IIR filter with order 4 from 0.3–30 Hz and then epoched between 0.5 s and 1.5 s with respect to the perturbation onset. An outlier rejection was done to exclude contaminated trials. On average 59 trials and 77 trials per participant were used for averaging. Finally, the EEG was averaged to retrieve the PEP. This was done for trials without as well as with reaction to the perturbation. The average reaction time was derived from the joystick data and the perturbation onset. To assess whether a movement-related cortical potential (MRCP) during the joystick movement occurred, the EEG was also averaged according to the joystick movement onset. The same analysis was done on the trials without joystick reaction, by using the same distribution of reaction times from the second part of the experiment.

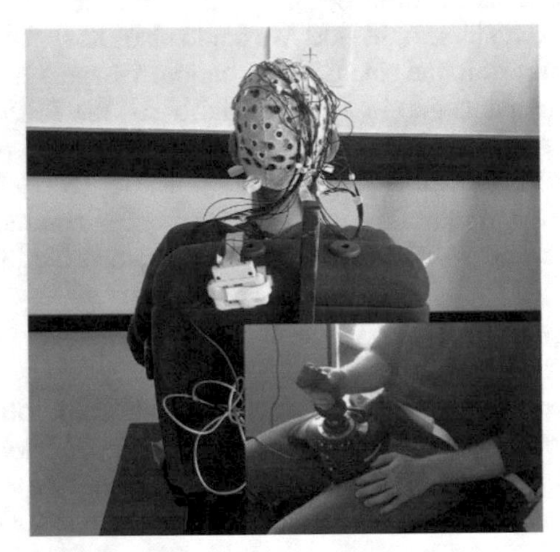

Fig. 1. Picture of the custom-build chair used to induce perturbations. The chair can be tilted using the handle seen behind the backrest. The amplifier is attached at the top of the backrest in order to record movements of the chair with the in-build accelerometer. In the second part of the study, participants used a joystick to "steer" to the opposite side.

4 Results

Neurophysiologically, the perturbation potential appears to be largest at electrode position Cz. The PEP was larger (N1) in the "reaction" condition than in the "no reaction" condition. Latencies of the N1 was almost identical in both conditions (see Fig. 2).

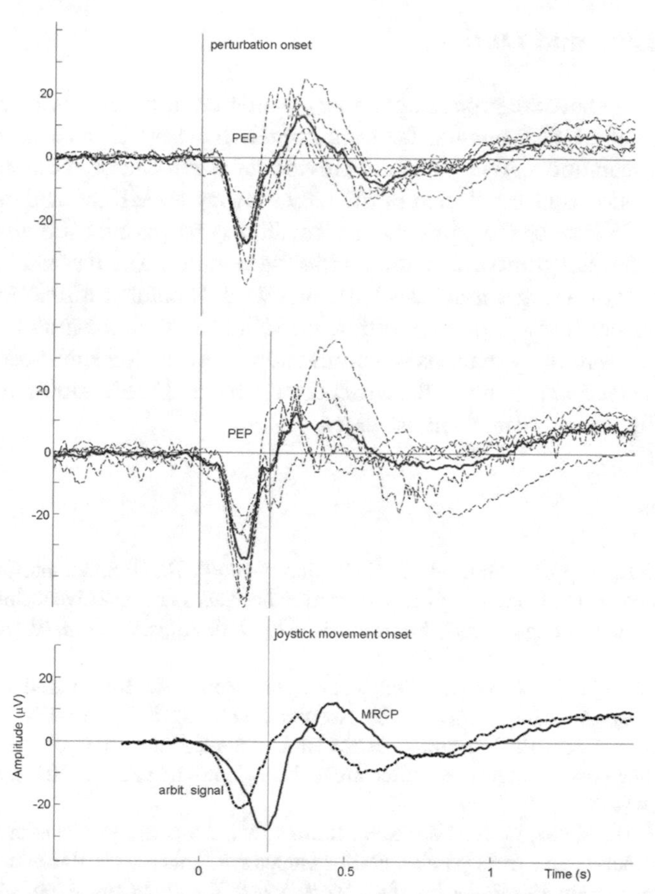

Fig. 2. Group responses of the perturbation at Cz. Upper panel: perturbation potential without reaction of the participants. Middle panel: PEP when participants reacted to the perturbation with movement of the joystick. The vertical red line indicates the mean reaction time. Lower panel: when averaging according to each single joystick movement (vertical line, reaction condition), a movement-related cortical potential (violet) appears. This does not happen, when the no-reaction data gets averaged (black dotted line).

From the behavioural point of view, participants needed 0.24 s ± 0.035 s to react on the perturbation and perform a joystick movement. The N1 amplitude was 8 μV ± 5 μV lager in the "reaction" condition, however this cannot be confirmed with statistics since the number of participants is too low. Also, N1 latencies were in the same range. Interestingly, the PEP had a slightly different shape and N1 amplitude in the "reaction" condition, so EEG was also investigated in relation to the joystick movement onset. Here, clearly an MRCP appears after averaging to individual movement onsets (see violet curve in Fig. 2, lower panel). By analysing the data from the "no reaction" condition, a smeared curve (arbitrary signal) of the perturbation potential appears (Fig. 2, lower panel black dottet line). This reflects, in contrast to the MRCP, no real physiological signal.

5 Conclusion and Outlook

With this work we have taken one step further towards a PEP-based pBCI. We could show that the shape of the PEP remains the same, although the N1 component was larger in the "reaction" condition. Also, we could provide some evidence that the motor reaction on the perturbation is also reflected in EEG, by showing an MRCP. The motor reaction modulates the PEP since the participants were already preparing ("Bereitschaftspotential") to react. Several points are of interest in a future study: (i) is it possible to detect the PEP in real-time on a single-trial basis? (ii) Since we did not find a difference in left/right PEPs in our recent study [10], is the difference visible through the motor reaction? (iii) How sensitive is our body in terms of perturbation (perturbation speed and tilt angle).

Our next intention is to move the current setup into a 2D scenario for the simulation and investigation in our described use cases.

References

1. Millán, J.d.R., Rupp, R., Müller-Putz, G.R., Murray-Smith, R., Giugliemma, C., Tangermann, M., Vidaurre, et al.: Combining brain–computer interfaces and assistive technologies: state-of-the-art and challenges. Front. Neurosci. **4**, 161 (2010). https://doi.org/10.3389/fnins.2010.00161
2. Müller-Putz, G.R., Scherer, R., Pfurtscheller, G., Rupp, R.: EEG-based neuroprosthesis control: a step towards clinical practice. Neurosci. Lett. **382**(1–2), 169–174 (2005)
3. Biasiucci, A., Leeb, R., Iturrate, I., et al.: Brain-actuated functional electrical stimulation elicits lasting arm motor recovery after stroke. Nat. Commun. **9**, 2421 (2018). https://doi.org/10.1038/s41467-018-04673-z
4. Zander, T. O., Kothe, C. A., Welke, S., Rötting, M.: Enhancing human–machine systems with secondary input from passive brain–computer interfaces. In: Proceedings of the 4th International Brain–Computer Interface Workshop & Training Course, pp. 144–149. Verlag der Technischen Universität Graz, Graz, Austria (2008)
5. Müller-Putz, G.R., Riedl, R., Wriessnegger, S.C.: Electroencephalography (EEG) as a research tool in the information systems discipline: foundations, measurement, and applications. CAIS **37**, 46 (2015)
6. Zander, T.O., Krol, L.R., Birbaumer, N.P., Gramann, K.: Neuroadaptive technology enables implicit cursor control based on medial prefrontal cortex activity. Proc. Natl. Acad. Sci. **113**(52), 14898–14903 (2016)
7. Lopes-Dias, C., Sburlea, A.I., Müller-Putz, G.R.: Online asynchronous decoding of error related potentials during the continuous control of a robot. Sci. Rep. **9**(1), 1–9 (2019)
8. Bos, D.P.O., Reuderink, B., van de Laar, B., Gürkök, H., Mühl, C., Poel, M., Nijholt, A., Heylen, D.: Brain-computer interfacing and games. In: Tan, D., Nijholt, A. (eds.) Brain-Computer Interfaces, pp. 149–178. Springer, London (2010)
9. Ditz, J.C., Müller-Putz, G.R.: Perturbation-evoked potentials: future usage in human-machine interaction. In: Davis F., Riedl R., vom Brocke J., Léger PM., Randolph A., Fischer T. (eds.) Information Systems and Neuroscience. Springer, Cham, pp. 271–277 (2019)
10. Ditz, J.C., Schwarz, A., Müller-Putz, G.R.: Perturbation-evoked potential can be classified from single-trial EEG. J. Neural Eng. (2020, accepted)
11. Dietz, V., Quintern, J., Berger, W.: Cerebral evoked potentials associated with the compensatory reactions following stance and gait perturbation. Neurosci. Lett. **50**(1–3), 181–186 (1984)

12. Duckrow, R.B., Abu-Hasaballah, K., Whipple, R., Wolfson, L.: Stance perturbation-evoked potentials in old people with poor gait and balance. Clin. Neurophysiol. **110**(12), 2026–2032 (1999)
13. Adkin, A.L., Quant, S., Maki, B.E., McIlroy, W.E.: Cortical responses associated with predictable and unpredictable compensatory balance reactions. Exp. Brain Res. **172**(1), 85 (2006)
14. Mochizuki, G., Sibley, K.M., Cheung, H.J., Camilleri, J.M., McIlroy, W.E.: Generalizability of perturbation-evoked cortical potentials: independence from sensory, motor and overall postural state. Neurosci. Lett. **451**(1), 40–44 (2009)
15. Varghese, J.P., McIlroy, R.E., Barnett-Cowan, M.: Perturbation-evoked potentials: significance and application in balance control research. Neurosci. Biobehav. Rev. **83**, 267–280 (2017)

Consumer-Grade EEG Instruments: Insights on the Measurement Quality Based on a Literature Review and Implications for NeuroIS Research

René Riedl[1,2](✉), Randall K. Minas[3], Alan R. Dennis[4], and Gernot R. Müller-Putz[5]

[1] University of Applied Sciences Upper Austria, Steyr, Austria
rene.riedl@fh-steyr.at
[2] Johannes Kepler University, Linz, Austria
[3] Shidler College of Business, University of Hawaii, Manoa, HI, USA
rminas@hawaii.edu
[4] Kelley School of Business, Indiana University, Bloomington, IN, USA
ardennis@iu.edu
[5] Graz University of Technology, Graz, Austria
gernot.mueller@tugraz.at

Abstract. Application of good methodological standards is critical in science because such standards constitute a precondition for high-quality research results. A fundamental question which has recently been raised in the NeuroIS literature is whether consumer-grade electroencephalography (EEG) instruments offer measurement quality that is comparable to research-grade instruments. Importantly, a notable number of EEG papers in the NeuroIS literature already used consumer-grade instruments, predominantly because such tools are typically portable, wireless, cheap, and easy to use. However, there is an ongoing discussion in the scientific community about these tools' measurement quality. To contribute to this discussion, we reviewed prior research to document major insights on the measurement quality of consumer-grade products. In essence, our results indicate that consumer-grade EEG instruments constitute a viable alternative to high-quality research tools. However, there are two important constraints on their use. First, as with any research, the use of consumer-grade systems is appropriate only when tied to the correct type of analysis. Second, in order to establish more definitive conclusions on consumer-grade systems' appropriateness, empirical validation studies are needed based on Information Systems (IS) tasks, paradigms, and types of analysis, and several other limiting factors have to be considered.

Keywords: Brain · Consumer-grade EEG · Electroencephalography · EEG · EPOC · NeuroIS · Research-grade EEG

F. D. Davis et al. (Eds.): NeuroIS 2020, LNISO 43, pp. 350–361, 2020.
https://doi.org/10.1007/978-3-030-60073-0_41

1 Introduction

Electroencephalography (EEG) measures electrical activity of neuronal networks in the brain. Using electrodes placed on the scalp, EEG captures with a very high temporal precision (milliseconds) the summation of synchronous postsynaptic potentials produced by a population of neurons. EEG is a widely used research tool in many scientific disciplines, including medicine, cognitive neuroscience, and psychology. In Neuro-Information-Systems (NeuroIS), EEG has been a subject of discussion since the genesis of the field (e.g., [1–5]), and the tool has also been applied frequently. A recent review indicates that EEG is even "the dominant tool in NeuroIS research" [6]. Specifically, it is reported that 37% of the empirical NeuroIS research applied EEG, followed by measurement of heart rate (22%), skin conductance (19%) and functional brain imaging (fMRI, 15%).

However, this review also points to an important detail of the EEG studies. Riedl et al. [6] indicate that several papers used instruments that were not developed for research purposes, they write: "In the last decade, companies with a primary focus on video gaming developed relatively inexpensive EEG-based headsets for players to control the game [...] Such EEG systems differ from established EEG research tools [...] this finding is an observation that deserves closer attention in future methodological papers". Examples for portable and wireless EEG tools which were not originally developed for research purposes are, among others, EPOC (Emotiv, San Franciso, USA), MUSE brain sensing technology (InteraXon, Toronto, Canada), and ThinkGear/MindWave (Neurosky, San Jose, USA). Motivated by this call for "closer attention," we conducted a literature review to identify and analyze scientific studies on the measurement quality of consumer-grade EEG instruments. Major results are reported in this paper.

Next, we briefly outline the methodology of the review (Sect. 2), followed by a description of major insights that we developed from the analyses of reviewed studies (Sect. 3). Finally, we report our conclusion (Sect. 4).

2 Methodology

Current best practice for research reviews is to include all relevant papers that are methodologically correct, regardless of source, because different sources have different biases [7]. For example, journal articles are less likely to publish papers that report no significant differences and those that challenge the status quo [7]. In February 2020, we used five different search strategies to identify relevant papers. First, we conducted several search queries in Web of Science, one of the worldwide largest databases with scientific articles, based on the following keywords: "EEG"/"Electroencephalography" (specification "Title") AND "reliability"/"validity"/"validation"/"sensitivity"/"diagnosticity"/"objectivity"/"accuracy"(specification "Title"). The terms related to measurement were derived from a well-cited paper on NeuroIS methodology [8]. The search yielded the following number of hits: reliability (32 EEG/2 Electroencephalography), validity (13/1), validation (26/1), sensitivity (23/1), diagnosticity (0/0), objectivity (0/0), accuracy (12/0). Second, we searched Google Scholar using the terms "EEG" and "reliability"/"validation" and "Emotiv" (the most popular consumer-grade EEG system found by our Web of Science searches). Third, we examined the Emotiv Web page devoted to research using their system. Fourth, we used forward and backward search on the papers we found. Fifth, we used personal communication in the scientific community in an attempt to identify further papers. Specifically, research related to the keywords "evaluation" and "classification" was added in this step. We ended up with a total of 16 relevant papers, which constitute the basis of our review.

3 Results

We present two tables. In Table 1, in ascending order based on publication date, we summarize the main findings of the 10 papers that sought to validate consumer-grade devices against research-grade devices. Presentation of the studies' main findings is based on literal citation of relevant text passages in order to avoid potential misinterpretations. We classify papers as using point-related analyses (e.g., Event-Related Potential, ERP) or time-frequency and spectral analyses (e.g., Event-Related Spectral Perturbation, ERSP).

In Table 2, we present the 6 studies that allow neurophysiological signals from consumer-grade devices to be classified using algorithms from research-grade devices. These studies include feature extraction studies, as well as ERP, and time-frequency analyses.

4 Conclusion

We reviewed 16 studies in which the measurement quality of consumer-grade EEG devices was assessed. The most commonly used device was the Emotiv EPOC. Table 3 summarizes the conclusions of the research studies in Tables 1 and 2. The majority of studies (14 out of 16) concluded that use of consumer-grade EEG was acceptable. The presented evidence suggests that reliability (assessed via test-retest: [18]), concurrent validity (assessed via benchmark to research-grade tools: [10, 11, 13–15, 19]), and comparative validity (assessed via predictable sensitivity of a measure to variations in psychological state: [10]) are acceptable. One of the ERP studies in our sample, however, reports less positive results. Duvinage et al. [12] write that "the Emotiv headset performs significantly worse" than the research-grade ANT system (p. 1). Based on this finding, Duvinage et al. [12] recommend that Emotiv should only be used "for non critical applications such as games" (p. 1).

When interpreting this finding, several *limiting factors* have to be kept in mind. *First*, no study in our sample used more complex tasks such as those typically used in IS research. We cannot conclude that results from less complex tasks (e.g., auditory tasks or basic visual tasks such as n-back) generalize or do not generalize to more complex tasks used in IS research (e.g., perception of complex visual stimuli such as websites), so we need more research comparing consumer-grade EEG instruments to research-grade tools. It is likely a question of *when* and *with what analyses* these consumer-grade systems are appropriate in the IS context. Thus, we call for future studies which examine the appropriateness of consumer-grade EEG for different tasks and environmental settings which are typical for IS research.

Second, as indicated in Tables 1 and 2, EPOC was the most frequently studied tool. This tool is a 14-channel instrument. Other tools such as ThinkGear (single-channel EEG), or Muse headband (4 channel tool) use fewer channels, while OpenBCI has

Table 1. Papers Validating Consumer-grade EEG against Research-grade Equipment

Source/Sample/Task	(1) Consumer-grade Equipment and (2) Research-grade Equipment, (3) Type of Analysis, (4) Finding
Alzu'bi et al. [9] N = 3 (adults) Visual and acoustic mental tasks	❶ EPOC (Emotiv, USA, www.emotiv.com) ❷ BrainAmp (Brain Products, Germany, www.brainproducts.com) ❸ Variety of feature extraction ❹ "The average 4-fold cross-validation accuracy over the three subjects were around 53% using PSD feature extraction and LDA classifier. Comparing this result to the one by […] which gains 58% accuracy by using the same feature extraction and classification methods but different EEG recording device (BrainAmp with Ag/Cl electrodes), it can be concluded that Emotiv can be used to gain comparable results to advanced EEG recording devices […] A comparison was attended to compare outcome accuracy of using inexpensive device and the accuracy [that] comes from advanced EEG recording device. The outcome results suggest that an inexpensive device, such as Emotiv, can gain comparable accuracy to the advanced devices, such as BrainAmp" (p. 100)
Johnstone et al. [10] N = 20 (study 1, adults) N = 23 (study 2, children) Eyes open (EO) and eyes closed (EC) resting conditions (study 1), EO and EC resting conditions and active conditions (relaxation, attention, cognitive load) (study 2)	❶ ThinkGear (Neurosky, USA, www.neurosky.com) ❷ Neuroscan (Compumedics, Australia, www.compumedicsneuroscan.com) ❸ Spectral (relative power) ❹ "The present study provides preliminary data concerning the validity of a new method of single-channel EEG measurement, examining concurrent validity via comparisons to a research system, and comparative validity via predictable sensitivity to variations in psychological state […] Overall, these results suggest acceptable validity for the new single-channel dry-sensor method, but of course there are trade-offs associated with the use of the headset. While the headset is convenient and wireless, and the dry-sensor takes less than 20 s to fit, there is a minor trade-off in terms of data quality compared to the research system considered here and a major trade-off in the number of recording locations" (p. 119)
Badcock et al. [11] N = 21 (adults) Measurement of auditory event-related potentials (ERPs), 566 standard (1000 Hz) and 100 deviant (1200 Hz) tones under passive (non-attended) and active (attended) conditions	❶ EPOC (Emotiv, USA) ❷ Neuroscan (Compumedics, Australia) ❸ ERP ❹ "Considered together, the results of this study suggest that the gaming EEG system compares well with the research EEG system for reliable auditory ERPs such as the P1, N1, P2, N2, and P3 measured at the frontal sites […] The apparent validity of the gaming EEG system for measuring reliable auditory ERPs, paired with its quick and clean set-up procedure and its portability, makes it a promising tool for measuring auditory processing in people from special populations who are unable or unwilling to be tested in an experimental laboratory" (p. 14)

(*continued*)

Table 1. (*continued*)

Source/Sample/Task	(1) Consumer-grade Equipment and (2) Research-grade Equipment, (3) Type of Analysis, (4) Finding
Duvinage et al. [12] N = 9 (adults) P300 paradigm in terms of recognition performance under walking and sitting conditions	❶ EPOC (Emotiv, USA) ❷ AdvancedNeuroTechnology acquisition system (ANT Neuro, Netherlands, www.ant-neuro.com) ❸ ERP ❹ "The Emotiv headset performs significantly worse than the medical device; observed effect sizes vary from medium to large. The Emotiv headset has higher relative operational and maintenance costs than its medical-grade competitor […] Although this low-cost headset is able to record EEG data in a satisfying manner, it should only be chosen for non critical applications such as games, communication systems, etc. […] the design of a specific low-cost EEG recording system for critical applications and research is still required" (p. 1)
Badcock et al. [13] N = 19 (children) Passive (P) and active (A) listening conditions. In P, children were instructed to watch a silent DVD and ignore 566 standard (1,000 Hz) and 100 deviant (1,200 Hz) tones. In A, they listened to the same stimuli, and were asked to count the number of "high" (i.e., deviant) tones	❶ EPOC (Emotiv, USA) ❷ Neuroscan (Compumedics, Australia) ❸ ERP ❹ "three key findings. First, whilst both EEG systems recorded a high proportion of accepted epochs, fewer were acceptable for EPOC […] may stem from reduced stability of EPOC's saline-soaked cotton sensors resting on the scalp, relative to the gel used with Neuroscan, which effectively glues to sensor the scalp with gel […] Second, the systems produced similar late auditory ERP [event related potential] and MMN [mismatch negativity] waveforms […] Third, there were only a few differences between the peak amplitude and latency measures produced by the EPOC and Neuroscan systems, which mostly related to delayed latencies for the EPOC system […] Overall, the findings of the present study paired with Badcock et al. (2013) suggest that EPOC compares well with Neuroscan for investigating late auditory ERPs in children." (p. 13/14)
De Lissa et al. [14] N = 13 (teenagers and adults) Face perception task (stimuli were upright and inverted gray-scale images of wrist-watches and emotionally-neutral Caucasian faces)	❶ EPOC (Emotiv, USA) ❷ Neuroscan (Compumedics, Australia) ❸ ERP ❹ "The N170 recorded through both the gaming EEG system and the research EEG system exhibited face-sensitivity […] The EPOC system produced very similar N170 ERPs to a research-grade Neuroscan system, and was capable of recording face-sensitivity in the N170, validating its use as research tool in this arena […] though with modifications tailored to specific research questions and methodologies. Thus, the use of such devices may prove a useful neuroscientific tool for investigating the neural correlates of face processing in populations of people who cannot attend, or cannot tolerate, ERP research laboratories" (p. 47/53)

(*continued*)

Table 1. *(continued)*

Source/Sample/Task	(1) Consumer-grade Equipment and (2) Research-grade Equipment, (3) Type of Analysis, (4) Finding
Wang et al. [15] N = 30 (adults) ERPs were induced by an auditory oddball task; amplitudes and latencies of N1, N2, and P3 were analyzed; reaction time (RT) and response accuracy were derived from synchronously recorded behavioral data	❶ EPOC (Emotiv, USA) ❷ Neuroscan (Compumedics, Australia) ❸ ERP ❹ "Electrophysiological analysis demonstrated that Emotiv system was able to detect component N1, N2, and P3. The peak magnitudes of these components recorded in the Emotiv group were remarkably lower. That might be explained by the different standard of amplifiers in two systems. Obviously, the amplifier for Neuroscan system is much better in sensitivity […] patterns of the amplitudes were similar which representing a same relationship, N1 < N2 < P3. These comparative results confirmed that Emotiv device could provide reliable auditory ERP signals, suggesting its potential for further development in medical applications" (p. 4)
Friedman et al. [16] N = 44 (adults) Video clips to elicit emotional responses. Positive and negative valence and arousal studied. Frontal wave asymmetry was examined	❶ EPOC (Emotiv, USA) ❷ Compared to Davidson et al. [17] ❸ ERP ❹ "Our results indicate that hemispheric asymmetry may be used as a good marker for emotional valence in EEG, even when using a consumer device. The analysis is mostly based on extreme segments, for which high arousal was reported, so it is likely that our results indeed related to the valence component of the emotional experience of our subjects. Our results are in line with the classic studies by Davidson et al. (1990) […] our conclusion is although the levels of noise and artifacts are extremely high it is nevertheless possible to extract useful information, in this case emotional valence, from such devices" (p. 935/936)
Rogers et al. [18] N = 59 (youth, adults, old adults) Conditions were eyes-open, eyes-closed, auditory oddball, and visual n-back Longitudinal data to calculate test-retest reliability (n, n + 1 day, n + 1 week, n + 1 month)	❶ ThinkGear (Neurosky, USA) ❷ not applicable (test-retest reliability of consumer-grade tool) ❸ ERP ❹ "Relative power (RP) of delta, theta, alpha, and beta frequency bands was derived from EEG data obtained from a single electrode over FP1 […] Intra-class correlations (ICCs) and Coefficients of Repeatability (CRs) were calculated from RP data re-collected one-day, one-week, and one-month later […] Eyes-closed resting EEG measurements using the portable device were reproducible (ICCs 0.76–0.85) at short and longer retest intervals in all three participant age groups. While still of at least fair reliability (ICCs 0.57–0.85), EEG obtained during eyes-open paradigms was less stable […] Combined with existing validity data (Johnstone et al. 2012), the current results suggests a portable device may provide a viable alternative to conventional lab-based recording systems for assessing changes in electrophysiological signals, and further application to the study of brain function using the system can be encouraged" (p. 87/95)

(continued)

Table 1. (*continued*)

Source/Sample/Task	(1) Consumer-grade Equipment and (2) Research-grade Equipment, (3) Type of Analysis, (4) Finding
Krigolson et al. [19] N = 120 (adults) Tasks: visual oddball paradigm, standard reward-learning task	❶ MUSE brain sensing technology headband (InteraXon, Toronto, Canada, www.choosemuse.com) ❷ Brain Vision Recorder (Brain Products, Germany) ❸ ERP ❹ "Our results demonstrate that we could observe and quantify the N200 and P300 ERP components in the visual oddball task and […] the reward-learning task […] 95% confidence intervals all statistically verified the existence of the N200, P300, and reward positivity in all analyses […] our work highlights that with a single computer and a portable EEG system such as the MUSE one can conduct ERP research with ease thus greatly extending the possible use of the ERP methodology to a variety of novel contexts" (p. 1)

Table 2. Papers using classification to validate consumer-grade devices

Source/Sample/Task	(1) Consumer-grade Equipment and (2) Research-grade Equipment, (3) Type of Analysis, (4) Finding
Ramírez-Cortes et al. [20] N = 8 (adults) P300 paradigm to attend to an object when it appears on the screen	❶ EPOC (Emotiv, USA) ❷ classification study ❸ ERP ❹ "The results presented in this paper are part of a project with the ultimate goal of designing and developing brain computer interface systems. These experiments support the feasibility to detect P300 events using the Emotiv headset, through an ANFIS approach, which can be used as information control for external devices in BCI applications" (p. 5)
Debener et al. [21] N = 16 (adults) Auditory oddball task while participants walked. P300 classification	❶ EPOC (Emotiv, USA) ❷ classification study ❸ ERP ❹ "We show that good quality EEG data can be obtained in such adverse recording conditions as naturally walking outdoors. All recorded trials were entered into the classification, after only moderate preprocessing that could be implemented online. Moreover, a chronological classification strategy that aimed at avoiding pitfalls in machine learning applications was chosen Lemm et al. [22]. The drop in classification performance from training (88%) to testing (73%) revealed nonstationarities in the data, a common problem in EEG-based BCIs. Advanced feature selection strategies implementing spatial filters may be needed to optimize classification performance. However, the good across-subjects and acrosstrials consistency, in combination with the excellent ERP test-retest reliability, suggests that our results are robust and should generalize to other real-world scenarios" (p. 1450)

(*continued*)

Table 2. (*continued*)

Source/Sample/Task	(1) Consumer-grade Equipment and (2) Research-grade Equipment, (3) Type of Analysis, (4) Finding
Rodriguez et al. [23] N = 16 (adults) IAPS pictures of positive and neutral valence. Alpha-band ERD study	❶ EPOC (Emotiv, USA) ❷ classification study ❸ Alpha-ERD (time-frequency analysis) ❹ "These results show evidence of a frontal asymmetry in the EEG of the participants during a positive induction, which is in accordance with science literature, that affirms that activation of left hemisphere is linked with positive emotion induction. This result supports the possibility of using the low-cost EEG devices, in particular Emotiv Epoc, as an emotional measuring tool in future studies" (p. 46)
Elsawy et al. [24] N = 3 (adults) Oddball P300 speller paradigm, using a grid of characters and identification of intensification of target character row or column	❶ EPOC (Emotiv, USA) ❷ classification study ❸ ERP ❹ "We examined the performance of the PCA ensemble classifier on data recorded using the Emotiv neuroheadset. We compared the performance of the method to that obtained using a concatenated feature vector-based classifier. Our results indicated that Emotiv neuroheadset can have acceptable results for P300 speller applications using the PCA ensemble classifier" (p. 5035)
Wang, Gwizdka et al. [25] N = 9 (adults) Time-frequency spectral analysis on an *n*-back task. Theta, alpha, and low gamma band corresponded to varying levels of workload level	❶ EPOC (Emotiv, USA) ❷ classification study ❸ Time-frequency analysis ❹ "The behavioral measures (increased RTs and decreased RAs) confirmed that different memory workload levels were experienced corresponding to different n-back levels. Different memory workload evoked associated EEG signal patterns that made it possible to classify the corresponding memory load levels. The change in signal power in the theta band (4–8 Hz) at frontal channels was found to be significant for distinguishing the lowest workload level (0-back) from the higher workload levels. The change in alpha band (9–13 Hz) and the low gamma band (30–40 Hz) were found to be useful for distinguishing memory workload levels between 1-, 2-, and 3 back levels" (p. 432/433)
Lakhan et al. [26] N = 43 (teenagers, adults) Feature extraction and classification of video clips by high or low level of valence and arousal	❶ OpenBCI (OpenBCI, USA, www.openbci.com) ❷ classification study ❸ Feature extraction/time-frequency analysis ❹ "To our knowledge, this study is the first to carry out an evaluation of OpenBCI applicability in the domain of emotion recognition. In comparison to medical grade EEG amplifiers with a greater number of electrodes and sampling frequencies, OpenBCI demonstrably could hold its own. A consumer grade, open-source device has a potential to be a real game changer for programmers or researchers on the quest for better emotion recognition tools. The device may facilitate further progress toward online applications since it is inexpensive and possibly affordable even to those with more limited purchasing power" (p.10)

Table 3. Number of Studies with Conclusions about Use of Consumer Grade Devices

Type of Analysis	Conclusion about Use of Consumer-Grade Systems		
	Acceptable	Mixed	Unacceptable
ERP	9	1	1
Time Frequency/Other	5	0	0

more (16 channels). The sole study examining use of the Muse system concluded that it was successful in measuring N200 and P300 using ERP analysis. The two studies examining ThinkGear found it to be adequate for spectral analyses, but fair or less stable for ERP analyses. Thus, if researchers use ThinkGear or Muse headband, for example, they should not use reliability or validity evidence from EPOC studies, because it is not appropriate to generalize validation results across different consumer-grade tools (in particular if the validation study assessed tools with more electrodes). Most of the consumer-grade tools have fixed electrode positions, which might not fit the research question under investigation, so researchers need to carefully select the most appropriate consumer-grade tool for their research.

Third, because much fewer electrodes are typically used in consumer-grade tools than in research-grade tools, various forms of EEG data analyses are hardly, or not at all, possible (e.g., coherence analysis or source localization, refer to [3] as a starting point). A similar issue exists if one wants to calculate the neural correlates of IS constructs based on complex EEG metrics (see Table 3 in [3]). Some researchers have expressed concerns that consumer-grade systems may have less precise time stamping of events in the data stream as recorded, which is a potential problem because ERPs are quite susceptible to even small variation in event locking, and measuring phase-locking can be highly dependent upon precise time stamps. However, there is no evidence of this problem in past research (see Tables 1 and 2). Nonetheless, given their generally lower precision compared to research-grade tools, consumer-grade tools may be best suited to spectral or time-frequency analyses that focus on larger brain regions over larger time periods and larger frequency ranges.

Fourth, NeuroIS researchers should be aware of several potential data quality issues related to those tools that could limit their use for certain types of research (e.g., see [19]), such as: (i) sampling rates (i.e., $\geq$250 Hz), (ii) possible artifacts, (iii) electrode type and quality, and (iv) event timing in ERP studies. Researchers should also consider that dry electrodes are much more susceptible to movement artifact, electromagnetic interference (50/60 Hz noise), and drift caused by sweating. This issue may be especially important when researchers are interested in detecting EEG components, such as P300, and differences in component amplitude (or spectral power) between experimental conditions. However, the empirical data in Tables 1 and 2 suggest this has not been an issue in past research.

Fifth, one drawback of some consumer-grade tools compared to research-grade tools is that access to the data stream is not well supported and can require more technical expertise to get into a form or output that is beyond what the built in algorithms provide. In some cases, these algorithms are not well-defined in literature or are considered proprietary. This is not an issue with the most popular tool used in past research (Emotiv), but we caution researchers to ensure the data produced by a consumer-grade tool is compatible with the analyses they wish to perform.

However, keeping these limiting factors and potential issues in mind, we conclude that consumer-grade EEG tools are a useful research tool. Advantages of consumer-grade products which are of particular relevance for NeuroIS researchers are the following: Subjects frequently report discomfort during the lengthy fitting and calibration procedures with research-grade products and hence fast participant preparation, as it is the case with consumer-grade tools, is beneficial. Consumer-grade recording systems are typically portable, easy to use, and can be used wirelessly (or in wired connections). Thus, they can be applied in naturalistic and authentic research settings, even application in field research is possible. Finally, the costs for consumer-grade tools make research feasible at far more universities. Low cost devices are available for less than US$ 500, while high-quality EEG research equipment may cost US$ 20,000 or even more. We note that many of the features of consumer-grade systems are not unique to them. Specifically, there are research-grade systems that are wireless, use dry electrodes, use sponge electrodes, etc. Further, there are research-grade systems that employ low-density setups as seen in the consumer-grade systems. Thus, when taking these factors into account, the primary benefit to consumer-grade systems is, ultimately, cost.

Finally, we emphasize that in order to establish more definitive conclusions about the measurement quality of consumer-grade EEG instruments, empirical validation studies are needed based on IS tasks and paradigms. Based on these future studies, it will be possible to better determine the research settings in which application of consumer-grade systems is appropriate. Specifically, it would be useful to examine the performance of research-grade and consumer-grade systems with core IS constructs, elucidating their relative performance with ERP and time-frequency or spectral analyses. Further insight into how these devices compare in measurement of more complex IS constructs will provide further insight into their appropriate use in NeuroIS. Nonetheless, based on the analyses of past research presented in this article, we conclude that consumer-grade EEG tools produce results similar to research-grade tools, thus indicating the potential for the application of consumer-grade systems in NeuroIS research.

References

1. Dimoka, A., Banker, R.D., Benbasat, I., Davis, F.D., Dennis, A.R., Gefen, D., Gupta, A., Ischebeck, A., Henning, P.H., Pavlou, P.A., Müller-Putz, G., Riedl, R., vom Brocke, J., Weber, B.: On the use of neurophysiological tools in IS research: developing a research agenda for NeuroIS. MIS Q. **36**, 679–702 (2012)
2. Dimoka, A., Pavlou, P.A., Davis, F.D.: NeuroIS: the potential of cognitive neuroscience for information systems research. Inf. Syst. Res. **22**, 687–702 (2011). https://doi.org/10.1287/isre.1100.0284
3. Müller-Putz, G.R., Riedl, R., Wriessnegger, S.C.: Electroencephalography (EEG) as a research tool in the information systems discipline: foundations, measurement, and applications. Commun. Assoc. Inf. Syst. **37**, 911–948 (2015)

4. Riedl, R., Banker, R.D., Benbasat, I., Davis, F.D., Dennis, A.R., Dimoka, A., Gefen, D., Gupta, A., Ischebeck, A., Kenning, P., Müller-Putz, G., Pavlou, P.A., Straub, D.W., vom Brocke, J., Weber, B.: On the foundations of NeuroIS: reflections on the gmunden retreat 2009. Commun. Assoc. Inf. Syst. **27**, 243–264 (2010). https://doi.org/10.17705/1CAIS.02715
5. Riedl, R., Léger, P.-M.: Fundamentals of NeuroIS – information systems and the brain. Springer, Berlin (2016)
6. Riedl, R., Fischer, T., Léger, P.-M., Davis, F.D.: A decade of NeuroIS research: progress, challenges, and future directions. Data Base Adv. Inf, Syst (2020)
7. Larsen, K.R., Hovorka, D.S., Dennis, A.R., West, J.D.: Understanding the elephant: the discourse approach to boundary identification and corpus construction for theory review articles. J. Assoc. Inf. Syst. **20**, 887–927 (2019). https://doi.org/10.17705/1jais.00556
8. Riedl, R., Davis, F.D., Hevner, A.R.: Towards a NeuroIS research methodology: intensifying: intensifying the discussion on methods, tools, and measurement. J. Assoc. Inf. Syst. **15**, i–xxxv (2014)
9. AlZu'bi, H.S., Al-Zubi, N.S., Al-Nuaimy, W.: Toward inexpensive and practical brain computer interface. In: 2011 Developments in E-systems Engineering. pp. 98–101. IEEE (2011). https://doi.org/10.1109/DeSE.2011.116
10. Johnstone, S.J., Blackman, R., Bruggemann, J.M.: EEG from a single-channel dry-sensor recording device. Clin. EEG Neurosci. **43**, 112–120 (2012). https://doi.org/10.1177/1550059411435857
11. Badcock, N.A., Mousikou, P., Mahajan, Y., de Lissa, P., Thie, J., McArthur, G.: Validation of the Emotiv EPOCR EEG gaming system for measuring research quality auditory ERPs. PeerJ. **1** (2013). https://doi.org/10.7717/peerj.38
12. Duvinage, M., Castermans, T., Petieau, M., Hoellinger, T., Cheron, G., Dutoit, T.: Performance of the Emotiv Epoc headset for P300-based applications. Biomed. Eng. Online **12**, 1–15 (2013). https://doi.org/10.1186/1475-925X-12-56
13. Badcock, N.A., Preece, K.A., de Wit, B., Glenn, K., Fieder, N., Thie, J., McArthur, G.: Validation of the Emotiv EPOC EEG system for research quality auditory event-related potentials in children. PeerJ. **3**, 1–17 (2015). https://doi.org/10.7717/peerj.907
14. de Lissa, P., Sörensen, S., Badcock, N., Thie, J., McArthur, G.: Measuring the face-sensitive N170 with a gaming EEG system: a validation study. J. Neurosci. Methods **253**, 47–54 (2015). https://doi.org/10.1016/j.jneumeth.2015.05.025
15. Wang, D., Chen, Z., Yang, C., Liu, J., Mo, F., Zhang, Y.: Validation of the mobile emotiv device using a neuroscan event-related potential system. J. Med. Imaging Heal. Inf. **5**, 1553–1557 (2015). https://doi.org/10.1166/jmihi.2015.1563
16. Friedman, D., Shapira, S., Jacobson, L., Gruberger, M.: A data-driven validation of frontal EEG asymmetry using a consumer device. In: 2015 International Conference on Affective Computing and Intelligent Interaction (ACII), pp. 930–937. IEEE (2015). https://doi.org/10.1109/ACII.2015.7344686
17. Davidson, R.J., Ekman, P., Saron, C.D., Senulis, J.A., Friesen, W.V.: Approach-withdrawal and cerebral asymmetry: emotional expression and brain physiology I. J. Pers. Soc. Psychol. **58**, 330–341 (1990). https://doi.org/10.1037/0022-3514.58.2.330
18. Rogers, J.M., Johnstone, S.J., Aminov, A., Donnelly, J., Wilson, P.H.: Test-retest reliability of a single-channel, wireless EEG system. Int. J. Psychophysiol. **106**, 87–96 (2016). https://doi.org/10.1016/j.ijpsycho.2016.06.006
19. Krigolson, O.E., Williams, C.C., Norton, A., Hassall, C.D., Colino, F.L.: Choosing MUSE: validation of a low-cost, portable EEG system for ERP research. Front. Neurosci. **11** (2017). https://doi.org/10.3389/fnins.2017.00109
20. Ramírez-Cortes, J.M., Alarcon-Aquino, V., Rosas-Cholula, G., Gomez-Gil, P., Escamilla-Ambrosio, J.: P-300 rhythm detection using ANFIS algorithm and wavelet feature extraction

in EEG signals. In: Proceedings of the World Congress on Engineering and Computer Science 2010, vol. 1, pp. 963–968 (2010)

21. Debener, S., Minow, F., Emkes, R., Gandras, K., de Vos, M.: How about taking a low-cost, small, and wireless EEG for a walk? Psychophysiology **49**, 1617–1621 (2012). https://doi.org/10.1111/j.1469-8986.2012.01471.x
22. Lemm, S., Blankertz, B., Dickhaus, T., Müller, K.-R.: Introduction to machine learning for brain imaging. Neuroimage **56**, 387–399 (2011). https://doi.org/10.1016/j.neuroimage.2010.11.004
23. Rodríguez, A., Rey, B., Alcañiz, M.: Validation of a low-cost EEG device for mood induction studies. Stud. Health Technol. Inform. **191**, 43–47 (2013)
24. Elsawy, A.S., Eldawlatly, S., Taher, M., Aly, G.M.: Performance analysis of a principal component analysis ensemble classifier for Emotiv headset P300 spellers. In: 36th Annual International Conference of the IEEE Engineering in Medicine and Biology Society, pp. 5032–5035. IEEE (2014). https://doi.org/10.1109/EMBC.2014.6944755
25. Wang, S., Gwizdka, J., Chaovalitwongse, W.A.: Using wireless EEG signals to assess memory workload in the n-back task. IEEE Trans. Human-Machine Syst. **46**, 424–435 (2015). https://doi.org/10.1109/THMS.2015.2476818
26. Lakhan, P., Banluesombatkul, N., Changniam, V., Dhithijaiyratn, R., Leelaarporn, P., Boonchieng, E., Hompoonsup, S., Wilaiprasitporn, T.: Consumer grade brain sensing for emotion recognition. IEEE Sens. J. **19**, 9896–9907 (2019). https://doi.org/10.1109/JSEN.2019.2928781

Author Index

F. D. Davis et al. (Eds.): NeuroIS 2020, LNISO 43, pp. 363–364, 2020.
https://doi.org/10.1007/978-3-030-60073-0

Printed in the United States
By Bookmasters